C0-BEH-908

TECHNOLOGY MANAGEMENT

CASE STUDIES IN INNOVATION

ROBERT SZAKONYI
Editor

ISBN 0-7913-1127-9

Printed in the United States of America

Auerbach Publications
210 South Street
Boston MA 02111 USA

CONTRIBUTORS

Ron K. Bhada

Department Head and Professor, Department of Chemical Engineering, New Mexico State University, Las Cruces

J.R. Champion

PhD, Consultant, Denver

Hector Cochrane

Technical Director, Cabot Corp, Tuscola IL

James E. Doherty

Vice-President, ITW/Magnaflux Co, Chicago

Thomas P. Fidelle

Manager, Applications Research, Great Lakes Chemical Corp, West Lafayette IN

Stephen J. Fraenkel

President, Technology Services Inc, Northfield IL

Alan L. Frohman

President, Frohman Associates Inc, Lexington MA

F.D. Galatola

President, MG Management Group, Valley Forge PA

J. Michael Grindel

Executive Director, Project Planning and Management Worldwide, RW Johnson Pharmaceutical Research Institute, Raritan NJ

William Guman

Director, Contract R&D Technology Development, Grumman Corp, Bethpage NY

Mark T. Hehnen

Director, Technology Commercialization, Weyerhaeuser Co, Tacoma WA

Stephen Hellman

Director, Technology Planning and Statistics, Consumer Products R&D Division, Warner Lambert Co, Morris Plains NJ

Donald F. Hoeg

Technology Management Consultant, Mount Prospect IL

Russell Horres

Biomedical Consultant, Del Mar CA

Kenneth Jarmolow

Former President, Martin Marietta Energy Systems, Oak Ridge TN

Norman E. Johnson

Vice-President, Corporate Research, Engineering and Technology Commercialization, Weyerhaeuser Co, Tacoma WA

James R. Key

Consultant, Hopkins MN

Colin MacPhee

Principal, Albion Enterprises, Scarborough, Ontario, Canada

Keith E. McKee

Director, Manufacturing Productivity Center, Illinois Institute of Technology Research Institute, Chicago

Yuv R. Mehra

President and COO, Advanced Extraction Technologies Inc, Houston

Louis A. Mischkind

Consultant, Development Department, IBM Corp, Purchase NY

Graham Mitchell

Director of Planning, GTE Labs, Waltham MA

Edward L. Soule

Vice-President, Corporate Research and Engineering, Weyerhaeuser Co, Tacoma WA

Robert Szakonyi

Director, Center on Technology Management, Illinois Institute of Technology Research Institute, Chicago

Tom N. Thiele

Former Director of Technology Application, Siemens Corporate Research, Princeton NJ

Adrian Timms

Associate Manager, Hershey Foods Corp Technical Center, Hershey PA

William D. Torregrossa

Director of R&D Planning and Analysis, Hershey Foods Corp Technical Center, Hershey PA

Robert J. Tuite

President, RJ Tuite Associates, Rochester NY

Thomas B. Turner

Certified Management Consultant, Great Falls MT

Myra Williams

Executive Director, Information Resources and Strategic Planning Group, Merck & Company Inc, Rahway NJ

CONTENTS

INTRODUCTION

The management of technology is important to industrial companies for two reasons. First, technology affects a company's competitiveness. It influences the quality of a company's products as well as the way those products are manufactured, the company services that can be provided, and the manner in which information is handled and communicated within a company. In short, technology affects most of a company's operations.

Second, although technology is important in itself, it cannot be considered in isolation. The development and use of technology must be managed effectively if the technology is to provide benefits. For example, most new-product development projects based on a new technology will flounder in the absence of clear technical and commercial objectives, coordination among technical and nontechnical managers, and the guidance of a skilled project manager.

Despite the importance of technology management, companies have made only moderate progress in building a body of knowledge in this field during the last 30 years. For example, Albert Rubenstein, who has conducted research in this field since the late 1950s, wrote in 1989 about the changes in technology management over this period:

> The art of managing technology has changed . . . but not always in the direction of improving the effectiveness of the R&D and innovation process and not always to the long-term benefit of the firm and to the national economy. Unlike progress in many fields of science and technology, progress in management of science and technology (i.e., R&D and innovation) does not occur in steadily positive increments. Some technology programs in some firms do not appear to be as well managed today as they were years or decades ago, and some are managed independent of the many lessons learned by leading companies over the past decades.[1]

Rubenstein's discussion of what needs to be done serves as the touchstone for this publication:

> All this suggests that we are still far from a testable, cumulative, comprehensive "science" of managing technology. . . . Effective management is still heavily dependent on the skills and insights of individual managers and the shared skills and experiences of the managers in a particular firm.[2]

The purpose of this publication is to use those individual managers' skills and insights to build up a body of knowledge concerning technology management. In this way, managers in other companies can share the skills and experiences of these individual managers. To achieve this purpose, many people with wide experience in technology management have contributed articles on diverse subjects. Almost

90% of the contributors are industrial managers or have spent most of their careers as industrial managers. The remaining contributors are people who, like the editor, have conducted research or consulted at more than 250 companies in the US, Canada, and Europe.

This book uses a case-study approach to channel the contributors' insights and experiences into useful articles. Most of the contributors use one or more cases from their experience to elucidate the lessons they have learned. Even those who do not draw on one or more cases provide numerous pragmatic examples to demonstrate their recommendations.

Because this publication emphasizes practicality and lessons learned in companies, it is expected that its readers will be not only R&D and engineering managers but any other managers involved in technology management. These others could be general managers or managers of new business development, manufacturing, marketing, quality assurance, or planning. All of these managers can benefit from this publication by gaining new insights and practical tools to help them manage technology more effectively in their own companies.

The individual sections in this publication address most of the issues that concern industry managers. The first section discusses the linking of technology to business goals. This subject is examined from three perspectives: selecting R&D to meet business goals, the role of technology planning in helping a company select R&D, and the selection of R&D for new businesses and services.

The second section covers the management of R&D. This subject is examined from four perspectives: managing, evaluating, and training scientists and engineers; improving the quality of R&D; managing computer and information resources for R&D; and coordinating R&D with the rest of the company.

The next three sections address different aspects of developing and commercializing products: project management and teamwork, new product development, and technology transfer. Finally, the last section addresses an important issue underlying effective technology management: the nurturing of innovation. This subject is examined both from the perspective of a small, start-up company and from that of a large, established company.

Notes

1. A.H. Rubenstein, *Managing Technology in the Decentralized Firm* (New York: Wiley-Interscience, 1989), p xvii.
2. Rubenstein, p xvii.

SECTION

1

LINKING TECHNOLOGY TO BUSINESS GOALS

Linking technology to business goals is perhaps the subject most discussed by technical and nontechnical managers. Every company can cite innumerable cases of R&D projects that were directed toward the wrong commercial goals or technical goals that were not supported sufficiently to accomplish the desired business goals.

In this section, the process of linking technology to business goals is examined from three perspectives: the selection of R&D projects, the role of technology planning in helping a company select better R&D, and the use of technology to develop new businesses and services.

First, Alan Frohman explains the market, technical, and business criteria that companies should use in their selection of R&D. Frohman states that companies, on the whole, should take a more systematic approach to evaluating R&D projects. In addition, Frohman illustrates how companies have made errors in this area and how these errors have been corrected.

Next, Hector Cochrane outlines the many issues that must be addressed in selecting R&D projects—from the necessary business considerations, to the definition of R&D projects, to the internal company process for selecting projects. Cochrane also analyzes several R&D projects that his company selected in the past.

Robert Szakonyi considers the advantages and limitations of various R&D selection techniques. In one chapter, Szakonyi points out that the R&D selection process should not be considered in a vacuum; the process is not an isolated one, and it should be recognized as comprising the whole chain of events, starting with the generation of a technical idea and ending with the completion or termination of an R&D project. In a second chapter, Szakonyi discusses an important issue that is often neglected: involving company non-R&D managers in the process of selecting R&D projects. Szakonyi recommends methods to help an R&D department and the non-R&D departments handle these issues more effectively.

In the chapters on technology planning, three industrial managers relate their companies' experiences in developing technology plans. R.J. Champion describes the system of strategic technology planning that he helped develop, enumerating each of the phases needed to implement such a system—from technology assessment, to program development, to documentation, to review and approval. In addition, Champion explains what is needed to make this system of strategic technology planning work effectively.

Kenneth Jarmolow describes the actions that he and his colleagues took to develop their company's technology plan. In this chapter, Jarmolow emphasizes the content of technology planning. Having pursued the information vital to company planning that had gone largely unrecorded by key personnel, he describes the results of his interviews with 200 scientific authorities in companies, universities, and government agencies around the world.

Adrian Timms and William Torregrossa write about the maturation of their thinking and practices as they have undertaken to develop an approach for technology planning, covering such issues as recognizing the need for support from senior management, eliciting R&D personnel's support and participation in technology planning, developing a balance between qualitative and quantitative analysis, and formulating principles for technology planning.

To illustrate the use of technology for developing new businesses and services, some industrial managers relate their companies' experiences in the remaining two chapters. Mark Hehnen, Norman Johnson, and Edward Soule describe Weyerhaeuser Co's efforts to commercialize technology from the corporate R&D department, explaining the rationale behind their company's methodology of identifying, developing, and exploiting its core technologies.

Graham Mitchell outlines another type of evolution related to commercializing technology: the shift from an emphasis on developing technology for products to a more balanced approach that equilaterally develops technology for both products and services. Mitchell explains the difference between product-based and service-based R&D and how these relate to each other.

MAKING BOOK IN R&D

Allocating Resources Effectively

Funding R&D projects is a gamble even if it is based on careful research and planning. Surprisingly, most R&D resources are still allocated according to a company's track record in particular markets and product lines, with little regard to changing market and business conditions. To ensure financial success, however, R&D projects must be funded according to their potential contributions to market, business, or technical development strategies.

Alan L. Frohman

During the past 20 years, I have worked with several dozen large, medium-sized, and small R&D organizations to examine how they allocate resources to R&D projects. My findings were startling and disappointing. The majority of these projects are justified and funded on the basis of past activity in the particular R&D area. Resources are allocated infrequently to products or processes according to their potential contribution to the bottom line; rather, allocations are based much more on history than on future potential.

The second startling finding was that less than one-third of the projects explicitly targeted factors that may be important to current or future markets. Allocations were based on a wide range of criteria—for example, a problem that a specific customer had, a complaint by senior management, or an area of interest to one of the technical people.

The third startling finding was that most of the projects did not take the corporate strategy into account; that is, no explicit link was made between R&D projects and support for the current business plan. This is not to say that R&D activities did not support the current business plan. If the support existed, however, it was on an informal, voluntary basis, led by the R&D managers, and may not have been communicated to or understood by the business leaders.

There is little logical defense for such a dire situation or for the way the majority of resources are allocated to R&D projects. This all-too-typical predicament has resulted from years of basing future budgets on extrapolations of past budgets. These findings also reflect the disproportionate influence short-term problems have on planning, which leads to efforts that do not have any broader-range, longer-term payoff to the corporation but are funded as if they do.

These findings are not surprising given the nature of human beings and organi-

---

ALAN L. FROHMAN, PhD, is president of Frohman Associates Inc. Lexington MA, a consulting firm specializing in R&D and technology management.

zations. All of us have interests and want to pursue them. We justify what we have done in the past by continuing to do it in the future. Furthermore, a change in direction may require skills, experience, and information that are not readily available. The problem this poses for allocating resources to R&D projects is one of balancing the criteria so that, when it comes time to decide which projects should be funded, those that deserve funding because of their contributions to the company's market position, overall technical strength, and bottom line are not squeezed out by less worthwhile projects. There simply are never enough resources to go around.

DEVELOPING APPROPRIATE CRITERIA FOR RESOURCE ALLOCATION

One particular approach to allocating resources to R&D projects works consistently well. It consists of applying three sets of criteria—market, technical, and business—to determine levels of funding for development projects. The benefits of using this approach are several:

- It is easy to use and readily understandable;
- It provides a common vocabulary for different functions to use in working together on R&D project planning, selection, and evaluation;
- It indicates explicitly for each funded R&D project its contribution to the business, the markets, and the overall technical stength of the organization.

In this chapter, three case studies are presented that show the application of the three sets of criteria to decisions regarding R&D project selection. In each case, my consulting firm was asked to determine how effectively each company was concentrating its technological resources on areas of strategic importance to the company. Inasmuch as the firm believes that this approach is extremely productive, it is not as important to use the particular criteria described as it is to have a common set of criteria and an established process for selecting R&D projects. These criteria should be thoroughly discussed, understood, and accepted by all relevant members of the organization.

Market Criteria

The first set of criteria for resource allocation to R&D efforts concerns the relevance of the project to the needs of the marketplace. Projects that are driven by a real market need are those that enhance an attribute of a product, process, or service that is known to be important to the customer. Reducing the noise made by a dishwasher, improving the quality of a drug, improving the taste of a cereal, or accelerating the speed of a machine—all exemplify projects that are linked to market need. Reducing the cost of manufacturing through modifying an energy-intensive process is another example of a market-driven project; in this case, R&D creates a new process that consumes less energy.

The only difference between these examples is that in the first set, the market is an external customer; in the second set, it is an internal customer—the company's own manufacturing department. Both types of customer are regarded as the market when applying these criteria to decisions about resource allocation. New products that are developed to gain entry in a new market or to render old products obsolete also fall under the category of R&D projects explicitly linked to market need. In all these cases, the market criteria specify an attribute to be achieved that will affect the customer.

this system is based on applying three sets of criteria—market, technical, and business—to determine funding levels for development projects

Technical Criteria

The second set of criteria concerns the contribution of the project to building the organization's technical competence in core areas. For example, the use of lithium catalysts was important in the manufacturing processes of a chemical company. The company supported a significant number of R&D projects aimed at understanding and characterizing the use of these catalysts in its various manufacturing environments. The company sought to maintain its expertise and leadership in the use of lithium catalysts in chemical processes through better resource management in R&D.

The projects undertaken by a defense contractor in the area of infrared sensors provide another example. The company justified its intensive efforts to understand infrared technology applications in the detection and identification of airplanes by claiming the importance of this technology to its business. The company developed a strong research base in infrared technology even though these projects did not apply to specific customer needs or problems.

It is especially critical to apply technical criteria (i.e., to ask how a project strengthens the company's technology base) when a company is contemplating new business development in an area spawned by a new technology. For example, with the shift from chemical to electronic imaging in photographic processing, major players in the field (e.g., Kodak and Polaroid) have had to undertake major efforts to build their electronics capabilities. Such large-scale projects most certainly require recruiting and hiring new technical personnel, building new facilities, purchasing new equipment, and possibly acquiring smaller companies to build the capability more rapidly. Taken together, such new business development efforts can be risky and expensive.

Projects that monitor technologies that are of interest to the company but that do not need a large investment may also satisfy the company's current needs. The company can keep in touch with technology developments simply by funding contracts at universities or by assigning one or two staff members to part-time technology development in this area. This level of support allows the company to watch the field as it develops and to be ready to enter it as soon as it would be appropriate or as resources become available.

By allocating resources to projects that strengthen its technical competence, a company can develop expertise in key technology areas that are vital to the company's future competitiveness. Such projects focus on building the strength of the organization itself, as distinct from providing products or solving problems in the market.

Many companies do not acknowledge the need to build technical competence when they allocate resources to R&D projects; there is no formal statement of this investment objective. These companies assume falsely that the technological capability will be there when it is needed or that the needed capability can be bought through acquisition or licensing. The experience of such companies as General Electric and DuPont indicates that some internal capability is necessary to pursue even this approach—that is, simply monitoring a technology or buying it outright is insufficient.

Business Criteria

The third and final set of criteria for project selection and funding concerns the company's explicit expectation of payoff from its R&D efforts. For example, companies considering a variety of new business opportunities now look more carefully for those that are environmentally safe. In addition, companies often have stringent return-on-investment requirements that must be satisfied.

In some companies, the hurdles that R&D projects must clear to receive funding are implicit. The resulting lack of clear, financial goals can lead to inefficient and sometimes wasted efforts when R&D managers try to get projects approved. For example, in a company seeking to create a development program for a new gauze fabric for the hospital market, the project's sponsors had to appear before corporate management six times before the program was finally approved. Approval hinged on the amount of capital investment the manufacturing facility would require. Corporate management was looking for a significant capital investment. (High start-up costs would indicate a barrier to entry for competitors, thus allowing the company more time to charge a premium price for its product). This was never clearly stated; the project's sponsors were not aware of this criterion and so worked arduously over several years to justify the project on the basis of lower start-up costs rather than competitive strategy. This rather extreme situation probably occurred because senior management was not explicit about what criteria it was applying in this decision. Reviewing this major project proposal forced these managers to develop the appropriate criteria for managing their business.

DETERMINING THE RIGHT CRITERIA FOR YOUR COMPANY

Developing the market, technical, and business criteria for allocating resources to R&D in a particular company should be a collective effort that involves organizational functions outside the R&D area. To understand what matters to customers in today's market and in the future, it is essential that R&D personnel talk to the marketing personnel involved. Better yet, R&D personnel should speak directly to customers whenever possible to find out what the customers' needs are; this is done regularly at a few companies, including 3M and Hewlett-Packard. Some of the most effective market research findings result from conversations between technical personnel and customers.

Those companies that do not maintain such dialogue with their customers are tempting fate. For example, the manufacturer of steam turbine engines for electric-power generating utilities, in an attempt to muscle its competition, put a massive R&D effort into reducing the size—and therefore the cost—of its turbines. As it turned out, the utility companies that bought these steam turbines were not interested in size or cost; they were concerned primarily with reliability. In fact, turbines that were cheaper and smaller often aroused their suspicion about the product's reliability. The company had poured tens of millions of dollars into R&D projects to improve features that the market was already satisfied with.

it is especially critical to apply technical criteria when a company is contemplating new business development in an area spawned by a new technology

Assessing a market's needs becomes even more difficult when the criteria for product purchases shift, as they did when the oil crisis of the 1970s resulted in a greater demand for fuel economy in cars. The R&D programs of the major US automakers were redirected toward reducing vehicle mass and weight and increasing the fuel efficiency of the engines. Funding had to come from other projects that concerned mechanical structures, wind noise, or other comfort features. The US car-buying public had shifted the predominant market criteria from comfort to fuel efficiency.

Defining technical criteria for R&D project funding falls clearly in the province of the R&D department. This department, perhaps better than any other, understands the maturity of the technologies the company is using and the potential of emerging technologies that may compete with current technologies or render them obsolete.

Business criteria for allocating resources to R&D projects need to be developed by senior management. If these criteria have not been established, the R&D department can make reasonable assumptions about them and present them to senior management before resources are actually allocated. Otherwise, much time and effort can be wasted by R&D in putting forth projects that are rejected for poorly understood reasons.

To determine funding criteria, it is almost always beneficial for management to establish work groups that involve representatives from all involved departments, including R&D, finance, marketing, sales, and manufacturing. The groups should have between 5 and 12 participants, depending on the size and organization of the company. In moderating such groups, my firm always starts with the technical criteria because it is often easier to reach agreement in this area. It is also usually the area in which the R&D representatives, who may otherwise feel defensive, receive the most recognition and support. In addition, after this portion of the exercise, all participants have a clearer understanding of R&D's contribution to the well-being of the business.

Next, the group should identify key market parameters that must be considered in R&D project selection. Market parameters should be established for current and future corporate needs on the assumption that customer needs and expectations change as products mature.

The business criteria are the last to be defined. They are often identified tentatively through a collective brainstorming exercise, after which one member of the group is charged with reviewing the results with senior management to determine whether these criteria should actually be used in allocating funds among competing R&D projects. After the criteria are established, it is useful to examine the current inventory of R&D projects to rate them against the resource allocation criteria. Usually, substantial reallocation of resources is necessary to meet the corporate objectives that have been clarified as a result of this exercise.

Rating Projects

Different kinds of systems have been used in different organizations to rate R&D projects against the established evaluation criteria. Most of the time, projects are simply ranked in importance against each other as high, medium, or low. In other organizations, a more quantitative system has been used—for example, a numerical scale from 1 to 5, with 1 meaning very low impact on technical, market, or business criteria and 5 meaning very high impact. For each project in a quantitive rating system, the highs, mediums, and lows are counted or the numerical ratings tallied, and the projects are then listed in order of their scores.

In general, the top 20 percent of projects, starting with those rated highest on the list, usually have the most significant impact on the organization; those may be called class A projects. It is important that these projects get whatever funding they need to progress toward their goals. The next 40 percent of the projects on the list are called class B projects. These are important to fund, but they should not be treated with the same urgency as class A projects. The bottom 40 percent of projects are ranked as class C; these should proceed only as funding becomes available.

Projects that are undertaken to comply with regulations, laws, and other requirements as well as those that counter significant competitive threats must receive funding no matter how they rate against the list of criteria. In all organizations, certain projects are nondiscretionary and take precedence over even class A projects. In funding such projects, however, it is important not to strip class A or B projects of all resources; and whenever possible, resources should be diverted from class C projects.

As a final check on the financial soundness of the rankings, the projects are charted in order of rank on a table that features three columns. The first of these columns indicates the dollar amount the company must invest before it will see a return. The second gives the expected sales five years into the future (if relevant). The third column shows the profit or return to the company, also for a five-year period. (A company could choose a different time horizon, depending on the cycles it uses for its business plan.)

HOW THE CRITERIA WORKED IN THREE CASES

In the following sections, three cases are presented that illustrate some fairly typical problems companies have with R&D resource allocation. These cases demonstrate the benefits that can be derived from defining and applying market, technical, and business criteria. The insights that can be gained just by analyzing problem situations using these criteria can be the difference between making and breaking a business.

the company can keep in touch with technology developments simply by funding contracts at universities or by assigning staff members to part-time technology development

The Loss of Core Technical Competency

This case involves a manufacturer of drill bits for oil-well drilling. Although not an expensive piece of equipment, drill bits are significant in oil drilling. They are the part of the oil rig apparatus that actually drills through such materials as hard rock, shale, sand, and mud. Drill bits endure tremendous pressure and high temperatures, and unexpected fractures can result. When this happens, the drilling must stop, the rig must be pulled up, and the bit replaced. These stoppages are very expensive because use of rigs can cost more than $100,000 a day; it is therefore important that the bits do not unexpectedly break during drilling.

Over the years, the company's R&D organization had gathered extensive practical experience in the design and manufacture of drill bits, and their products were recognized worldwide as best in the field. Experienced senior engineers, who were the mainstay of the R&D organization, had lots of hands-on, trial-and-error experience developing designs for drill bits that could withstand certain kinds of stresses, pressures, and tears. They also learned how to alloy metals so that the drill bits would be able to endure a variety of forces below the earth's surface. The manufacture of these bits was still very much an art that only a few engineers understood and practiced.

Customers would contact senior engineers directly with requests for the specific bits they needed. The senior engineers would design the bits knowing what worked in the past, and the specifications would then go to manufacturing, where years of experience resulted in flawlessly produced, reliable bits. This pattern was repeated for so many years that the R&D engineers' expertise and competence was now taken for granted by the entire company. No one had systematically examined the R&D department's contribution to this company's success.

The situation was upset when a new head of R&D was hired. A senior scientist with strong technical experience, he was interested in replacing the more experienced, senior R&D engineers in the organization with several recent doctoral graduates. Attractive early retirement programs and some pressure induced many of the senior engineers to leave.

After a year, the new head of R&D started having problems. My firm was asked to help the company locate the source of these problems and to evaluate the company's new technology strategy. We started by identifying product criteria of a technical nature that were important to the company's markets. From discussions with customers, we found that they evaluated drill bits on consistency, average lifespan, and speed. The consistency of the bit was determined by how predictably the bit performed over time. Near the end of its expected life, the drill bit would be examined and possibly replaced to prevent cracking below ground. The second criteria, average lifespan, referred to how long the bit was expected to last. The longer a bit could drill, the more efficient the use of the oil rig was. The last factor found to be important to customers was the speed at which the bit rotated; the faster a bit goes, the more progress it makes in a day.

The departed senior engineers understood that these performance requirements were important to customers, and they therefore focused R&D efforts on meeting them. Without the benefit of these engineers' judgment, however, R&D efforts took on a whole new direction as projects were undertaken to investigate turbulence and to apply computer modeling, simulation, and mathematical analysis in the design of drill bits. In the long term, these projects might have paid off if they had been aimed at performance factors that were important to the customer.

It was evident from our analysis that the R&D organization, by retiring its experienced engineers, had ushered out its technological competitivenes. The exercise of identifying the company's core areas of technical competence allowed the new manager to recognize his mistake in recommending early retirement for the senior engineers.

The next step in our evaluation process was to identify business criteria for technology resource allocation. Competitive advantage was determined to be a key rationale for funding R&D projects—in short, any effort that would make the company's product better than other bits currently on the market could be justified. It was also determined that all funded projects must have a champion, someone willing to put the work and effort into seeing a project through from development to delivery to the marketplace.

Through our analysis, we discovered that most of the bits on the market had reached the mature stage of the product cycle and that changes to these products, if any, were minor. As a result, the third business criterion for resource allocation to R&D projects was the development of new bits. An additional criterion was established for projects aimed at developing a new bit—namely, that it be usable for drilling other kinds of underground metal and minerals, even though its initial, primary application would still be oil drilling. The company correctly foresaw decreasing future demand for oil, and it recognized the need to use its expertise in developing drilling bits for other underground deposits.

Several important conclusions were drawn from this analysis. The first was that the focus of projects undertaken had to shift to product parameters that were important to customers. Then, recognizing the strategic contribution of the departed senior engineers, the company resolved to dedicate resources to rebuilding and sustaining the technical competence these senior engineers once represented. Efforts that had begun to bring in the recent crop of doctoral graduates were scaled back in favor of a much more gradual hiring program.

The process of identifying market factors was useful in building a communication bridge between the R&D and marketing functions. This relationship had not existed previously; in the past, the marketing department resented the engineers for dealing directly with customers, and the engineers perceived the marketing personnel who lacked technical expertise as unsophisticated and unhelpful in selling the world-class products they developed. In the end, however, each was able to acknowledge the other's role in and contribution to the company's success.

a lack of clear financial goals can lead to inefficient and sometimes wasted efforts trying to get projects approved

The new spirit of cooperation between these groups came into play in resolving problems with product quality. In the past, there had been major battles over drill bits that fractured prematurely. R&D personnel blamed the marketing department for selling the wrong product; marketing personnel blamed the R&D department for not developing an effective product. After working together to establish product criteria based on customer needs, the participants came to perceive quality as a joint responsibility. The marketing and R&D functions pooled their talents by forming a new product quality unit with the task of examining product failures and providing the customer with a more satisfactory product.

Missing the Market

The subject in this case—a manufacturer of engines for cars, boats, and other vehicles—had a sizable R&D department with a wide range of technical skills. Over the years, R&D had become an island unto itself in the organization. The R&D department undertook projects mainly on the basis of technical criteria that were determined within the department itself; the development effort was concerned with using technology for its own sake.

The chief engineer was the most powerful person reporting to the president of the company. As time passed, the relevance of the R&D department's projects to market needs became more and more remote. Rapid turnover of senior management had allowed the R&D department to grow increasingly more autonomous. No recent company president had possessed the incentive or time to attempt the battles necessary to review the projects favored by top R&D engineers. The company had a sufficiently dominant market position, however, to survive this critical problem.

Fortunately for the company, a powerful new chief engineer took over the reins of the R&D department and decided to make its efforts relevant to the overall business plan. My consulting firm was asked to assist in this process and did so by presenting a workshop on R&D resource allocation to the more than 100 senior engineers in the company. We started the workshop with a discussion about recent competition the company faced from Japan. We offered data showing how Japan was selling superior products on the basis of cost, quality, and performance features. This introduction had a purpose: the chief engineer wanted the R&D department to take responsibility for making the company more competitive. Although many engineers found this a difficult challenge, they understood that their new leader was telling them that priorities were going to change.

Although a great deal of data was available concerning the reasons customers bought one type of engine instead of another, it had never been used explicitly in the R&D project selection and funding process. Customers' principal criteria for purchase were dependability, product quality, and quality of labor. The engineers could easily see how they could have a direct impact on each of these factors.

The perceived dependability of an engine is very much determined by its ability to start the first time the customer tries it. Quality is affected by the use of precision parts to minimize noise, leaks, and breaks. Quality of labor is affected by the layout of the engine (i.e., Is it pleasing to the eye? Does it look neat and not crowded?). Although these three factors could be translated easily into product specifications that the engineering organization could control, these factors had never been considered. In fact, the list of current R&D projects showed that projects tended to be weighted according to whether they incorporated new technology—yet new technology was low on the customers' list of important criteria.

The two core areas in which the company sought to establish technological competence were the application of electronics (i.e., integrated circuits) to the control of engines, and engine design for the optimum configuration of parts. We found that the organization had many designers with the skills required for engine design but that it lacked personnel skilled in electronics engineering and computer technology. After identifying electronics engineering as a target area of technical competence, the R&D department recognized that it needed to hire staff experienced in applying electronics to engine control.

Among the business criteria we identified, competitive advantage came first; all funded R&D projects had to contribute to making an engine that is superior to others in its class. The second business criterion stated that the customer had to be able to see and benefit from technical features added to the engine. Compliance with regulatory and emission standards was the third criterion.

After reviewing the three criteria in the workshop, the engineers formed subgroups and rated all the projects that the R&D organization was supporting. The results showed the need to undertake a massive redistribution of resources, taking away from projects that were justified mainly for technology's sake and giving to projects that focused more on market-oriented factors.

During the next several years, engineers were reassigned to projects that were relevant to market needs. To develop the organization's competence in electronic engine controls, new personnel with electrical engineering and computer technology backgrounds were hired. Within three years, fully two-thirds of R&D resources were reallocated. The payback from this effort soon became clear. After three years, there were twice the number of new products and innovations with a demonstrable impact on the market than there had been in the entire history of the company. The engineering unit, which had been seen as uninspired and unresponsive to the market, was now regarded as an innovator of new products and features that perceptibly enhanced the product's value to the customers.

Too Many Projects

In this case, the client company is a contract R&D organization whose final product is usually a study or a report. This firm was interested in expanding its business to include commercialization of some of its project work. With approximately $15 million in contracts, mostly from the US government, the business had undertaken a long list of projects, many very small in size. For several of these projects, the company had invested additional resources to investigate their potential for commercialization. It was very clear at the outset that, if even a few of the projects had commercial potential, the resources would not be available to pursue them. It was therefore necessary to decide which projects deserved further commercial development.

one company poured tens of millions of dollars into R&D projects to improve features that the market was already satisfied with

The company's core areas of technological competence were relatively easy to identify because they were the reason most customers sought its services. One area was electrocatalyst gas-diffusion electrodes, and the other was periodic-current processes. All personnel in the organization were involved in some way with these two electrochemical technologies. The major market for products that could be spun off from research consisted of large power-generation equipment manufacturers. This class of capital equipment is very expensive to purchase and maintain, and the cost of the products the company would have developed was minor compared to the cost of the overall system.

The cost of the entire system was also a major concern for customers, so one of the market criteria for funding commercial projects was that they reduce the purchaser's capital cost. The other market criterion was that the projects contribute to enhancing the performance of the entire system, which meant providing better output or yield from the same input of materials. Although such other factors as the product's cost, ease of use, and reliability were important, they were not considered significant compared to the benefits of a product that would reduce the cost of the system and enhance its yield.

R&D projects aimed at the development of products with those advantages would much more likely result in products with the promise of commercial payoff. This realization was important, because few of the organization's projects were focused on these market criteria. (This was understandable given that the company was a contract R&D operation, and the criteria that customers used to award contracts differed completely from those applied to the commercialization of products.)

Several business criteria were established for selecting projects for commercialization. First, there had to be a strong conviction that the R&D contract could lead to a commercial product. Second, there had to be a clear path toward commercialization (i.e., a partner who would work with the R&D organization to commercialize the product had to be identified). The third criterion was a 15% return on company funds invested in commercialization, and the fourth was that the effort be financially self-sustaining within three years. Because it was small, the company did not want to undertake projects that did not pay off in the short term and that would thus drain its resources. It wanted to be able to cut its losses before they became too large, even if doing so meant missing out on a project that would succeed over a longer period of time.

On the basis of these criteria, the company's commercially promising projects were ranked. Only three projects satisfied the criteria and received the time, attention, and effort required for commercialization; another 10 were discontinued. The exercise of establishing criteria helped the company to understand and focus clearly on the work that it had to do to make itself successful by a whole new set of standards.

ACHIEVING STRATEGIC RESOURCE ALLOCATION

In all three cases presented, the benefits of applying a systematic approach to R&D resource allocation were evident. In each case, this approach resulted in clear and universally accepted criteria for assigning priorities to R&D projects, which in turn helped bridge differences between R&D and other departments in the companies cited. It identified the technological areas in which the R&D departments needed to build or maintain strength in order to respond to the marketplace. It also explicitly linked the R&D function with both established and potential customer needs and resulted in increased innovation. The R&D departments focused their efforts directly on supporting the overall business strategy. Finally, it clarified priorities for resource allocation so that every one involved understood why some projects got more resources and others got less. This shifted the nature of decisions regarding resource allocation from subjective and personal to objective and strategic.

MODERN-DAY ALCHEMY?

Selecting R&D Projects

The R&D project selection process is facilitated by thorough project proposals reflecting complete understanding of the company's goals, competition, technologies, and customers. Teamwork and careful assessment of project risk are fundamental to the development and evaluation of these proposals.

Hector Cochrane

Many R&D managers claim that selecting successful R&D projects is more of an art than a science. Although this task involves a high degree of risk and uncertainty, it is possible to define a structure and to detail guidelines for successfully managing the selection of R&D projects. This chapter discusses four important stages in the R&D project selection process: business considerations, definition of possible R&D projects, the selection process, and project monitoring. The chapter is based on the author's experience as technical director of a medium-sized specialty chemical company whose products are sold globally to numerous customers in diverse markets.

STAGE 1: BUSINESS CONSIDERATIONS

During this stage of the R&D project selection process, technology managers need to identify their customers, establish their business goals, and evaluate the competitive situation. These tasks are described in the following sections.

Identifying and Understanding Customers

R&D projects are usually designed to satisfy a customer's need—but the technology manager must first define who the customer is. Customers can be classified as external or internal. External customers include a company or market to which the technology manager's company provides products or services or other divisions within the company. Internal customers are within one division of the company—for example, the sales and marketing department, production, or quality assurance.

An R&D project can be successful only if the customer's expectations have been satisfied. For example, a well-known specialty chemical company spent considerable time and effort developing a new additive to improve the thickening properties of fumed silica in epoxy resin systems. The company tested the product under room temperature conditions but neglected to test the product as the industry would—that is, after accelerated aging and under high-temperature cure conditions. Under these conditions, the product was ineffective and therefore a com-

---

HECTOR COCHRANE, PhD, MBA, has more than 18 years of experience in various aspects of new-product development, including 11 years as technical director of the Cab-O-Sil Division of Cabot Corp, Tuscola IL.

mercial failure. Developing a close relationship with customers and a clear understanding of customer needs is critical to the success of a company's R&D projects.

Establishing Business Goals

In general, many more R&D projects are proposed than can be supported by the available resources. To select the projects that support the company's business objectives, the technology manager must understand the organization's business goals and markets.

First, the technology manager must consider what the company's business goals are, over various time periods:

- Short term (i.e., less than one year).
- Intermediate term (i.e., one to three years).
- Long term (i.e., more than three years).

Second, the technology manager must know the company's financial goals. Typical expectations are a return on assets of greater than 20%, a 20% profit before taxes, and 25% of profits derived from products developed in the previous five years. Senior management will have little interest in a significant new product development project if, for example, the projected market price cannot meet the current financial goals for capital expenditure or if it cannot produce acceptable profits based on projected manufacturing costs.

Third, the technology manager must understand the company's strategic and tactical business goals. For example, it seems that Honda Motor Co's goal is to become a world leader in automobiles. Its strategy is to develop high-quality, medium-priced cars manufactured in plants around the world. Its tactics are to build plants in the US and Europe and to gain market share in the luxury automotive market by building new models such as the Acura to compete with the lower-priced Mercedes-Benz and BMW models.

Next, the technology manager must assess the status of the markets that the company is servicing. For example, the market could be growing (e.g., the civil aviation market), mature (e.g., the chemical industry), or dying (e.g., the US shoe manufacturing industry).

Finally, the technology manager must have information regarding the company's current global positioning, its share of existing markets, its entry into new markets with existing or new products, and other market factors. The technology manager must know whether the markets are very price conscious (e.g., commodity markets, such as household paints) or less price-sensitive (e.g., specialty markets, such as pharmaceuticals).

Knowing the Company's Competitive Situation

The technology manager should also understand what the company's current competitive situation for its products and services in their current and newly developing markets. Michael Porter's model of the forces driving industry competition can be used to analyze the effect of product competition (see Exhibit 1).[1]

The technology manager should ask many questions:

- How many competitors does the company face? What is their size and position in the marketplace? What are their strengths and weaknesses?
- Are they basic in (i.e., do they produce) the raw materials required to make the product?
- Do they have complementary products?
- Is the company a market leader or a follower?
- Are new technologies being developed that will make the company's current product or service obsolete?
- Is there new federal or state legislation that will affect the company's current products or markets (e.g., low volatile organic content for coatings)?
- What products does the company anticipate its customers will want in the next five years?
- How will the company's competitors react?

STAGE 2: DEFINITION OF POSSIBLE R&D PROJECTS

During this stage, the technology manager should monitor competitive technology, obtain input for R&D projects, compare the risks of various R&D projects, and develop new technologies. These tasks are discussed in the following sections.

Exhibit 1
Forces Driving Industry Competition

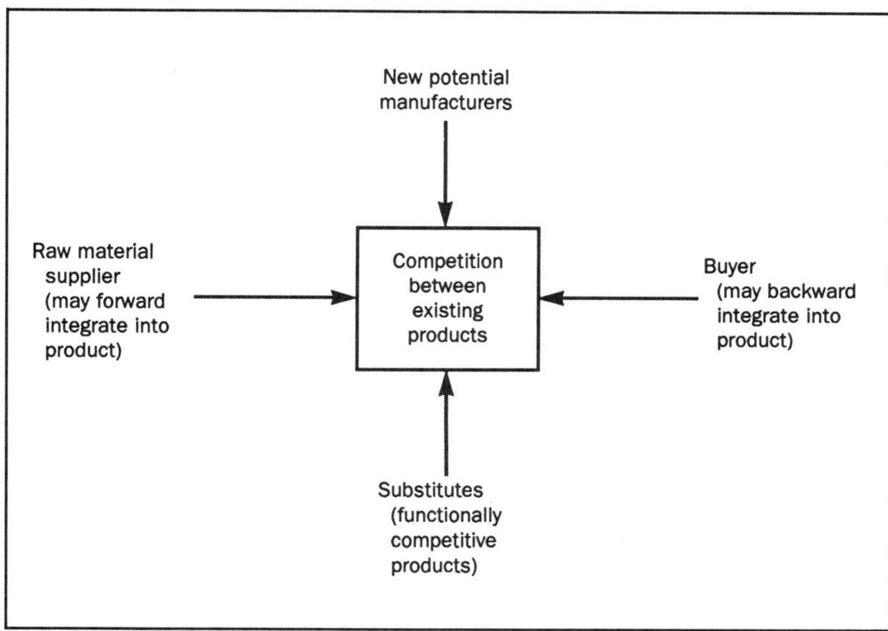

Monitoring Competitive Technology

When planning R&D projects, it is very important for the technology manager to develop a clear understanding of the status of current and newly developing products or services in the marketplace. This may be achieved by:

- Monitoring patent literature, trade and academic publications, and competitors' literature and by attending seminars and meetings.
- Obtaining customer feedback.
- Conducting an in-house evaluation of competitive products to assess performance and uniformity.

It is the responsibility of the R&D group to be the company's technical eyes and ears. The group should understand all major technical aspects of the business and should constantly assess the technical capabilities of the company's competitors. In addition, the group should be aware of alternative technologies to manufacture competitive products and major changes in customers' technologies that may affect the technical utility of the current product being sold to the customer.

Obtaining Input for R&D Projects

For R&D projects that are being developed, the technology manager should obtain input from the following groups:

- Customers—Customers can provide input on the performance of current and competitive products as well as on their future needs. A successful technical relationship involves the open exchange of nonconfidential information between the technical personnel in the supplier and user organizations. On occasion, this may result in joint product or application development programs.
- Marketing and sales—Marketing and sales can provide input regarding improved products, new products, new applications, and new customer needs.
- Production—Production can identify new or improved process controls.
- Quality assurance—Quality assurance can facilitate R&D efforts by developing new test protocols that are faster, less variable, and more meaningful.
- R&D—R&D staff may contribute new technology, products, applications, and tests.
- Senior management—On the basis of corporate strategic objectives, senior management can provide input on new exploratory directions.

It is critical for the technology manager to clearly understand the objectives of R&D projects suggested by the people in all of the groups cited as sources of input. For all groups except customers, this is best achieved by asking these people to make a written request to initiate the project. The request should include an unambiguous statement of their needs, the time they need the project completed, and the benefit to the company (e.g., generation of extra business). When the request is received, a member of the R&D department who is knowledgeable in the area of technology in which the project falls should be assigned to work with the individual. This process should ensure that the R&D department fully understands the project requirements; then, the project request should be converted into the standard R&D project proposal format, including supporting documentation.

Comparing Types of R&D Projects and Their Associated Risks

R&D projects can be categorized as:

- Basic technology projects.
- Product-related projects.
- Application-related (i.e., market-related) projects.
- Test development projects.

Projects can vary from small improvements to completely new developments. The probable success level of an R&D project is related to the size of the shift the project requires from the existing status quo with respect to technology, product properties, and end-use market. For example, projects that require only small deviations from the current status quo have a much higher probability of success (i.e., a low risk of failure).

It is very useful for technology managers to analyze proposed R&D projects according to risk level. Exhibit 2 lists eight possible R&D projects, constituting different combinations of these three parameters (i.e., the product, the technology to make the product, and the market for or application of the product) and indicating whether these parameters are existing or new. Examples are also provided.

The project with the lowest probability of success (i.e., with the highest risk factor) is R&D Project 1, in which all parameters are new. An example of this is the development of the computer chip during the early 1970s.

Exhibit 2
Analysis of Eight Proposed R&D Projects

R&D Project	Product	Technology to Make Product	Market or Application	Example
1	New	New	New	Computer chip
2	New	New	Existing	Optical fiber
3	New	Existing	New	Videocassette rental
4	New	Existing	Existing	Oat bran cereal
5	Existing	New	New	Car phone
6	Existing	Existing	New	Graphite fibers for golf clubs
7	Existing	New	Existing	Plain-paper copiers
8	Existing	Existing	Existing	Improved-quality product

Exhibit 3
Project Risk Assessment

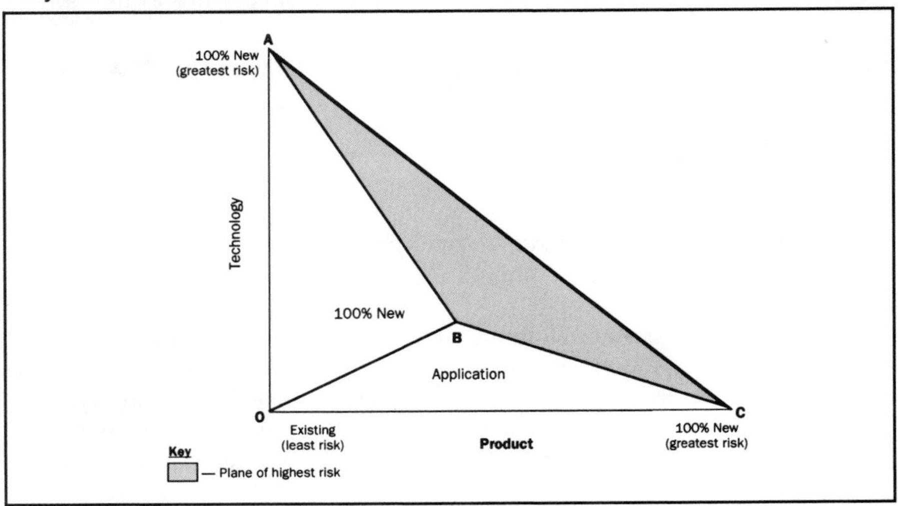

The project with the highest probability of success (i.e., with the lowest risk factor) is Project 8, in which the three parameters are already existing. An example of this case would be a project to improve a product's lot-to-lot uniformity.

Of course, not all projects involve extensive change with respect to product properties, technology to make the product, and product application; often, the projects involve small or partial changes. In these cases, technology managers can analyze the magnitude of the approximate risk of a potential R&D project by using the diagram in Exhibit 3. The three axes indicate the level of technology change required to make the product, the changes in the product's properties, and the new applications for the product. The least risk in carrying out the project (i.e., the highest probability of success) occurs when the three parameters are in a plane close to the origin. The highest risk (i.e., the lowest probability of success) occurs in plane ABC.

The least successful commercialized R&D projects are those that have been driven by technology or new products. The most successful projects have been market driven (i.e., they have been developed for an existing market).

This discussion attempts only to crudely compare the probable success of various projects, not to assess the financial benefits of low-risk versus high-risk projects. Obviously, major technological breakthroughs are made because companies are prepared to spend significant resources to work on projects close to the plane of highest risk. Often, these risks are accompanied by large financial rewards after the technology, if successful, has been widely commercialized.

Developing New Technologies

Often, market information suggests that a company develop new products using new technology or enter new markets with existing products. The technology manager must carefully assess how to approach such R&D projects. Several approaches can be used to tackle this problem.

the technology manager must consider such factors as the company's current global positioning, its share of existing markets, and its entry into new markets with existing or new products

One approach is to develop in-house technology. This is the least desirable method, however, because of the time required to develop the necessary in-house expertise. Another approach is to attempt a joint development program with a leading potential user. This approach can have limitations if the potential user is unwilling to share market and technical information.

The third approach is to hire a consultant. An effective consultant should be able to provide an accurate market overview, determine potential growth, assess the competition, and identify the technical leaders. The consultant should also understand the potential utility of the company's existing product or projected new product in the end-use application. The same market and business input will be required in the R&D project proposal to judge whether to continue to fund the project or terminate it. If the decision is made to continue the project, the best approach may then be to attempt joint development with a potential user.

STAGE 3: R&D PROJECT SELECTION

During this stage, the technology manager needs to form the R&D project selection team and handle any problems that arise during the project selection process. The technology manager also should balance short-term programs with medium- and long-term programs.

Forming the R&D Project Selection Team

During the R&D project selection process, the technology manager should form a team composed of technical personnel and representatives from production, sales and marketing, quality assurance, and senior management. The technology manager should choose the team members with great care and should give primary consideration to how well they know the needs of their portions of the business and the company's overall business objectives. It is essential that they be team players and have some technical background. The technical director should be the team leader, but other technical personnel with different areas of expertise should be part of the team.

The project selection team is thus composed not only of the group carrying out the projects (i.e., the R&D group) but of their internal customers for the projects—with the exception of the external customers (i.e., companies buying the products). Hence, the internal customers are involved in both the data collection for the R&D project proposal and in the final decision-making process determining which projects will be carried out. This should result in project proposals that are better defined and in better future cooperation between the various departments (e.g., to support the projects and their future commercialization).

An alternative approach that is more traditional and easier but less desirable is for the technical department alone to develop the R&D program and budget on the basis of input from its external and internal customers and then to select the R&D projects. The R&D program and budget is then submitted to the general manager and the sales, marketing, production, and quality assurance managers for their comments.

Handling Problem Areas in R&D Project Selection

When selecting R&D projects, technology managers often make the following errors:

- Choosing projects that lie outside the company's business goals.
- Selecting non–market-driven projects (i.e., projects for which the market is not defined).
- Failing to correctly define the problem to be solved.
- Failing to obtain the necesary support for the project from other groups.
- Underestimating the time, labor, or money required, the technological difficulties of the project, or the customer cooperation required to complete the project.

Most of these errors are related to incomplete knowledge of the project proposal. The project proposal is usually prepared by technical professionals or their supervisors. If the company's products are employed in several end-use applications, it is wise for the company to hire technical personnel who are familiar with the technology associated with the company's major markets. These should be well-trained, knowledgeable people who have frequent contact with their technical peers in the marketplace and who are cognizant of the company's future technical needs. They should also be familiar with the company's general business goals and the competitive situation in the market.

The company should foster an easy exchange of technical information and trust between the technical staff and the internal customers of the R&D programs (i.e., sales and marketing, production, and quality assurance). This information exchange should reduce misunderstandings when R&D projects are proposed by these groups and should make it easier to obtain such information as market or production savings data to support the project initiation.

Exhibit 4 provides a checklist for many of the items that should be included in the R&D project proposal form. Although this list is not exhaustive, it indicates the amount of information that should be collected.

Finally, the technology manager should find that the formation of the R&D project selection team with representatives of the project's internal customers should help the team reach consensus on those projects that are most important to the company. In addition, it is essential that R&D personnel be included in teams that review projects from the marketing, production, and quality assurance departments.

Exhibit 4
Checklist for R&D Project Proposals

- ☐ Name of the customer requesting the project.
- ☐ Description of the problem to be solved or new product to be developed.
- ☐ Description of the current market and products in use.
- ☐ Competitive analysis.
- ☐ Approaches to solving the problem (e.g., better product uniformity, different mixing equipment, or new product).
- ☐ For a new product, a definition of properties and determination of whether to modify existing technology or use new technology.
- ☐ If technology changes, information on:
  - — Core technology or new technology.
  - — New capital.
  - — Any hazards or safety or premanufacturing notices.
  - — Additional manufacturing costs.
  - — Packaging.
- ☐ Projected manufacturing cost and sale price.
- ☐ Distribution channels.
- ☐ Projected market size and growth during the next five years.
- ☐ Any changes projected in base market technology.
- ☐ Steps and time to bring the product to market.
- ☐ Tentative commercial development plan.
- ☐ Statement of how the project meets the company's long-term business goals.
- ☐ Statement of how the project meets the company's financial goals.

Balancing Long-, Medium-, and Short-Range Programs

In developing an R&D program, the technology manager should develop a balance between the efforts being given to short-term projects (i.e., less than one year), medium-term projects (i.e., one to three years), and long-term projects (i.e., more than three years in the future). The length of time given to a project usually relates to the urgency or complexity of the problem to be solved. For example, short-term projects are usually simple projects that have a high probability of success or are initial screening projects to determine whether more effort should be spent on the subject in the future.

Medium-term projects are more complex and may relate to small changes in product properties, new applications development, or complex new test development. Long-term projects relate to major core technology activities or to the development of significantly new products or market applications.

A company's actual priorities determine the time spent on each type of project. Ideally, the balance should be approximately 40% spent on short-term projects, 40% on medium-term projects, and 20% on long-term projects. There are several reasons for this balance. Short-term projects have a higher chance of success and provide satisfaction sooner to internal and external customers. The company, of course, needs satisfied customers for continued future support and finances, and success is important for the morale of the R&D group. Medium- and long-term projects are also necessary, however, for the company to continue to develop core technologies, to provide customers with new products, and to develop radically new markets. Unfortunately, they also have more setbacks, which offset some of the success of the short-term projects.

STAGE 4: MONITORING R&D PROJECTS

Successful R&D projects are not only selected correctly but accomplished quickly and effectively. To achieve this, it is important for management to monitor the progress of the project and to give it reasonable additional support to ensure that it stays on track.

Projects can be monitored through routine monthly progress reports and informal meetings between the project leader and the project selection team. Because the project selection team is broad based, it should be able to provide any additional support the project may need.

CASE STUDIES

To illustrate the importance of making the correct R&D selection decisions, five case studies are provided. Each case study reviews the major points affecting whether the research project was successful.

Case A: Developing a Technology-Driven Project that Failed Commercially

The Cabot Corp, a world leader in the manufacture of carbon black and silicon dioxide through high-temperature flame processes, initiated an R&D project to examine the feasibility of manufacturing a new high–surface area metal oxide using a new vaporizable metal chloride that had become commercially available. The project was successful and a unique new metal oxide was created. Although no significant market had been identified for the product, Cabot decided to commercialize it, and the company pursued significant market development activities. Although several small markets were found for the product, after a few years, none of the markets had developed to any significant extent. Financial analysis indicated the continued unprofitability of the product, and its manufacture was discontinued.

This is a typical example of a technology-driven project (see Exhibit 2, Project 1). A new product was developed using new technology (i.e., the flame hydrolysis of the new metal chloride); unfortunately, no major market need for the product was identified initially and none was found subsequently. Hence the product was a significant technological success, but a commercial failure.

Case B: Developing a Technology-Driven Product that Succeeded Commercially

Cabot initiated an exploratory R&D project to develop a new hydrophobic fumed silica using a silicone fluid as a treating aid. The project was a technical success: a new product was developed using new technology. However, the product was not initially a commercial success until a market for the product was found. The new product solved two problems that the existing product had suffered from. The new product is now a great commercial success.

a successful technical relationship involves open dialogue of nonconfidential information between technical personnel in the supplier and user organizations

This project started out as an example of Project 1 in Exhibit 2: a technical success without a market. It became an example of Project 2 in Exhibit 2 only when the commercial market for it was identified.

Case C: Developing a Product to Meet Customer Needs

One of Cabot's major US customers obtained from its foreign affiliate new product formulations that were based on a grade of fumed silica that a competitor manufactured and for which Cabot did not have an equivalent grade. Cabot initiated an R&D project to match the competitive grade. The experiments were conducted and were a success. The new product was manufactured, the customer's needs were satisfied, and the R&D project was a commercial success.

This is an example of the development of a new product using known technology for an established market—that is, an example of Project 4 in Exhibit 2.

Case D: Developing a Product to Meet Market Needs

Several years ago, Cabot was approached by a company that had a new, inexpensive chlorosilane by-product. Cabot initiated an R&D project to evaluate whether this raw material could be used to match the properties of Cabot's existing fumed silica products. After extensive studies, the correct flame conditions were found to make fumed silica products with properties equivalent to Cabot's current grades. The project has been a great commercial success.

This is an example of the production of an existing product using new technology (i.e., the new feedstock) to service the needs of an existing market. Thus, it is an example of Project 7 in Exhibit 2.

Case E: Improving a Product for an Existing Market

Aqueous silica sols are used commercially to polish silicon wafers before making computer chips. Cabot initiated an R&D project to develop an aqueous dispersion of fumed silica as an alternative wafer polishing system. To assist in this project, Cabot decided to retain as a consultant a company that serviced the supply needs of the wafer polishing industry and that could carry out studies of wafer polishing. Cabot projected that a major potential advantage of the fumed silica dispersions would be their much higher purity compared with the silica sols. The presence of ionic metallic impurities in the silica sols was expected to cause problems in chip manufacture as the circuits became more miniaturized.

The experiments were successful, and several wafer polishing formulations were developed. The products were commercialized, but extensive customer trials found that the dispersions gave slower polishing rates than the silica sols in the initial rough polishing step; however, the fumed silica dispersions gave better surface properties in the final wafer polishing stage. Recently, a potential new large polishing market for the aqueous fumed silica dispersions has developed. The Cabot systems have given superior polishing rates compared with the competitive silica sols, and the higher purity of the fumed silica dispersion is an added benefit.

The initial R&D project was a partial commercial success. There is great hope that the modified dispersions will be a much greater commercial success in the newly emerging market. This is an example of the development of a new product using new technology for an existing market (i.e., the initial silicon wafer application). It therefore is an example of Project 2 in Exhibit 2. This is also an excellent example of the use of a consultant or product distributor to help develop a new product in an application or market in which the company has no expertise.

CONCLUSION

It has been stated that the hallmarks of business excellence are high quality of decisions, programs, and plans; high commitment; and high speed.[2] These criteria should also apply to excellence in the selection of R&D projects to maximize the overall benefit to the company.

This chapter has emphasized that to make the best project selections, the technology manager must start with well-prepared project proposals (see Exhibit 4). Such proposals require that the individual preparing the proposals clearly understand the nature of the business, the company's short- and long-term goals, the competitive market situation, the technology in which the project lies, and the specific needs of the customer for whom the work is being planned.

Notes

1. M.E. Porter, *Competitive Strategy: Techniques for Analyzing Industries and Competitors* (New York: The Free Press, 1980), p 4.
2. R. Richardson, "Organizational Culture: What It Is and How to Change It," Adhesive and Sealant Council Summer Conference (June 1990).

Acknowledgments

The author wishes to thank Richard Geiger for useful discussions, Roberta Porter for typing the manuscript, and the Cabot Corp for its permission to publish the information in this chapter.

SO MANY PROJECTS, SO LITTLE TIME

Improving the Selection of R&D Projects

Although various techniques can be useful for selecting R&D projects, effective management of the R&D selection process requires a broader perspective that includes not only careful review of the techniques currently in use but improvements in the generation of new ideas and timely termination of ongoing, unproductive projects.

Robert Szakonyi

One of the great ironies in R&D management is that the issue that concerns R&D managers most is also the issue on which little progress has been made during the past few decades: the selection of R&D projects. Naturally, R&D managers are concerned about other issues, such as managing R&D projects, transferring technology, and commercializing new products. They recognize intuitively, however, that if their staffs are not working on the right R&D projects, it is not very important how well they manage or transfer these projects.

Because of the significance of this issue, researchers on R&D management (e.g., academicians and consultants) have tried for decades to develop techniques for improving the selection of R&D projects. Very few of the thousands of R&D selection techniques that have been developed over the last three or four decades, however, have been used by R&D managers in industrial companies. According to Albert H. Rubenstein, "The current state of practice in many large firms, as well as in most small and medium-sized ones, is not much more advanced than it was two or three decades ago, even though the theoretical and academic state of the art has progressed rapidly, with increasing mathematical sophistication and increased potential for computer-based systems."[1]

A few fairly rudimentary R&D selection techniques, however, are widely used by industrial R&D managers. For example, checklists and scoring methods are used by R&D managers to evaluate research projects and many development projects. Discounted cash flow analysis is also used by R&D managers for evaluating a development project that is moving toward commercialization.

These techniques have the advantage of being simple; R&D managers can easily learn how to use them. In addition, the type of information needed for evaluating R&D projects with these techniques can be gained without too much difficulty. Checklists, scoring methods, and discounted cash flow analysis, however, have been available to R&D managers for at least three decades. In other words, R&D managers have been using the same rudimentary techniques for more than 30 years to deal with the issue that concerns them the most: the selection of R&D projects.

The development of R&D selection techniques has not solved the problem of

ROBERT SZAKONYI, PhD, is the director of the Center on Technology Management at IIT Research Institute, Chicago. He has performed consulting work for many companies in a variety of industries and has written two books and more than 40 articles on technology management.

how to select R&D projects. To find a solution, R&D managers must first realize that they cannot use a technique to solve what is essentially a management problem (although R&D selection techniques do still play a role in R&D project selection). Instead, managers should look at the broader management issues related to how a company as a whole evaluates R&D projects.

Specifically, R&D managers must address two critical management issues. First, they should look at the entire process of evaluating and selecting R&D projects—from generating technical ideas to terminating unproductive ongoing R&D projects. Second, when selecting R&D projects, they should involve non-R&D managers in the company (e.g., the general manager and the marketing and manufacturing managers).

This chapter first reviews six commonly used techniques for selecting R&D projects and then addresses these two management issues. The aim is to identify actions that an R&D department can take to improve R&D project selection. Many company examples are used to describe the actions that R&D departments and companies can take.

TECHNIQUES FOR SELECTING R&D PROJECTS

There are thousands of R&D selection techniques;[2] they can be classified in terms of various types:[3]

- Checklists.
- Project profiles.
- Scoring.
- Analytic hierarchy process.
- Scoring with mathematical programming.
- Financial methods.
- Financial methods with risk analysis.
- Decision trees.
- Goal programming.
- Scoring with simulation.
- Mathematical programming.
- Decision trees with mathematical programming.
- Multiattribute utility theory.
- Simulation.
- Stochastic goal programming.

No consensus has emerged among R&D managers or academicians, however, regarding which R&D selection technique is the most effective.[4]

Six techniques are fairly representative of the various types available. This section describes the advantages and disadvantages of these six. The purpose is not to decide which R&D selection technique is the best or whether R&D managers should use any of these techniques. The purpose is to help R&D managers decide if and when they want to use any particular technique.

scoring methods allow managers to compare projects and combine quantitative and qualitative assessments in one value

A second purpose is to sensitize R&D managers to some of the key issues that underlie the use of any R&D selection technique. These issues specifically concern the quality of the information used and the essential questions that must be addressed when a manager is selecting R&D projects, regardless of whether R&D managers use any particular technique.

Advantages and Disadvantages of R&D Selection Techniques

Six types of R&D selection techniques are evaluated: checklists, scoring methods, discounted cash flow analysis, decision and risk analysis, linear programming, and dynamic programming. These techniques vary in two respects: the degree of mathematical rigor and the type of information required to use them. The mathematical rigor of the technique increases enormously from checklists through to dynamic programming, but so does the requirement for precise and comprehensive information.

Checklists. Checklists help R&D managers evaluate an R&D project in terms of such criteria as cost, likelihood of technical success, profitability, size of potential market, development time, patent status, and market share. A checklist could include as many criteria as an R&D manager wants. An R&D manager may evaluate a project in terms of descriptive rankings (e.g., high or low) or in terms of numeric rankings (e.g., 1, 2, 3, 4, 5) regarding various criteria.

The advantage of checklists is that they are simple and easy to use. They usually require information that can be obtained reasonably easily. For example, checklists do not require an exact prediction of future sales, profits, or investments. Moreover, the results of an evaluation gained by using a checklist can be communicated easily. Therefore, checklists make it possible for an R&D manager to quickly identify the glaring strengths or weaknesses of a possible R&D project.

One of the biggest disadvantages of checklists is that an R&D manager cannot compare two R&D projects with high rankings on different criteria. For example, a checklist does not help the R&D manager compare the merits of a project with high potential profitability but low likelihood of technical success with a project with low potential profitability but a high likelihood of technical success. Checklists also are based on subjective ratings and therefore may not yield accurate judgments. Finally, because checklists do not yield estimated financial payoffs, they cannot be used to analyze how productive a project would be at different funding levels.

Scoring Methods. Scoring methods build on checklists by incorporating weights for the various criteria. For example, if an R&D manager used the two different projects just described, and if twice as much weight was given to the likelihood of technical success as to potential profitability, the R&D manager would select the second project. Although both projects have rankings of 1 and 5, the second project has the 5 ranking on the criterion that has twice the normal weight, whereas the first project has a 5 ranking on the criterion with normal weight. Therefore, scoring methods allow R&D managers to obtain a single value for each of several projects so that they can compare projects. Scoring methods also allow quantitative and qualitative assessments to be combined in one value, which may be either strong or weak. For example, an assessment of cost figures can be combined with an assessment of the status of patents.

Scoring methods also have major disadvantages, however. First, the weights assigned to various criteria are based on subjective judgments. Second, company managers often differ in how they determine the weights for the criteria. Third, there usually are difficulties in identifying mutually exclusive criteria. For example, although both potential market share and potential profitability may seem to be important criteria, they are not mutually exclusive. The size of a company's market share affects its profitability. The value of a project that was based on using both of these criteria would therefore be biased. Finally, like checklists, scoring methods do not yield the estimated financial payoffs of an R&D project.

Discounted Cash Flow Analysis. This R&D selection technique is easily understood and does yield the estimated financial payoffs of an R&D project. Discounted cash flow analysis (also known as cost/benefit analysis or net present value calculation) consists of calculating project costs and benefits in terms of the appropriate discounted rate for the value of cash in the future. One advantage of this technique is that it is used in other areas of business in a company: thus, it appeals to financially oriented business managers. In addition, this technique has the advantage of explicitly recognizing R&D expenditures as an investment.

This technique also has several weaknesses, however. First, it does not include such noneconomic considerations as social or environmental factors. Second, it does not take into account the time required to complete a project. Third, it can be used only when the costs and benefits can be estimated with certainty, which means that its use is usually restricted to development projects that are near commercialization. Finally, because it discounts the cash benefits of a project the further out they occur in the future, it favors investments that are aimed toward short-term financial payoffs.

Decision and Risk Analysis. This technique allows R&D managers to evaluate the costs and benefits of an R&D project (or of a sequence of R&D projects) with greater depth than is possible with discounted cash flow analysis. By using both decision analysis and risk analysis, the most likely potential consequences of an R&D project can be evaluated.

Decision analysis uses decision trees to portray the many possible outcomes of a project at various stages (or the many possible outcomes of a sequence of projects). Decision analysis also portrays the relative probabilities of the outcomes at these various stages. By establishing the estimated financial payoffs with each outcome, the R&D manager can determine the most probable benefit of an R&D project.

Risk analysis is used to evaluate the probability distribution of various outcomes of different projects. For example, an R&D manager can compare the breadth of the range of various probabilities for an outcome from one project with that of an outcome from another project. The R&D manager might decide to select a project whose outcome is not as high if the probability distribution of this outcome is narrower (i.e., there is less risk of the outcome not occurring), rather than a project whose outcome is higher, if the probability distribution of this outcome is very broad (i.e., it is less likely to occur). The advantage of this technique is that it enables R&D managers to portray accurately the decision options that would arise as an R&D project is carried out.

regardless of which technique managers use to select projects, they must be concerned with the quality of the information they use

The disadvantage of this technique is that it requires a great deal of precise information. Not only do the costs and benefits need to be estimated precisely, the R&D manager needs to estimate probability distributions for each stage of a project (or for each project in a sequence). In addition, even though this technique helps R&D managers evaluate a project (or sequence of projects) with great rigor, it does not take into account tradeoffs between projects. In other words, it does not allow R&D managers to deal with resource constraints that occur when they must evaluate a portfolio of projects.

Linear Programming. This technique is derived from operations research: R&D managers use various sets of equations to determine the optimal use of limited resources. As a technique for selecting R&D projects, it can optimize the benefits of a portfolio of R&D projects by taking into account the limits on the resources that could be used to carry out these projects.

The advantage of linear programming is that it allows R&D managers to go beyond the artificial constraint of evaluating a project (or a sequence of projects) as if it were the only project to select. In addition, because R&D managers always face limits in their resources, another advantage is that linear programming allows them to deal with alternative levels of funding for any project.

Linear programming has several disadvantages, however. First, it requires an enormous amount of precise information. The expected benefits of each project must be quantified in a way that is consistent with the objective functions contained in the equations. Second, R&D managers must precisely define each project's requirements with regard to resources and the limits on resource availability. Two other disadvantages are that this technique does not deal either with uncertainty or with interdependencies between projects. Finally, in analyzing resources, linear programming treats the benefits stemming from the incremental use of resources in a linear fashion (i.e., the 100th unit of a resource produces as much benefit as the first unit of that resource), which is unrealistic.

Dynamic Programming. Dynamic programming is more advanced than linear programming; it overcomes the disadvantage of treating benefits stemming from the use of resources in a linear fashion. To gain this advantage, however, dynamic programming has even greater information requirements than linear programming does. Another disadvantage is that the models of dynamic programming are difficult to solve because of the complexity of the equations. Finally, dynamic programming requires R&D managers to determine the appropriate probability of technical success as a function of an R&D department's past and current R&D spending—and figuring out such a relationship is extraordinarily complex.

Applications for the Six Techniques

As can be seen, all of these R&D selection techniques have advantages and disadvantages. The simpler techniques could be used for evaluating an R&D project that has great uncertainties and about which little precise information can be obtained—for example, when an exploratory project is being evaluated. The more quantitative techniques could be used for evaluating an R&D project when the R&D manager has much more accurate information covering costs and benefits—for example, when a project that is moving toward commercialization is being evaluated.

Garbage In . . .

Regardless of which technique R&D managers employ to select R&D projects, they must be concerned with the quality of the information they use. This information could be flawed. Estimates of the probability of technical success, sales volume, and profitability are, for example, particularly unreliable. Even the costs of an R&D project are difficult to estimate accurately. According to one study, the actual costs of R&D projects usually turn out to be two or three times larger than the initial estimates.[5] Consequently, the study concluded that resource allocation decisions involving the return factors of an R&D project and based on such inaccurate estimates can only be regarded as worthless.[6]

The quality of the information used in selecting R&D projects can be improved in several ways. First, more attention could be paid to the level of expertise of an evaluator of an R&D project. Second, R&D people could be trained to improve their probability assessments. Third, records could be kept of evaluations of R&D projects so that biases might be identified.

The most important means of improving the quality of information, however, is to concentrate on answering the essential questions involved in evaluating and selecting R&D projects, including:

- What criteria should be used in evaluating an R&D project?
- Which criteria are the most important?
- What are the benefits of a project?
- What are the costs of a project?
- How much risk is acceptable?
- What are the key uncertainties related to a project?

- What is the quality of the information being used?
- How quantifiable is this information?
- What key factors cannot be quantified?
- What information is not available?
- Are there interdependencies among projects?
- What are the resource constraints in terms of funding, facilities, and staffing?
- What would the results of a project be if it were funded at alternative levels?
- What decisions about projects should be first?
- Which members of a company should participate in selecting R&D projects?

to improve project selection, managers must balance the need to foster creativity with the importance of critically reviewing ongoing projects

When senior managers have answered these questions thoroughly, they will use any R&D selection technique more competently. They will also be more capable of dealing with any inadequacies in the information they use.

GENERATING IDEAS FOR R&D PROJECTS AND TERMINATING UNPRODUCTIVE PROJECTS

To improve the process of selecting R&D projects, R&D managers must look for ways to improve the ideas that serve as the basis for new R&D projects. Some surveys indicate that R&D managers consider finding good R&D proposals to evaluate to be a bigger problem than actually selecting projects from these proposals.[7] In other words, many R&D managers recognize a great need for fostering creativity in their staff.

R&D managers must also consider whether to continue funding an ongoing R&D project as part of the selection process. This is because 85% to 90% of the projects in an average R&D department's portfolio consist of these ongoing projects.[8] R&D managers want to find ways to terminate unproductive R&D projects early so that the resources from those projects can be reallocated to more promising projects. R&D managers must therefore instill greater discipline in the R&D project evaluation process. Many R&D managers, however, do not recognize the problem of cutting unproductive ongoing R&D projects:

> Many projects continue from year to year or even from decade to decade without a clear link to the changing needs and objectives of the firm or the new opportunities and threats posed by outside forces in the marketplace and the economy. . . . Sometimes the momentum for continuing irrelevant or unpromising projects and programs comes from the interests and skills of the R&D staff. Often, however, it comes from the poor linkage between business planning and strategic technology planning.[9]

Numerous managers of companies find that the toughest thing to do in their company's laboratory is to terminate an ongoing R&D project.

Consequently, to improve R&D project selection, R&D managers must take seemingly contradictory actions. On one hand, they need to foster greater creativity, which means they should encourage an environment in which R&D staff members are not too critical of ideas that are in the embryonic stage. On the other hand, R&D managers must establish much more critical reviews of ongoing R&D projects. These problems are discussed in the following sections.

Generating Better Technical Ideas

The first consideration for generating better technical ideas is the type of environment that should be encouraged. In this environment, R&D staff members should suspend judgment when considering new ideas until further investigations can be conducted. To create such an environment, the overly critical attitude that many scientists and engineers have in evaluating new ideas must be changed:

> Successful scientists and engineers are often their own worst enemies in an idea generating session. Their training and prior success may compel them to blot out things that seem silly or that smack of irrational thinking and faithful excursions . . . [10]

After establishing the right environment, R&D managers can take five actions to generate better technical ideas. These actions are discussed in the following sections.

Investigating Ideas from Various Sources. Ideas for R&D projects can come from many sources: customers, competitors, suppliers, government agencies, and technical and trade journals. Ideas can also come from marketing, manufacturing, purchasing, quality assurance, and other departments in a company. Many R&D departments are not adept at exploiting the ideas from these other sources.

Using Idea-Generation Techniques. A variety of techniques can be used to stimulate thinking in new areas. One technique is brainstorming—generating as many ideas as possible during a short period of time. During this period, no one in a group is allowed to criticize ideas.

Another technique is synectics, which involves looking for a creative solution to a problem by taking a two-step approach. First, group members are required to imagine new ways of ordering things. Then, they are required to force-fit their possible solutions to the problem. By doing this, they can obtain new ideas or a new perspective on a problem.

A third technique is morphological analysis, which requires a group to examine all possible combinations of factors for solving a problem. By evaluating many combinations of factors that would not otherwise have been considered, the group can gain new ideas.

Establishing an Idea Bank. Ideas in companies often need to be generated a second, third, or fourth time because no record is kept of them. Consequently, just keeping a record of these ideas would save a great deal of time and effort. For example, a materials company maintains an idea bank, with contributions from members of all of the company's functions—particularly those functions that interact frequently with customers, such as marketing, technical services, and R&D.

understanding the strengths and weaknesses of a company's technologies enables managers to define appropriate criteria for evaluating R&D projects

Comprehensively Screening Ideas. For example, a chemical company screened R&D ideas annually. First, a planning group compiled a list from a wide range of sources (e.g., customers, technical journals) of the 150 critical ideas or issues that affected the company. Next, key managers pared the list down to the 20 most important ideas or issues. (This step was the most difficult because it forced the managers to resolve their differences of opinion about what the company's main problems and opportunities were.) Afterward, 20 employees were assigned to investigate each idea or issue by themselves. Last, the key managers used these investigations to decide on the merits of any R&D projects that were based on those 20 ideas or issues.

A glass company conducted a similar screening to select new product ideas. In the first stage, key managers reviewed approximately 500 ideas. If any of these managers liked an idea, that idea was considered in the second stage. About 50 of the 500 ideas reached this second stage, in which the managers defined the critical questions regarding an idea. They also took limited actions, such as conducting preliminary laboratory tests or making a business evaluation, to investigate the merits of an idea. If an idea was still considered excellent, it was taken to the third stage. Only 5 to 10 of the original 50 ideas made it to this stage, in which the company conducted a detailed venture analysis that covered the technical, marketing, manufacturing, financial, and organizational aspects of a possible new product development. Finally, if an idea still seemed worthwhile to pursue, it was brought to a fourth stage. Only one or two of the original 500 ideas made it this far. At this stage, senior company managers decided whether to approve a new product or new business development.[11]

Exposing R&D Staff Members to Other Parts of the Company. R&D staff members who have been exposed to other parts of the company—such as marketing or manufacturing—are more likely to generate promising technical ideas.[12] Because they have seen how technical ideas must relate to marketing and manufacturing requirements, they are more likely to be able to identify promising opportunities.

Terminating Unproductive R&D Projects

Just as proposals for new R&D projects should be evaluated and ranked, so should ongoing R&D projects. Obviously, if some ongoing R&D projects are reviewed regularly and are meeting expectations, they can be handled in a cursory fashion. Ongoing R&D projects that are proceeding poorly, however, should be singled out for very critical review.

For example, the R&D managers of a chemical company required all R&D staff members to identify their projects that ranked in the lowest 20% in terms of technical merits and potential payoffs for the company. This enabled the R&D managers to concentrate their attention on ongoing projects that deserved a critical review.

For a company to follow this approach, it must have clear guidelines on how ongoing R&D projects should be evaluated. It must spell out the criteria concerning how technical progress should be evaluated and how market potential should be measured. In addition, the R&D staff should recognize that R&D projects that are terminated are not necessarily stopped forever. New technical evidence, changing market needs, new government regulations, and other factors can create the need to revive an R&D project.

FOUR MANAGEMENT APPROACHES FOR IMPROVING R&D PROJECT SELECTION

On its own, an R&D department can improve R&D selection by:

- Systematically evaluating the company's technologies.
- Evaluating potential applications of the R&D project.
- Critically reviewing the types of R&D that are usually selected to discover possible blind spots.
- Seeking new ways to tap the skills of middle-level R&D staff.

These methods are described in the following sections.

Evaluating the Company's Technologies

To evaluate R&D proposals effectively, R&D managers must first systematically evaluate the strengths and weaknesses of the company's technologies. This knowledge enables R&D managers to define the criteria for evaluating R&D projects. For example, if R&D managers know that many of their company's core technologies are mature, they would think very differently about such criteria as technical potential and technology applications than they would if they knew that their company's core technologies could be advanced much further. With mature technologies, R&D managers would place a high priority on finding novel technical ideas in other disciplines. With growth technologies, R&D managers would rely more on their traditional disciplines. The following paragraphs describe six approaches to evaluating a company's technologies and provides examples from actual companies.

Developing a Technology Plan. The R&D managers of an instrument company analyzed their technologies as part of the development of a technology plan. They identified technologies that should be supported wholeheartedly; technologies that should receive some support but should mostly be just monitored; and technologies that should not be supported.

Analyzing the Company's Technologies in Relation to Its Products and Markets. The R&D (and non-R&D) managers of an electronics company first analyzed the company's 28 products in relation to its 32 markets. All of the company's products and markets were evaluated in terms of how effectively the company differentiated itself from competitors. The product/market combinations with the greatest differentiation were considered the company's strongest. Afterward, the R&D managers evaluated the strength of the company's technologies that served those product/market combinations to identify the company's most pressing R&D needs.

R&D personnel often shortchange consideration of downstream issues when evaluating projects

Comparing the Company's Scientific Boundaries and Business Boundaries. The R&D managers of a pharmaceutical company identified technological needs by comparing their areas of scientific strengths with their areas of business needs. First, they determined the boundaries of the company's various technical disciplines; these scientific boundaries showed the company's distinctive technical competencies. Afterward, the R&D managers determined the boundaries of technical needs of the company's businesses. By examining these two boundaries together, the R&D managers could identify needs within the businesses that the company's distinctive technical competencies did not serve as well as areas within the company's distinctive technical competencies that did not serve the business.

Analyzing Technologies in Relation to Business Opportunities. The R&D managers of an aerospace company analyzed their technologies in relation to their potential for gaining a major contract with the US Department of Defense. The R&D managers asked themselves the following questions:

- How strong is each of the company's technologies?
- What is the probability that the DoD will solicit bids in the business areas in which these technologies could be applied?
- What is the probability that the company will win a contract in these business areas, given the strength of the company's technologies?

Auditing the Company's Manufacturing Plants. The R&D managers of a chemical company audited the adequacy of the technologies used in all of their company's manufacturing plants. The R&D managers assessed whether each plant's technologies met standards related to the use of energy, yields, conversion efficiency, and the amount of maintenance required.

Assessing the Company's Technologies in Relation to Competitors' Technologies. The R&D managers of a food-processing company did this in three ways:

- By comparing R&D projects within a business area.
- By analyzing competing product lines in terms of how the products were produced (e.g., what raw materials were used).

- By analyzing how their competing technologies served similar functions.

Evaluating Potential Applications

Many of the important criteria related to evaluating and selecting R&D projects concern marketing, manufacturing, purchasing, sales, distribution, and financial considerations. In other words, they concern downstream issues relating to how R&D would be applied. Unfortunately, many R&D staff members shortchange these considerations when evaluating R&D projects:

> Programs that reach the stage of being put on a development time schedule rarely fail because the technical objectives are unattainable. Rather, other factors, such as time to completion, size of market, customer acceptance, cost, fit with the resources of the company, or manufacturability, prove decisive. . . . Unfortunately, projects sometimes fail because the drama of attacking spectacular technical problems [that] prove [to be] solvable drowns out adequate attention to other data [that] could have demonstrated that the project should not be pursued.[13]

R&D staff members often become engrossed in technical issues and neglect many nontechnical issues related to evaluating R&D projects. For example, R&D staff members often exhibit narrow thinking in the following ways:

- Continually resisting analyzing the benefits from their R&D projects.
- Having extremely limited knowledge about their company's existing products.
- Resisting working on R&D projects with excellent commercial payoffs.
- Ranking the various features of the company's products in the exact opposite order of the way in which customers value these features.

To combat this narrow thinking, R&D managers must provide their staffs with work assignments and training to expose them to all of the many nontechnical issues that must be considered and taken seriously when the department is evaluating and selecting R&D projects. Many of the criteria that are frequently omitted from the selection process concern nontechnical issues.

An example concerning an R&D department of a power equipment company illustrates the importance of using the right set of criteria. For many years, this R&D department failed to develop a new product. Finally, the R&D managers realized that their department's failure stemmed from the way it selected its projects: the department had been slighting nontechnical considerations.

The R&D managers at first decided to continue pursuing the best of their ongoing R&D projects, along with the new projects they had recently selected by using nontechnical as well as technical criteria. Six months later, however, they compared the merits of both sets of projects and concluded that they had to terminate all of the old projects. From the broader perspective they had gained by using the new set of criteria, they realized that the old projects would never pay off. Eventually, by concentrating on the new projects, they did make their R&D department much more productive.

incorrect assumptions about the relative value of different types of R&D may cause neglect of projects that offer tremendous potential benefits for a company

Identifying Blind Spots

Another way to improve the evaluation and selection of R&D projects is for R&D managers to look critically at the types of R&D projects that are usually selected. This may help them discover possible blind spots that stem from certain assumptions that R&D managers can make, such as an assumption that one type of R&D is more valuable than another. If R&D managers do make such incorrect assumptions, they usually neglect a type of R&D that could help their company greatly. The following sections describe six types of blind spots that R&D managers of various companies have experienced. As these examples show, R&D managers can make the wrong assumptions about the type of R&D that is more valuable. When they do emphasize the wrong type of R&D, they also select many inappropriate R&D projects. Consequently, R&D managers should continually ask themselves critical questions about the overall direction of their R&D efforts.

Manufacturing R&D. An R&D manager of an electronics company admitted that his company had traditionally neglected manufacturing R&D, which caused the company's businesses to suffer. For example, the company was frequently the first to develop a new product. It then captured a large market share and made great profits for the first few years. Because the company neglected manufacturing R&D (and manufacturing in general), however, product sales usually slipped after a while. Competitors eventually figured out how to make the same products more economically; they then captured most of this company's market share.

Product R&D. An R&D manager of a chemical company admitted that his R&D department began product R&D much too late. Traditionally, this company produced and sold mainly chemical commodities, and the R&D department therefore focused mostly on how to improve the production of these commodities. When the market for chemical commodities began weakening, the R&D department believed that it could help the company remain profitable by continuing to advance the state of the art of manufacturing R&D and by licensing its manufacturing technology. After more than five years of failure, however, the R&D department finally began conducting the R&D that it really needed; product R&D aimed at developing specialty chemicals.

Software R&D. The R&D managers of a computer company whose products had slipped in the marketplace recognized that their R&D department traditionally had focused completely on R&D to improve hardware, at the expense of software R&D. When these R&D managers recognized how weak their capabilities in this area were, they recruited many software specialists and retrained members of their R&D staff in the technical disciplines related to software.

Research of Developmental Problems. After a novel development effort was plagued with problems, the R&D managers of a pharmaceutical company were forced to recognize that trailblazing research had to be supported throughout development, not just during the discovery stage. A development effort must deal not only with the performance of a drug, but with safety, reliability, purity, and ease of use.

Product Maintenance R&D. The R&D managers of a food-processing company did not recognize this blind spot until the company's existing products suffered. Because these R&D managers worked in a corporate R&D department, they concentrated solely on R&D that would lead to new products. Consequently, they seldom talked with the marketing managers who were responsible for existing products. Eventually, however, the technical problems with the existing products became so serious that these marketing managers pressured the R&D managers for technical support. The R&D managers finally recognized the importance of maintaining the strength of existing products.

Cost-Cutting Opportunities. One of the primary assignments of an R&D department of a rubber products company was to help cut costs of the company's products. For years, the R&D department's goal for product R&D was to develop less costly materials. Unfortunately, it was never successful and was therefore criticized by the company's business managers. When one of the R&D managers was later transferred to manage the central engineering group, which conducted process R&D, this manager realized that the R&D department could have reduced product costs by improving how the products were manufactured. Instead of focusing on product R&D all along, the R&D department should have concentrated on process R&D.

Seeking New Ways to Tap the Skills of Middle-Level R&D Staff

Bench-level scientists and engineers and senior R&D managers have reasonably clear roles in the process of evaluating and selecting R&D projects. Scientists and engineers are expected to submit the majority of proposals for R&D projects. Senior R&D managers have the ultimate say regarding the R&D projects selected.

The role of middle-level R&D staff in this process, however, is often unclear. This group consists of first-line supervisors and other middle-level R&D managers (depending on how large an R&D department is) as well as scientists and engineers who have progressed beyond the bench level (assuming that an R&D department has a technical ladder). By finding new ways of tapping the skills of middle-level R&D staff, an R&D department can improve its process of evaluating and selecting R&D projects and can provide these R&D staff members with constructive roles in an important R&D activity. The following sections describe six ways R&D managers can tap the skills of middle-level R&D staff members.

Using First-Line Supervisors as Linking Pins Between Bench-Level Scientists and Engineers and Other R&D Managers. First-line supervisors are at a critical level in the organization of an R&D department because they manage scientists and engineers directly. Therefore, they should stimulate scientists and engineers to generate new technical ideas, and they should explain to scientists and engineers the guidelines of senior R&D managers regarding the overall direction for R&D.[14]

first-line supervisors can play a vital role in communicating a company's objectives to scientists and engineers

Involving Middle-Level R&D Managers in the Evaluation of New Proposals for R&D Projects and in Ongoing R&D Projects. Middle-level R&D managers should play a major role in evaluating R&D. An example from a pharmaceutical company shows what middle-level managers can contribute. The senior R&D managers were busy with many activities and therefore could not evaluate R&D projects thoroughly. Because they recognized that the R&D selection process would suffer, senior R&D management upgraded the role of middle-level R&D managers in the R&D selection process. To ensure that middle-level managers were capable of carrying out their new role, the senior R&D managers discussed new business opportunities with them and trained them in strategic planning. The senior R&D managers also encouraged middle-level R&D managers to propose new R&D projects.

Using Scientists and Engineers on a Technical Ladder to Evaluate New Technical Ideas. The R&D department of a building materials company did this. Because these scientists and engineers had great technical experience, they offered many insights on how to improve new technical ideas.

The technical experts chosen to make such technical assessments should be generalists, however. Specialized technical experts fail to identify the merits of technical ideas outside their technical area.[15]

Assigning a Broad-Gauged Middle-Level Technical Specialist to Coordinate Various R&D Efforts Aimed at Serving One Business. The R&D managers of a chemical company did this. The technical specialist chosen not only helped tie together product R&D, marketing, and the manufacturing plants but also helped coordinate the selection of R&D projects in an area.

Forming Technical Councils of Middle-Level Technical Specialists to Focus on the Potential of R&D in Their Technical Area. A large automotive company formed such technical councils in 10 technical areas, including forging, computer-integrated manufacturing, and specialty materials. Although the mission of these technical councils was to help transfer technology throughout the company and to advise on acquiring technology from outside the company, these technical councils played a role in evaluating and selecting R&D projects.Even though they should not be given the actual responsibility for selecting R&D projects (because the members of a technical council would come from different R&D groups and possibly report to different business units), these technical councils could provide advice on R&D projects in their technical area.

Establishing Centers of Excellence of Diverse Middle-Level R&D Staff Members Whose Technical Work Is Directed Toward Common Problems. The R&D managers of a food-processing company established a center of excellence for the technologies underlying a key flavor in the company's products. All of the middle-level R&D staff members who made up this center worked on common problems. Before the center was created, the members had seldom communicated with one another because they came from different technical specialties. By being brought together, these technical specialists realized that their R&D projects overlapped. Their new associations with one another also affected how they evaluated and selected their own R&D projects.

CONCLUSION

R&D selection techniques can be useful in selecting R&D projects. The solution to improving R&D project selection, however, is to look beyond R&D selection techniques and to deal with the entire problem—that is, how a company as a whole evaluates and selects R&D projects. Specifically, this means that R&D managers should review the R&D selection techniques they are using, improve how technical ideas are generated, terminate unproductive ongoing R&D projects so that resources can be reallocated to more promising projects, and improve their approach to R&D evaluation processes. R&D managers should also involve non-R&D managers in selecting R&D projects; this partnership is discussed in the next chapter.

Notes

1. A.H. Rubenstein, *Managing Technology in the Decentralized Firm* (New York: John Wiley and Sons, 1989), p 291.
2. W.E. Souder and T. Mandakovic, "R&D Project Selection Models," *Research-Technology Management* (July–August 1986), p 36.
3. P. Fahrni and M. Spatig, "An Application Oriented Guide to R&D Project Selection and Evaluation Methods," *R&D Management* 20, no 2 (1990), p 160.
4. D.H. Hall and A. Nauda, "An Interactive Approach for Selecting IR&D Projects," *IEEE Transactions on Engineering Management* 37 (May 1990), pp 126–131.
5. H. Thomas, "Some Evidence on the Accuracy of Forecasts in R&D Projects," *R&D Management* 1, no 2 (1971), p 59.
6. Thomas, p 69.
7. Rubenstein, pp 291–292.
8. L.W. Steele, "Selecting R&D Programs and Objectives," *Research-Technology Management* (March–April 1988), p 25.
9. Rubinstein, pp 291–292.
10. W.E. Souder, *Project Selection and Evaluations*, (New York: Van Nostrand Reinhold, 1984), p 19.

11. J.W. Muncaster, "Picking New Product Opportunities," *Research Management* (July 1981), pp 26–29.
12. Souder, p 24.
13. L.W. Steele, *Managing Technology: The Strategic View* (New York: McGraw-Hill, 1989), p 106.
14. Souder, p 21.
15. P.V. Rzasa, T.W. Faulkner, and N.L. Sousa, "Analyzing R&D Portfolios at Eastman Kodak," *Research-Technology Management* (January–February 1990), p 28.

PARTNERS IN PROJECT SELECTION

Forming Partnerships with Non-R&D Departments

Robert Szakonyi

Improvements in the process of evaluating and selecting R&D projects require collaboration between managers of R&D and other departments. Together, these managers can establish criteria for evaluating R&D projects and define arrangements for sharing responsibility in the selection process.

It is not possible for R&D managers to operate effectively in isolation from other company functions. Ensuring that development efforts are aligned with company objectives and customer requirements necessitates input from various sources. This chapter describes how R&D managers should work with managers from other departments to improve the process of evaluating and selecting R&D projects. (The preceding chapter describes several internal issues R&D managers must address to improve the process of evaluating and selecting R&D projects, including careful evaluation of the selection techniques currently in use and assessment of the methods used to generate ideas for new projects and terminate ongoing projects.)

Because long-term R&D projects are usually not the first priority of non-R&D departments, this chapter discusses steps R&D managers can take to ensure that long-term R&D is not neglected. This chapter also describes how R&D and non-R&D departments can share responsibility for the project selection process.

PROTECTING LONG-TERM R&D

Many R&D departments face a common dilemma: how to establish a partnership with non-R&D departments to select R&D projects without sacrificing long-term R&D. This partnership is needed because the information required to evaluate a new product development is so varied. For example, the Industrial Research Institute lists the following R&D project evaluation criteria:[1]

- Cost.
- Likelihood of technical success.
- Profitability.
- Size of potential market.
- Development time.
- Fit with overall objectives and strategy.
- Capability to market product.
- Market trend and growth.

ROBERT SZAKONYI, PhD, is the director of the Center on Technology Management at IIT Research Institute, Chicago. He has performed consulting work for many companies in a variety of industries and has written two books and more than 40 articles on technology management.

- Capability of manufacturing product.
- Market share.
- Patent status.
- Potential product liability.
- Capital investment required.

An R&D department can answer only one item on this checklist by itself: the likelihood of technical success. To answer the other items, an R&D department must work with marketing, manufacturing, legal, and other departments, as well as with the company's general manager.

Many R&D managers recognize that non-R&D departments should participate in selecting R&D projects, but they often fear that the short-term pressures facing non-R&D managers will make them neglect long-term R&D. This fear among R&D managers exists across all industries, including the eight diverse examples profiled in the following sections.

Threats to Long-Term R&D

The role of R&D as described in each of the eight industries highlights different aspects of the difficulties long-term R&D faces. Each industry example draws on the experiences of at least three but often as many as five or six companies from an industry.

Long-term R&D is given somewhat more support in the chemical, electronics, aerospace, and pharmaceutical industries than in the pulp and paper, food-processing, building materials, and oil industries. There are some companies in the last four industries described that have a long and venerable tradition of supporting long-term R&D. Nevertheless, even in these companies, long-term R&D is generally not exploited as well as it could be. In many other companies in these industries, long-term R&D may be no more valued than it is in the first four industries described.

Pulp and Paper. The pulp and paper industry might be expected to support less long-term R&D because the industry is mature and capital-intensive. R&D managers in pulp and paper companies find that business managers as a whole are interested mainly in existing businesses. When R&D managers in this industry propose long-term R&D, they are often asked whether any other company is currently conducting this R&D. If other companies are, R&D managers are more likely to get such R&D funded. If this is the way R&D projects are selected, however, such companies will not be leaders in R&D. They also probably will not fund much long-term R&D.

Food Processing. Food-processing companies do not fund much long-term R&D. On the whole, these companies are driven by the marketing department, which is almost always oriented toward short-term goals. Any long-term R&D is funded and conducted mainly for defensive purposes. When corporate researchers in food-processing companies perform long-term R&D, they often have difficulty getting adequate direction from their technical counterparts in the operating divisions—the people in product development groups.

> **although R&D departments in pharmaceutical companies often are encouraged to pursue long-term projects, they still face great pressure to support short-term marketing strategies**

Building Materials. R&D staff members in building materials companies usually find that they are given little discretion to choose long-term R&D projects. The marketing managers in these companies tend to specify in great detail how they want R&D resources used. R&D people are therefore the implementers of the marketing managers' strategies.

Oil. At first glance, companies in the oil industry might seem to be significantly different from companies in the industries mentioned so far. For example, in contrast to professionals in other industries, many business managers in oil companies are former engineers. Nevertheless, although it might be expected that these business managers would therefore support long-term R&D, in actuality they view technology as a means of fixing a problem, not as a tool with which to create growth opportunities. Consequently, in oil companies, long-term R&D typically plays a minor role.

Chemicals. Many chemical companies have a history of being process oriented—that is, most of their R&D was directed toward improving processes rather than developing new products. In these cases, the R&D department was set up mainly to provide technical services to the plants. As more chemical companies have become oriented toward producing and selling specialty chemicals, however, R&D departments in this industry have shifted their direction. R&D departments are becoming more attuned to responding to the marketing department's requirements. In both situations, R&D staff members in chemical companies have often considered their real bosses to be the business managers rather than the R&D manager to whom they report. In these cases, long-term R&D has almost always been shortchanged.

Electronics. Despite the aura of being in a high-tech business, business managers of electronics companies usually focus on quarterly sales and profit goals as much as business managers in other industries do. Even though electronics R&D often pays off only in 5 to 10 years, electronics companies are not necessarily directed more toward long-term goals than are other types of companies. Many R&D managers in electronics companies are frustrated by their lack of communication with business managers about strategic business goals.

Aerospace. In aerospace companies, engineers typically make up at least 40% of the company's entire staff. Also, the general managers in an aerospace company often are former design engineers. Despite the important role that engineers play in aerospace companies, however, the status of long-term R&D in aerospace companies is frequently not very high. Although the R&D staff members may be engineers, they are different from the design and manufacturing engineers, who outnumber R&D staff by a factor of 10. The primary responsibility of design and manufacturing engineers is to support the businesses with existing technologies. R&D staff members, on the other hand, are responsible for developing new technology that often takes 10 years or more to become commercially feasible. R&D professionals frequently find that design and manufacturing engineers are not interested in this new technology. Instead, many design and manufacturing engineers are more comfortable exploiting the technology that they already understand. Consequently, despite the sizable funding for long-term R&D that exists in aerospace companies, R&D staff members in this industry often have difficulty transferring new technology to business units.

Pharmaceuticals. Because long-term R&D is so vital to the success of a pharmaceutical company, R&D managers in pharmaceutical companies have the most freedom of any R&D managers to pursue long-term R&D. They do face great pressure, however, from marketing and business managers to channel much of the R&D toward supporting short- or medium-term marketing strategies. In many cases, R&D managers in pharmaceutical companies are able to resist this pressure. Nevertheless, they still have difficulty working with marketing managers in directing long-term R&D toward strategic business goals, which are often not clearly thought out.

Protecting Long-Term R&D from Neglect

Although R&D managers might wish that the business managers (and other non-R&D managers) in their company were not under pressure for short-term results, they should be realistic and recognize that things are not going to change. Given this situation, R&D managers can keep long-term R&D from being neglected by using the tactics described in the following sections.

Adding Value to Existing Products. The R&D department of a consumer products company preserved the importance of long-term R&D by deciding that none of its company's products should become cash cows (i.e., products that support the growth of other businesses). Instead, this R&D department continually carried out long-term R&D that would pay off by adding value to existing products.

Exploiting Opportunities to Provide Technical Services to Customers. Rather then grudgingly carry out fire-fighting missions, the R&D staff of a chemical company used these opportunities to learn what the customers' more serious problems were. For example, the R&D staff might ask, Why are these technical problems continually occurring in the customer's facilities? By identifying more fundamental problems, this R&D staff was able to focus its long-term R&D projects productively.

Providing the R&D Staff Discretionary Time. Although some R&D managers give their staff members the opportunity to work a small part of their time (e.g.,

10%) on R&D that interests them personally, many R&D professionals do not take advantage of this time. Frequently, they feel pressured to meet the short-term needs of the businesses, and they therefore use their discretionary time to work on their regular technical assignments.

R&D staff members should be encouraged to use discretionary time to explore their own technical ideas

An R&D manager of a materials company found an effective way to combat this problem. He personally coached many younger R&D staff members and encouraged them to explore some of the technical ideas that occurred to them. This R&D manager was in his 50s and realized that he had reached the height of his technical accomplishments. Rather than being primarily a principal investigator or project manager, he became a coach and elder statesperson within the laboratory. In his new role, he was able to guide R&D professionals in how to look beyond their regular technical assignments for opportunities to carry out long-term R&D.

Separating Research from Development. Even though research and development should be closely linked, sometimes it makes sense to separate them, to counter a tendency to emphasize development at the expense of research. In many companies, the term *R&D* means, in practice, mostly development. By separating research from development in terms of funding or organizationally (if the company is large enough), R&D managers can ensure that research is given a higher priority.

Finding Customers Interested in the Results of Long-Term R&D. The R&D managers of an aerospace company found that the best way to interest business managers in any long-term R&D was to work circuitously rather than to approach them directly. These R&D managers found a customer who appreciated the value of long-term R&D: a group within the US Department of Defense. By working closely with the customer, these R&D managers carried out exploratory R&D aimed at testing the merits of this line of inquiry. When they arrived at more conclusive results, these R&D managers approached the business managers with their results and with the information that a customer was interested in the long-term R&D.

WORKING WITH NON-R&D DEPARTMENTS

In addition to the concern regarding long-term R&D, the fundamental problem in the partnership between R&D and non-R&D departments concerns the difficulty that both sides have in sharing responsibility. This difficulty arises from the complexities involved in determining responsibility for R&D project selection.

For example, although an R&D department is considered responsible for managing R&D resources, the R&D resources are, in reality, company resources that are provided to the R&D department to serve the company's purposes. Furthermore, for R&D results to take shape, the whole company must be involved. To ensure that R&D resources are managed effectively, however, a company's senior business managers usually hold the R&D department accountable. Defining responsibility for how effectively a company's R&D resources are used is therefore not simple.

Similarly, it is not simple to isolate the effects of R&D from the business it serves:

> The influence of an innovation, be it success or failure, spreads throughout the company, and in the final analysis it becomes an integral part of the business. It affects not just one department in isolation (i.e., not just the R&D department) but several elements of the dynamic system called the business.[2]

In most companies, the R&D department selects projects by itself and involves non-R&D departments only if a project has particular significance or entails a major investment. Some R&D directors prefer this arrangement because they believe the selection of R&D projects is a technical matter and non-R&D departments should become involved with an R&D project only after it proves to be technically feasible.[3]

These two points are the heart of the issue:

- Are R&D projects concerned only with technical matters?
- Should non-R&D departments remain uninvolved until an R&D project proves to be technically feasible?

The answers to these two questions not only define how industrial R&D should be managed but convey ideas about what teamwork means.

According to some R&D directors, teamwork means that each department determines its own activities and then involves other departments when its activities prove successful. There is another definition of teamwork, however. This definition addresses the fact that successful activities involving a company as a whole—such as those in new-product development—require that all departments that eventually must be involved help guide those activities at an early stage. This definition of teamwork implies that R&D staff members must consider such criteria as potential market and new-product manufacturability when evaluating the technical merits of the R&D project to develop this new product.

For such teamwork to actually exist, however, requires more than just early evaluation of these criteria. It also requires that marketing professionals take the lead in evaluating the potential market and that manufacturing staff members take the lead in evaluating manufacturability. In other words, having the R&D staff evaluate the potential market or the manufacturability of a proposed new product is insufficient.

Sharing responsibility for evaluating the merits of an R&D project or a new product development is not easy. The departments must decide, for example, how much weight should be given to the opinions of the marketing department's evaluation of the technical potential of a proposed R&D project, or how much weight should be given to the opinions of R&D regarding the potential market of a new product that would stem from an R&D project? These issues are not easy to decide. Nevertheless, few correct decisions will be made regarding the competitive potential of an R&D project if either R&D or marketing is excluded from the discussion.

R&D resources are, in reality, company resources provided to the R&D department to serve the company's purposes

Furthermore, other, often delicate issues arise when an R&D department works with non-R&D departments to select R&D projects. R&D directors and their staffs could be anxious about sharing responsibility for selecting R&D projects and may feel that their status has been reduced. Sharing responsibility for selecting R&D projects highlights the dilemma that all members of an R&D department face: the desire to carry out challenging R&D, but to also make this R&D result in commercial payoffs. The dilemma is inherent in industrial R&D, as evidenced by R&D project checklists that include marketing, manufacturing, and financial considerations.

In other words, sharing responsibility for selecting R&D projects does not create this dilemma. Instead, it provides a solution for industrial R&D staff members trying to deal with this dilemma.

The next three sections provide examples concerning the issues that need to be addressed by R&D and non-R&D departments in their efforts to share responsibility for selecting R&D projects. These issues deal with the questions that need to be considered when R&D projects are being evaluated, the arrangements that R&D and non-R&D departments can develop for working together, and the technical or business goals that R&D and non-R&D departments can formulate to support a constructive relationship.

Questions to Consider During R&D Project Evaluation

R&D and non-R&D managers need to address together the following questions to foster a common understanding of how R&D projects should be evaluated and selected:

- What criteria should be used in evaluating R&D projects? Which criteria are the most important? How much accuracy in the data about possible R&D projects is required?
- How should the likelihood of technical success be evaluated (e.g., with three-point estimates, a range of probabilities, or a one-point or best estimate)?
- How important is it to have a balanced portfolio of R&D projects? How should the term *balanced portfolio* be defined?
- How important are R&D projects that build up a company's technology base—that is, projects that are valuable to a company even though they do not provide any direct forseeable benefits?
- What rule of thumb should be used for hurdle rates in evaluating R&D projects? How much flexibility should be allowed in the use of hurdle rates?
- What commercial calculations should and should not be made in evaluating exploratory R&D? For example, should the attractiveness of a potential market be used, rather than a projected return on investment?

- How is commercial success defined with regard to marketing considerations? For example, how important are three ways of measuring commercial success: sales volume, profitability, and market share?
- Which issues related to company capabilities and resources should be taken into account in evaluating how well a company would be able to handle the result of an R&D project if that project were to become truly successful?

How R&D and Non-R&D Departments Can Arrange Their Partnership

R&D and non-R&D managers need to discuss more than just the criteria used in evaluating R&D projects. They also need to discuss how they can work together. There are three possible partnership arrangements for R&D and non-R&D departments. Although all three have merit, none have ever been adopted because of various organizational problems in companies. There are few companies in which the R&D/non-R&D partnership was clearly defined. Moreover, many R&D and non-R&D managers have described problems coordinating R&D with the rest of the company that have stemmed from their company not having such arrangements.

One possible arrangement was developed by a marketing manager of a very large division in a consumer products company. This marketing manager outlined the responsibilities for both the corporate R&D department and the operating division in selecting R&D projects. The manager also contrasted this desirable arrangement with the existing relations between the corporate R&D department and the operating division.

The marketing manager specified six types of R&D in the company:

- R&D for other divisions.
- New product development in totally new areas for the company.
- Long-range R&D.
- Applied research (e.g., concerning possible changes in the mix of raw materials used in a product line).
- New product development for this operating division implemented at the division's initiative.
- New product development for this operating division implemented at the corporate R&D department's initiative.

The marketing manager then indicated how much influence the corporate R&D department and the operating division currently had in deciding on projects in these various types of R&D and how much influence each department should have in these areas. The corporate R&D department had total influence over the first two types of R&D (i.e., R&D for other divisions and new product development in new areas). The marketing manager felt that this influence was appropriate.

From the manager's point of view, however, the corporate R&D department and the operating division shared control appropriately in only one of the remaining four types of R&D. This was in the area of new product development for the operating division that was implemented at the division's initiative. The division, appropriately, had a tremendous amount of control.

Over long-term R&D, the division had almost no influence, when it should have had at least some degree of influence. Over applied research that directly affected its businesses, the division had only a moderate degree of influence, whereas it should have had a high degree of influence. Over new product development for the division implemented at the initiative of the R&D department, the division had only a slight degree of influence, when it should have had a moderate degree of influence. Although all managers may not agree with the categories or judgments of this marketing manager, the scheme could have facilitated useful discussions between the corporate R&D department and the marketing manager's operating division.

various organizational problems have prevented adoption of effective partnership arrangements between R&D and other departments

The second arrangement between R&D and non-R&D is illustrated by a household products company. Several R&D managers identified two main areas in which both the R&D department and the rest of the company should improve so that future efforts in selecting R&D would be more productive. The R&D department was supposed to terminate a sizable percentage (i.e., 25%) of its ongoing projects because too many projects were being pursued that were underfunded and understaffed. R&D was also supposed to define the tasks, resources, and time constraints of its projects more accurately. In turn, the rest of the company was supposed to define more effectively the criteria for the use of R&D resources and the real goals of the businesses.

The third arrangement is illustrated by an electronics company. The manufacturing manager laid out new guidelines by which the R&D department and the business groups should work in the future. Until then, the business groups had provided 95% of the direction for selecting R&D projects. This manufacturing manager wanted to change the existing situation radically. Specifically, he wanted the R&D department to learn much more about the company's businesses, provide much greater strategic direction to the company, and develop new system concepts for integrating products in the company's businesses.

Technical or Business Goals

The R&D and non-R&D departments should have technical and business goals to help guide their selection of R&D projects. To work constructively together, they need a common set of strategic goals, which serve as guideposts for evaluating and selecting R&D projects.

Three types of technical or business goals are illustrated: goals related to how R&D projects are directed, strategic concepts that involve both technical and business aspects, and goals related to the requirements of exploiting new business opportunities. In some of the examples, described in the following sections, the companies did not achieve these goals. Usually, inertia within a company prevented the more visionary managers from succeeding. Every technical or business goal had merit, however, and could have served as a guidepost when the company evaluated and selected R&D projects.

Goals Related to How R&D Projects Are Directed. The assumptions involved in how a company directs its R&D projects are usually not recognized until the marketplace changes radically. Often, company managers realize that they need to drop their old assumptions and replace them with new ones.

For example, after an electronics company lost much of its share in many markets, the company managers analyzed all of their operations as part of a strategic planning session. They discovered that the parameters of their businesses had changed. There were fewer market segments, R&D projects were more complicated, and more complex manufacturing operations were required. When they recognized this, the company managers began concentrating on fewer markets and fewer—but more significant—technologies. In developing these new technologies, the R&D managers also had to carry out much larger R&D projects that were also more interdisciplinary than previous R&D projects.

Another example is a machinery company that needed to redirect its R&D projects. Previously, this company had been the leader in its markets. Because of its technical strength, it relied on its ability to develop the new generation of a product. After many years, however, it lost its technical predominance, and competitors began taking over its market share. Given these new difficulties, this company had to think more systematically about how it would regain market share. It began planning a technical strategy for developing sequential generations of products. No longer would R&D projects be selected and carried out to address just the problems of developing the new generation of a product. They also would be directed toward providing a platform on which future generations of a product could be built.

Strategic Concepts. Strategic concepts involve both technical and business considerations. They provide clear direction to many departments within a company regarding how the departments must operate to exploit a new business opportunity.

For example, a dental products company successfully pursued the strategic concept of developing transparent braces for teeth. In carrying out the idea, the company ran into many technical challenges, such as developing a transparent material that was as strong and reliable as the material being used in existing products. There were also business challenges because dentists and the users of braces had to be persuaded that the performance of the transparent braces was satisfactory and that the value of these braces justified a higher price. Because the company developed a strategic concept, however, the R&D and non-R&D departments worked in tandem to successfully implement it.

Another strategic concept, which was not implemented, was developed by R&D managers in a metals and equipment company. This company provided metal for dental fillings to dental laboratories and also provided the computer hardware needed to design the fillings accurately. Some R&D managers identified a new business opportunity for their company: developing the software for the hardware that the company sold. The idea seemed sound: not only would this company provide a needed service, but it would offer a total package of products and services to dental laboratories. The full recognition of this business opportunity could have helped direct many R&D projects productively. Although this company had the capabilities to move in this direction, however, other R&D managers who were oriented toward the traditional businesses of the company prevented this strategic concept from being tested.

strategic concepts provide clear direction to many departments within a company regarding how they need to operate to exploit new business opportunities

Goals Related to the Requirements of Exploiting New Business Opportunities. When company managers understand their company's strengths and weaknesses, they can correctly define the requirements for exploiting new business opportunities successfully. By defining these requirements clearly, they also can provide guidelines for the type of R&D project that should be selected.

For example, after trying to develop many new products—some successful and some unsuccessful—the managers of a household products company recognized for the first time what their company's strengths and weaknesses were. Its strengths were in distribution and in producing specialty products that were consumed as if they were commodities (i.e., very frequently). Although their company was quite proficient in developing products that combined three or four chemicals, it was poor at developing products with more complicated formulations. Products with complicated formulations did not use the strengths of the company's distribution system, because these products were more likely to have stability problems while being delivered and because they tended to be specialty products that were not consumed frequently. When the company managers, including the R&D managers, recognized this, the R&D managers were able to direct R&D projects toward goals that fit the company's capabilities.

In contrast, another example illustrates how the requirements of exploiting new business opportunities were not understood. This occurred in a chemical company that traditionally produced commodity chemicals. The company's business managers were accustomed to assessing such large investments as a $1-billion plant. When the company had to concentrate more on developing specialty chemicals, more than just the product lines had to change; the philosophy of doing business also had to change. Specialty chemicals are very different from commodity chemicals. For example, an investment in a plant to produce specialty chemicals may be much smaller—only a $2 million plant may be needed. To succeed in this business, a company often should take small steps—that is, make a little, sell a little. In other words, the company could not succeed with specialty chemicals simply by lowering the costs of a standard product below those of its competitors. Because the business managers in this company had difficulty making this transition, they also had difficulty making investment decisions regarding specialty chemicals. Without clear direction on how to exploit new business opportunities in specialty chemicals, the R&D department, in turn, had difficulty in selecting R&D projects.

BENEFITS OF SHARING RESPONSIBILITY

An R&D department benefits in at least three major ways by sharing responsibility with non-R&D departments for selecting R&D projects. One benefit is a more focused portfolio of R&D projects. One of the endemic problems of R&D organizations is a tendency to spread themselves too thin.[4] Although one reason for this is that R&D managers sometimes have difficulty setting priorities, another reason is that the business groups themselves do not set priorities. Because of either of these circumstances, many R&D departments have pursued many R&D projects that one person would work on one or two days a month. On the other hand, if R&D and non-R&D departments are required to work closely to scrutinize and select R&D projects, both sides are more likely to eliminate peripheral R&D projects.

A second benefit of sharing responsibility is that better R&D projects are selected. For example, the R&D department in an appliance company had always selected R&D projects in isolation, and the results from most of these projects were not used. After the business groups in the company became involved in selecting R&D projects, everything changed. Marketing managers started providing direction regarding how they wanted the products improved. For example, they specified that they wanted to sell washing machines that washed linens significantly cleaner and made less noise. R&D staff members were delighted to be provided with clearer direction from marketing. From then on, they also worked on better R&D projects.

A third benefit of sharing responsibility is that non-R&D departments gain a sense of ownership of the R&D project. Unless non-R&D departments help shape the R&D product that they are eventually going to use, they will not be fully committed to using it.

This is illustrated by a problem that R&D departments continually encounter: the unity of purpose with which an R&D project is pursued frequently gets lost when the technology is transferred to business operations. For example, the researchers in one chemical company had almost single-handedly developed a novel technology. The project floundered, however, just at the point at which a combined developmental and business strategy needed to be formulated. The nonresearch departments had not shaped this technology and were unsure about what to do with it. Fortunately, the problem was eventually solved by creating an interdisciplinary team to manage the project. If such a team had been created at the beginning of the project, however, this problem would not have occurred.

MAKING THE TRANSITION TO SHARING RESPONSIBILITY

It usually takes a sustained effort by both the R&D and non-R&D departments to make the transition from not sharing responsibility for selecting R&D projects to doing so. The following stumbling blocks often prevent such a transition:

- R&D staff members do not think of ways in which the non-R&D staff can provide guidance on R&D projects.

- R&D staff members are forbidden by R&D managers to interact closely with people in marketing or manufacturing.
- Non-R&D managers are not interested in meeting with R&D managers.
- Business managers do not want to be tied down to a clear set of priorities.
- Functional departments of all kinds do not want to disturb a system in which each function totally controls its own activities.

one of the endemic problems of R&D organizations is a tendency to spread themselves too thin

To break logjams like this takes sustained effort. One materials company took five years to develop coordination between the R&D and non-R&D departments for the purpose of selecting R&D projects. The R&D manager worked with both sides. He first helped the R&D department so it could generate better technical ideas and defend the role of R&D in the company. He then demonstrated to non-R&D managers that their businesses were being threatened. He also persuaded them to work with R&D.

In companies with less extreme problems, it will surely take less time than it took this company. Nevertheless, both an R&D department and non-R&D departments must be fully committed to making such a transition.

CONCLUSION

The most important improvements in the process of evaluating and selecting R&D projects can be made only by R&D managers working in conjunction with non-R&D managers. Both sides together need to develop the criteria for evaluating R&D projects, the arrangements for sharing responsibility, and the company's common technical and business goals. By working closely with non-R&D managers and by improving the entire process of how R&D projects are evaluated, R&D managers can succeed in selecting better R&D projects.

Notes

1. R.H. Becker, "Project Selection Checklists for Research, Product Development, and Process Development," *Research Management* (September 1980), pp 34–36.
2. B.C. Twiss, *Managing Technological Innovation* (London: Longman, 1974), pp 142–143.
3. Twiss.
4. L.W. Steele, *Managing Technology: The Strategic View* (New York: McGraw-Hill, 1989), p 112.

IF IT AIN'T WRIT, IT AIN'T THUNK

Developing a Corporate Technology Plan

One of the best ways to receive continuous corporate support for R&D projects is to coordinate annual technology development plans with the general business development plans that are followed by senior management. This plan should be based on the same data, time frame, and business goals as the general business development plan and should be written for distribution, keeping management and R&D personnel informed of the long-term goals of the R&D group.

J.R. Champion

Effective R&D managers are constantly concerned with technology management issues that are strategic to the company's overall objectives. These objectives can best be achieved by using a formal system for identifying, collecting, analyzing, organizing, documenting, and presenting the information required to make R&D's case to management. In other words, what is needed is a strategic technology planning system. To structure such a system four questions should be foremost in the R&D manager's mind: Which of the company's business needs and opportunities require R&D support? Is R&D supporting those needs? If R&D is supporting them, according to what priorities should this support continue? and How can R&D convince the company's management that it has the right answers to these questions?

As logical as these questions are, answering them intelligently takes a great deal more than logic. It requires continuous analysis of the company's technology needs and opportunities in coordination with continuous self-assessment by R&D staff of the department's goals and capabilities. It also requires that R&D managers acquire the skill of communicating the results of these analyses clearly and persuasively to the company's decision makers.

A strategic technology planning system is primarily a tool for technology planning; it defines the company's technological objectives (i.e., the areas of technology to be worked on). This is distinct from project planning, which is concerned with how these objectives are attained. Second, the word *strategic* implies an intimate connection between the technological and the strategic business objectives of the company. The coordination of both sets of objectives involves analyzing all pertinent technological objectives and selecting those that are deemed best able to further the company's goals.

J.R. CHAMPION, PhD, recently retired from his position at Manville Corp, Denver, where he most recently served as senior director of corporate research. Previously, he had served as the company's director of technology planning and as division general manager.

This chapter describes the corporate technology planning system that was designed for and used successfully by a Fortune 500 company to significantly improve the company's use of R&D resources and the effectiveness of R&D performance.

THE CORPORATE TECHNOLOGY PLANNING SYSTEM

The corporate technology planning system is based on the premise that R&D's mission is to lead and guide the company technologically. This role requires that the R&D department assume the responsibility for and take the initiative in developing, documenting, and communicating the corporate technology plans to the corporation's various departments. In addition, it requires securing the participation of all the relevant functions (e.g., marketing, accounting, and sales) in formulating the plans.

Organizations that Would Benefit from the System

The corporate technology planning system can be used in a company of any size with any number of strategic business units (SBUs) and with centralized or decentralized operations. The company may be based on young, middle-aged, or mature technologies. The company for which the system was designed is a multibillion-dollar corporation engaged in a broad range of commodity businesses that were based on relatively mature technologies. At the time of the system's introduction, these businesses were divided into 48 SBUs under the responsibility of 13 operating divisions.

The company's R&D operations were completely centralized and were composed of 360 professionals, technicians, and administrative or clerical personnel, all organized into 10 technical departments that reported to a vice-president of research and development. The 10 departments were subdivided into 27 sections.

The departments and sections were technology specific, and each provided technological suport to any division or business unit that needed the technology in which the department or section specialized. The areas in which technological support was provided were typical for most industrial research organizations—that is, new product and process development, existing product and process modification, general technology research and development, and technical service support to marketing and manufacturing. In recent years, the company has changed its mix of businesses and has decentralized a large part of its R&D operations; however, many of the decentralized R&D departments continue to use the corporate technology planning system.

How the Planning Process Works

The corporate technology planning process, as shown in Exhibit 1, requires R&D management to first perform three separate technology assessments to identify the company's needs and opportunities. It then compares these needs and opportunities with the goals toward which R&D is working; revisions are made to the existing technology program as needed, the cost and benefits of the new program

Exhibit 1
The Corporate Technology Planning Process

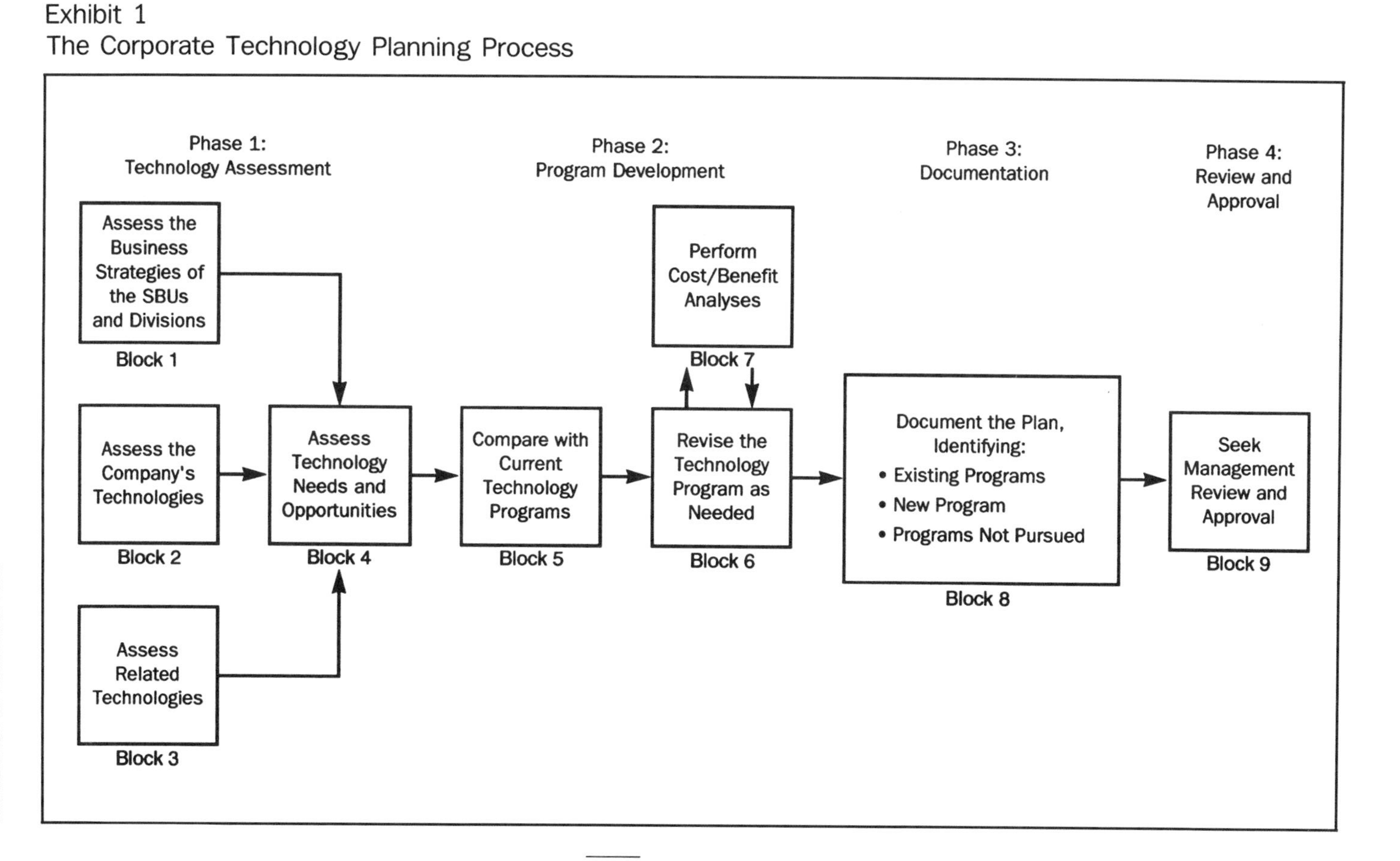

are calculated, and the plan is documented. This document becomes the basic technology plan that is then subject to review, negotiation, and approval by management on various levels—departmental, SBU, divisional, and corporate. If the planning process involved R&D's internal clients from the beginning, as it should have, the division- and corporate-level reviews should produce only relatively minor changes to the plan.

Phase 1: Technology Assessment

The technology assessment phase encompasses blocks 1 through 4 of the flow diagram shown in Exhibit 1. The three technology assessments the R&D area needs to perform are identified in blocks 1, 2, and 3.

Assessing Business Strategies. Block 1 requires each R&D group and subgroup to understand as thoroughly as possible the business strategies of the SBUs and the divisions that they support and to analyze these strategies for their technological implications. The information obtained for the assessment of business strategies will necessarily be strongly market oriented and will tend to be short term in its applicability. It will reflect only a part, albeit an important one, of the company's technology needs. If the corporate business strategy encompasses more than the sum of the divisions' strategies, the R&D department also has responsibility for analyzing that strategy to determine what must be accomplished technologically to ensure the success of a project.

Ideally, the R&D function's analysis of business strategies should be conducted at the same time that the SBU, division, and corporate managers are developing their strategies. In the real world, of course, such coordination is not always feasible. Some business managers are laggard in developing their strategies; others may be reluctant or may simply forget to invite their R&D colleagues to their planning sessions.

Using the corporate technology planning system, R&D managers should invite themselves to such sessions. For business units that have not completely formulated their strategies, R&D managers must simply make the most probable assessments of those strategies on the basis of existing information and their own discussions with the appropriate managers. In such cases, R&D managers should confirm that current research programs still address the company's business needs. If a business manager is unable to provide such confirmation, the R&D manager should proceed as planned and advise the business manager accordingly. This course of action will not seem improper if it is executed diplomatically and professionally. Experience with the corporate technology planning system has shown that when business managers were slow in developing strategies, the R&D manager's requests for information often served to prompt these managers to act.

In traditional, passive R&D organizations (i.e., those that wait for others to initiate contact), such initiative is seldom taken for fear of strong negative reactions on the part of R&D's internal clients. If the R&D manager handles interdepartmental situations properly, however, such reactions probably will not occur.

the R&D manager should take responsibility and initiative for developing, documenting, and communicating the corporate technology plans

To properly handle these situations, R&D managers must change their attitudes toward themselves; they must have the self-confidence that comes from knowing that the R&D organization performs a function that is critical to the success of the business. Although this is a basic truth, many R&D managers don't recognize it. Second, R&D managers must demonstrate that supporting the business managers' objectives shows that they are as concerned with the success of the business as they are with the success of the R&D programs. Third, they must show that they are professional and responsible managers who organize R&D operations with careful planning. Almost all business managers will respond positively to this approach; they will see in the R&D manager someone who understands the realities of the business world and who is concerned with the bottom line. In the company for which the corporate technology planning system was developed, it was the author's experience that the R&D organization was held in higher esteem by the rest of the company as a result of its initiative and strategic management.

Assessing Company Technologies. As shown in block 2 of Exhibit 1, another area subject to analysis by the R&D organization under the corporate technology planning system is the company's technological position in relation to its competition and the state of the technology. This assessment identifies longer-term needs and opportunities that must be incorporated into plans so that the company can conduct project planning beyond the visible technological horizon and achieve the appropriate balance between short-term and long-term R&D funds allocation. The R&D organization is required to make these assessments for all the businesses it supports and to document the results in the technology plan.

Assessing Related Technologies. The third part of the technology assessment phase, shown in block 3 of Exhibit 1, requires the R&D department to look beyond the specific technologies in which the company is engaged to developments in related technologies that could have an indirect impact, whether adverse or favorable, on the company's businesses. One example of such a development is the potential impact of high-temperature ceramics on automobile engine and engine component manufacturers. Because industrial R&D practitioners tend to focus only on the company's existing technologies, R&D managers should make specific assignments with respect to related technologies. In the company for which the corporate technology planning system was designed, the senior scientist (who was at the top of the technical career ladder) was responsible for defining related technologies with which the company should be concerned and for ensuring that assignments were made within the R&D organization to monitor these technologies.

Identifying Technology Needs and Opportunities. The final step in the technology assessment phase is to identify and substantiate the needs and opportunities the R&D department believes the company must address. This R&D priority list should then be presented in whatever form is most appropriate to management before the program development phase begins.

Phase 2: Program Development

This phase is represented by blocks 5, 6, and 7 in Exhibit 1.

Comparing with Current Technology Programs. As shown in block 5 of Exhibit 1, the first of these steps requires that the needs and opportunities identified in the technology assessment phase be compared with the slate of projects on which R&D personnel are currently working. This enables R&D managers to determine what changes need to be made to the plan.

Revising the Technology Program. The next step, as shown in block 6, stipulates that needed changes be made. If R&D has been doing strategic technology planning all along, most of the changes will consist of updates—that is, changes resulting from modifications in business strategies or from new technological insights. If strategic technology planning is just being introduced, however, the R&D organization probably will discover that it is dedicating resources to some unworthy programs at the expense of worthy ones. For example, the company for which the corporate technology planning system was developed decided after the first year of using the system to cancel 20% of their R&D projects and to introduce a few high-potential programs that should have begun several years earlier.

Performing Cost/Benefit Analyses. Cost/benefit analyses, as shown in block 7, must be performed for every single project in the revised technology program identified in block 6. These analyses are aggregated at departmental, SBU, divisional, and other appropriate organizational levels. R&D expenses and capital expenditures are estimated for each year of each project's duration; if the development effort is successful, additional estimates are provided for capital (i.e., plant and equipment costs) and marketing or sales expenses.

It should be evident that the R&D department must work closely with the engineering, manufacturing, and marketing departments to develop these estimates. It should also be clear that R&D must do some reasonably detailed project planning to be able to forecast each project's duration and cost.

Benefit estimates are stated in two ways: in dollars of additional sales (or sales losses avoided) for the first year of commercialization and four years beyond, and in dollars of gross earnings resulting from these sales. In the case of cost reduction and productivity improvement projects, benefits are stated in dollars of additional gross earnings. Here again, R&D management must work very closely with the appropriate non-R&D functions to ensure that those groups in the company that are most competent to make these estimates agree to them.

Although the company in this case study usually carries individual business results down to the net earnings level, it decided to use gross earnings to estimate the benefits of R&D efforts because the R&D function has the most direct impact on that level of earnings. In addition, all benefit estimates were made on the assumption that the development and commercialization phases of the projects would be completely successful.

the R&D function's analysis of business strategies should be done at the same time the SBU, division, and corporate managers are developing their strategies

Working with these premises has significant advantages. First, it gives management a feel for the total commercial potential of the project, which is an important parameter in decisions vis-à-vis project funding. Second, it sidesteps the problem of estimating a project's probability of success, a subject on which everyone has a different opinion and for which estimates change as the project progresses. Even when 100% success is assumed in the plan, everyone is still entitled to a personal estimate. And although the benefit dollars stated in the plan assume a 100% success rate, the document also indicates for each project a high, medium, or low probability of success so that the decision makers are not left completely without guidelines.

All cost/benefit estimates should be stated in this year's dollars, and the plan should clearly state this. The rationale for using constant dollars here is much the same as it was for assuming a 100% rate of success for R&D projects—namely, that the nonproductive disagreements that arise from inevitable differences of opinion about inflation rates over the coming years will be avoided.

Although the cost/benefit analyses are a lot of work, they are an essential part of responsible technology planning, and R&D managers must perform them. After the painful process has been gone through once, carefully and thoroughly, subsequent cost/benefit analyses will be much easier because the plan will require updating for the most part and because the R&D staff and management will have learned from the first experience.

Phase 3: Documentation

Although a great deal of preliminary documentation is generated during the program development phase, it is necessary to prepare the written plan in its final or near-final form.

Documenting the Plan. This step is indicated in block 8 of Exhibit 1. The final corporate technology plan really consists of a nested system of documents. These documents, which are described later in some detail, identify the following:

- The programs R&D plans to continue, and reasons for doing so.
- The new programs R&D plans to undertake and why.
- The technology needs and opportunities that are not addressed and the reasons for not pursuing them.

This last item is important because calling special attention to these issues underlines their importance and gives management an opportunity to reconsider important strategic decisions. Each new edition of the plan will repeat these items until management makes a firm decision to either proceed with them or drop them.

At the company for which the corporate technology planning system was developed, R&D managers were considerably reluctant to thoroughly document their plans at the outset. This was understandable but unacceptable. Clear and complete documentation is a sine qua non of effective planning for these reasons:

- If it is not written down, it is not thought out.
- If it is written down, it is there for future reference.
- If it is written down clearly, it will communicate clearly to those who will be affected by the work proposed in the plan.
- If it is written down, it need be prepared only once—it can be reproduced and communicated uniformly any number of times
- If the plan is written down, preparing it each year is a relatively simple matter of updating.

For these reasons, great emphasis is put on complete and accurate documentation in the corporate technology planning system. R&D managers can achieve this objective by developing detailed and standardized contents and formats for plan documents. Standardization of format not only eases the documentation process but also helps users of the document find specific items of information.

Phase 4: Review and Approval

Seeking Management Review and Approval. The last step in the planning process, as shown in block 9 of Exhibit 1, is the final review and approval that immediately precede publication of the plan. It is not the only review that occurs as the process evolves toward a final plan document. It should be evident that approval should be sought throughout the corporate technology planning process, particularly during the program development phase.

The final review should be performed by all managers affected by the plan, including various levels of R&D and operating management and appropriate members of senior management. The review can be performed jointly, with R&D and operating management working together, or it can be done by each function independently. If the planning process has proceeded with the required close coordination between R&D managers and operating management, the final review will be a quick look at the plans to which all parties have previously agreed. If the required coordination has been lacking, however, the final review will be a disaster.

THE PLANNING CYCLE

The corporate technology planning system calls for the annual production and publication of the many component plans that constitute the corporate technology plan. The technology planning cycle should be timed to coincide with the corporate business strategy planning cycle to allow the business planners to have their input to the technology planning process and to allow the technology planners to have their say in the business planning process. Publication of the corporate technology plan would then occur at about the same time as the business and corporate strategic plans are released.

R&D managers must demonstrate that they are as concerned with the success of the business as they are with the success of the R&D programs

Although the documentation of the corporate technology plan is published only once a year, the planning process is continuous. R&D programs and projects should be added, deleted, and modified during the year as the needs of the business and the company change. Unless a complete redesign of a particular technology plan is required, changes that take place during the year should be reflected in the next annual plan's documentation.

A NESTED SYSTEM OF PLANS

The corporate technology planning system comprises a nested system of plans, as illustrated in Exhibit 2. The basic building blocks of the system are the business unit technology plan and the general technology program plan. The business unit technology plan defines the technology strategy and program for a single business unit, and one such plan is prepared for each unit in the company. General technology program plans cover single projects or groups of projects that involve a technology that supports more than one business unit or division as well as project work in areas not related to existing business units. These documents are prepared by the appropriate R&D departments and distributed to all individuals and organizations with a direct interest in the plans.

The building blocks can be incorporated into any number of plan documents corresponding to various organizational or technological groupings in the company. For example, Exhibit 2 shows aggregations at several levels: R&D departments and divisions, operating divisions and subsidiaries, and the entire corporation. Fewer or more aggregations are possible, depending on the needs of the company.

In addition to producing business unit technology plans and general technology program plans, the R&D function must prepare its own departmental plan comprising all the business unit technology plans and general technology program plans for which it is responsible. In addition, the director of technology planning for the company may want to compile higher-level plans that summarize and consolidate the plans prepared by others in the R&D organization. The plans should cover a span of five years to conform to the planning cycles used for strategic business plans and enable easier cost and benefit calculations. The plans must include all programs, however, regardless of their duration.

The Business Unit Technology Plan

This plan is the cornerstone of the whole corporate technology planning system of plans. It brings together all the information gathered and processed in the technology assessment and program development phases of the planning process and identifies what is to be done and why, by whom, for whom, by when, at what

Exhibit 2
The Corporate Technology Planning System of Plans

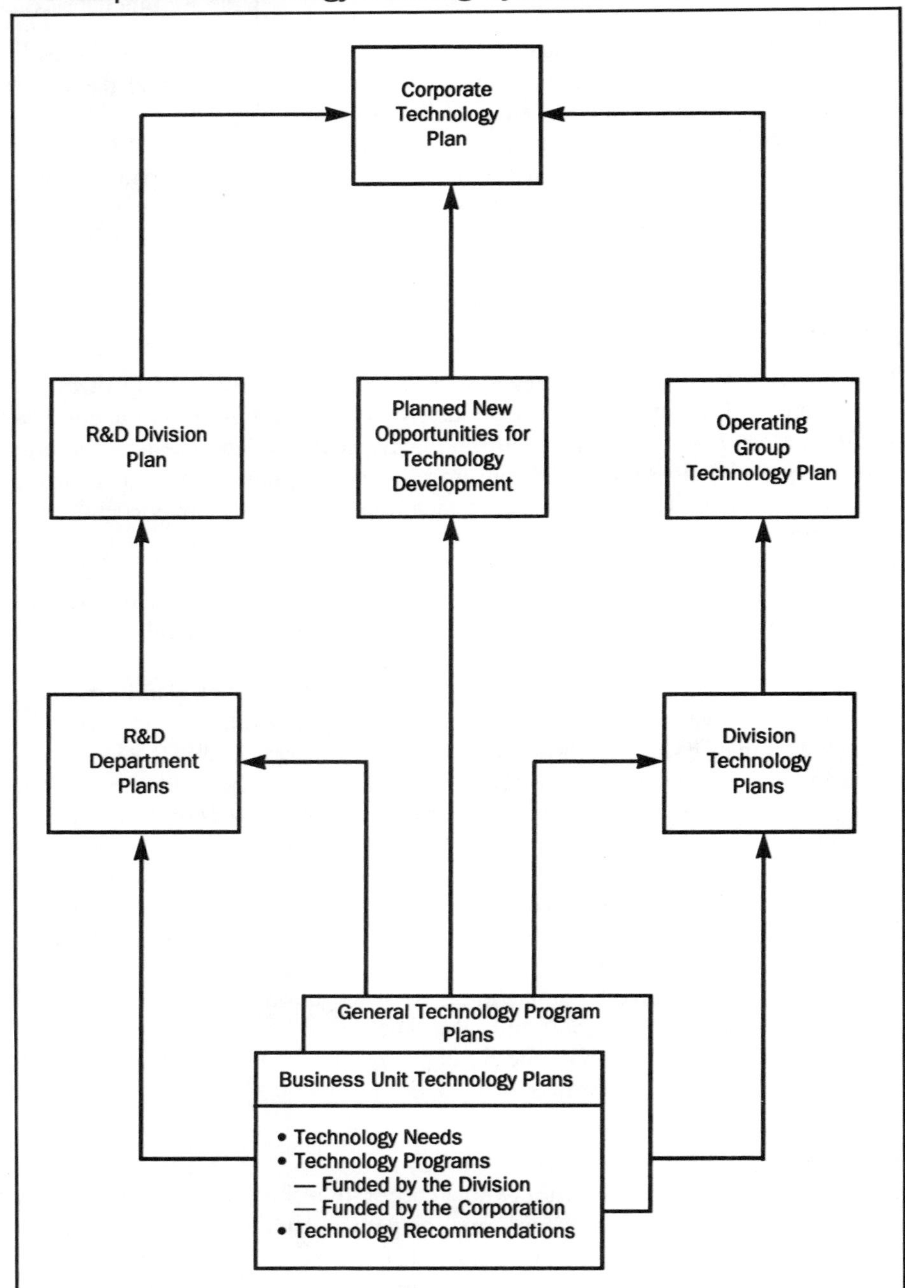

cost, and with what benefits. Because it is such a critical document, its content and rationale are worth looking at in greater detail. Exhibit 3 shows a sample business unit technology plan document in topical outline form. The following paragraphs explain the rationale for each section of that outline.

Section A: Executive Summary. This is a brief statement of objectives and conclusions.

Section B: Characteristics of the Business. The primary purpose of this section is to help those in the R&D organization who prepare the business unit technology plan understand the business they support well enough to be able to describe its salient characteristics in their own words. A secondary benefit is that it defines for the readers of the plan the nature of the business that is the subject of the document.

The subsections of this part of the plan describe the characteristics of the business, including its major product lines, applications, and markets; the nature of the products (i.e., commodity or specialty); past, present, and projected market growth, company market share, company sales and gross earnings (extending five years in both directions for past and future data); major competitors and their share; patent and licensing status; and the nature of any non-US business.

Exhibit 3

Sample Outline of a Business Unit Technology Plan

A. Executive Summary

B. Characteristics of the Business
   1. The Nature of the Business
   2. The Nature of the Products
   3. The Scope of the Business
   4. Major Competitors and Their Share of the Market
   5. Patent and Licensing Status
   6. International Aspects

C. Technology Assessment
   1. General Discussion
   2. Company's Current Technological Strengths
   3. Technological Weaknesses and Threats
   4. Development Potential of the Company's Current Technology
   5. Achieving and Maintaining Technological Leadership
   6. Technological Opportunities

D. Business Strategy and Required Technological Support

E. The Technology Program
   1. Summary of Required Technological Support
   2. Currently Approved Program Versus Required Support
   3. Estimated Benefits and Costs
   4. Project Lists

Section C: Technology Assessment. This section outlines for each business unit the strengths, weaknesses, problems, threats, and opportunities in the company's technology as compared with its competitors and with the state of the art. Such an analysis serves to identify what needs to be done to strengthen or optimize the company's position. This assessment applies to products, processes, and, when appropriate, underlying technologies.

As the sample outline indicates, the technology assessment section has several subsections. The subsection dealing with the company's current technological strengths lists the major strengths and identifies what needs to be done to maintain them. This information requirement is included in the plan format to guard against the very natural and common tendency to take for granted the company's strengths and thus ignore possible threats to the prevailing corporate well-being in terms of growth, development, and profitability.

In the subsection on technological weaknesses and threats, weaknesses are defined as current problems, and threats as problems that may arise later. R&D planners must identify significant problems and threats and state what the company needs to do to solve or avoid them.

The subsection on development potentials of the company's current technology requires the R&D planner to consider the gap between the current state of the company's technology and the maximum potential of that technology. This analysis can be done for products as well as processes and can be very critical in indicating whether efforts should be devoted to improving the current technology or shifting to a new technology.

The next subsection, on achieving and maintaining technological leadership, answers the following questions:

- Which companies are considered the technological leaders in the industry?
- What does each of those companies do or have that gives it a leadership position (e.g., product design, processes, R&D, technical service, number of patents)?
- Which technological programs should the company undertake to become the technological leader or to maintain its position if it is already the leader?

The subsection devoted to technological opportunities is included to prevent planners from overlooking favorable opportunities by focusing too narrowly on problems and threats. Here, the R&D planner must identify technology options that, if properly exploited, could give the company an advantage over its competitors or that could lead to profitable additions or modifications to the business unit's products or processes.

Section D: Business Strategy and Required Technological Support. This section of the plan identifies elements of the business unit strategy that have technological implications as well as technology development work that must be undertaken to support that strategy. The main objective here is to formalize in writing any agreements between the business manager and the R&D planner to ensure that both functions are operating with the same goals.

although cost/benefit analyses are a lot of work, they are an absolutely essential part of responsible technology planning

In developing this section of the business unit technology plan, R&D planners cannot allow the absence of a clear-cut business strategy to deter them from preparing this portion of the plan. Rather, the planner should take the initiative and outline for business managers who are still developing their strategies the assumptions the planner is making about that business strategy as well as any technology development work that is planned to support that strategy. Such a move usually results in some clarification from the business managers. The planner can proceed only on the basis of companywide assumptions for the balance of the business strategy, because it usually doesn't make sense to terminate technological support until the entire strategy has been finalized.

Section E: The Technology Program. This section of the plan indicates what will be done to meet the technology requirements identified in the earlier sections of the plan, and it provides broad benefit and cost estimates. The section has several subsections.

The subsection on summary of required technological support lists all major areas of required work, not only those that have been approved. In the subsection devoted to currently approved program versus required support, the plan's authors indicate whether the currently approved program covers all the technological requirements identified in the previous subsection. Currently approved programs are those that are underway or that have been approved to begin at any time during the five-year planning period. If a currently approved program does not address all the requirements, the planners must identify the exceptions and the reasons for not addressing them. Part of the planners' responsibility as technical advisers to the company is to speak up when they believe the company is missing a significant opportunity by not funding certain R&D programs. This portion of the business unit technology plan gives planners the chance to say so—firmly, clearly, and diplomatically, in the form of a recommendation.

The subsection on estimated benefits and costs presents the summary cost/benefit analysis for the entire R&D program outlined in the plan, including programs that are proposed but not yet approved. It sums up benefit and cost estimates for each of the individual projects identified in the next subsection devoted to project lists. Benefits are stated in dollars of additional sales and sales losses avoided as well as additional gross earnings resulting from these sales and from cost-reduction and productivity improvement programs. The sales projections and gross earnings ratios used to calculate benefits must be obtained from and accepted by the responsible business managers. Cost estimates are provided for development (direct and capital expenses), technical service, and capital expenses required to implement R&D program results. The project lists constitute the final section of the business unit technology plan. These lists itemize the individual projects that, taken together, constitute the R&D program in support of the business unit.

MAKING THE SYSTEM WORK

As important as these skills are, however, the corporate technology planning system primarily needs strong leadership in the technical as well as nontechnical departments. The system works most effecftively under R&D management that believes that R&D can best support the company through initiative and strategic management. R&D managers must believe in the benefits of planning and accept this responsibility as an important activity to which time and effort must be devoted. By recognizing that R&D is responsible for involving the operating side of the company in technology planning, those technology managers who are able to see the world through the eyes of the operating managers and who can talk the same language will fare best in getting others to participate in the planning process.

Business managers generally believe that financial benefits are possible from R&D work, and when they work jointly with R&D to develop and substantiate these benefits, that belief is strengthened. For their part, these managers can contribute to the planning process by developing clear business strategies and making the effort and time to interact with R&D as an essential and equal partner in the corporate enterprise.

It is common for systems such as the one described in this chapter to work better on paper than in the real world; the corporate technology planning system is no exception. Despite considerable success in bringing about the conditions required to make the system work, some R&D departments and some of the business groups fared better than others. However, all R&D departments, all business groups, and the company as a whole derived significant benefits from working with the system, including the following:

- R&D's effectiveness and efficiency improved markedly because the right projects (i.e., projects that met the company's goals) were high priorities—Because there was a plan against which to measure and adjust progress, R&D could continue its level of effectiveness.
- Both the R&D and operating groups were able to manage R&D's support capabilities much more strategically—The R&D agenda became an integrated set of programs clearly linked to the company's broader business objectives rather than a heterogeneous collection of individual, loosely related projects.
- R&D rose to the challenge of adopting a proactive operations role rather than a passive support role and thus applied its special and unique talents on the planning and implementation of the company's strategies and operations.
- In their negotiations and deliberations with operating business managers, R&D managers became far more able to state their case with clarity and forcefulness because the planning system had forced them to think issues through completely.
- The rest of the company, which tended to denigrate R&D before the introduction of the corporate technology planning system, gained a significant understanding of and appreciation and respect for R&D and began to deal with R&D personnel as equals.
- As a result of acquiring increased respect in the company, the R&D group enjoyed greatly improved morale and, along with it, improved productivity.
- The corporation as a whole made more effective use of its R&D resources, which must have had a beneficial impact on the bottom line.

The most significant conclusion that can be drawn from the corporate technology planning experience is that R&D managers should devote as much effort to planning what they should be working on as they dedicate to planning individual projects. If they do so, even imperfectly, such planning will yield many benefits for them as well as for their companies.

if the required interdepartment coordination has been lacking, the final review will be a disaster

R&D SLEUTH-SAYERS

Developing a Strategic R&D Plan

Long-term R&D planning requires insight into three factors: the corporate charter, the customer, and the potential of new technologies. A structured program for investigating these three areas can yield significant information on which to build future technical programs.

Kenneth Jarmolow

Many R&D organizations fail to grasp the basic rule of their operation: The customer must be satisfied. R&D's most steady customer is none other than senior management, and the best way to please management is to provide R&D results that will support the company's products and the company itself currently and in the future. When an R&D organization is considering a long-term plan (i.e., 5- to 10-year), it must make crucial decisions on which products to support and how to support them even though these products haven't been developed yet, and it must do so with consistency and an acceptable degree of accuracy in forecasting.

As a rule, this problem has not been addressed seriously because customary, straightforward approaches have seemed sufficient. After all, R&D planners have reasoned that they know the business of the company and are certainly well versed in the science and technology supporting it. Their solution is to simply extend last year's plan, which was based on the plan from the previous year. In this way, the R&D department hopes to bank on experience and move into new areas gradually and with little risk.

This approach fails because everything changes—the market, the competition, and the progress of science, especially overseas. Senior management must ensure that the R&D organization not only accommodates these changes but, preferably, provides products clearly superior to the competition's. This allows the rest of the company to get on with engineering, manufacturing, and marketing, with the best possible basis for commercial success. Such markedly improved products are rarely based on mere extensions of current technology.

Much of the technology underlying the products that will be created during the next 10 years is being developed now—and probably not in industrial laboratories. Technology developed in industrial laboratories, although worthwhile for other reasons, is not usually the basic technology that other companies will use to develop new products. An R&D organization's challenge is to know its company—its character, capabilities, and market—and to identify the technologies that will enhance its future. R&D must fit the technology to the company.

During 1987 and 1988, my company required a lot of information available neither in the library nor in the laboratory. The company needed to determine

---

KENNETH JARMOLOW **recently retired from the presidency of Martin Marietta Energy Systems Inc, where he was responsible for the research and production activities of the Department of Energy at Oak Ridge TN. Just before retiring, he was vice-president for technical planning for the corporation. Previously, he was vice-president and director of research and development for Martin Marietta Corp, Bethesda MD, for 20 years.**

and understand the latest technology—basic and applied—that might affect its future. To evaluate that impact, it was necessary for R&D management to understand the company better—particularly the plans, expectations, and hopes of senior management as well as the company's customers and competitors.

FINDING THE CORPORATE CHARACTER

To achieve this objective, the company's R&D department conducted an inquiry that was the analog of a scientific investigation. Extensive interviews were held with the experts, and their responses to R&D investigators' questions are summarized in the following sections.

The Senior Management Perspective

Senior management's input generally indicated that only the technology that the company would attempt to develop into products is worth pursuing. To understand what these products are, it is necessary to understand the company's gestalt (i.e., its driving force), long-range plans and strategy, and corporate statements.

Such pronouncements frequently describe only objectives; they do not provide the means to achieve them. They are more a vision of ideal management than a plan for running the enterprise. The R&D department's interviewers (who should be experienced, senior-level managers) must discern and dissect the real meaning of such comments. Corporate management is predominantly concerned with shareholder expectations, return on investment, return on equity, stock values, the views of the board of directors, and the current balance sheet and income statement, but not with long-term research and development. Interviewers must ask questions incisively to get the answers they need.

The Marketing Team Perspective

The ideal marketing team combines a deep understanding of corporate management with extensive knowledge of the industry, the competition, and customers. Such a team can be a useful resource in determining areas of research that will be attractive to corporate management and that will fall within the corporation's capabilities to develop and in supporting the development of competitive new products.

The Federal Agency and Think Tank Perspective

Organizations that analyze global science and technology opportunities—especially as they affect such organizations as the National Aeronautics and Space Agency (NASA) and the Department of Defense (DoD)—must analyze R&D opportunities in much the same way as the R&D organization must for its company. Their vast resources and connections are well worth enlisting, and they are willing to help. Although such institutions are not sources for corporate information, they can provide insight into emerging technologies and their 5- to 10-year potential. Because they are neither in the research trenches battling the mysteries of science nor in the marketplace, however, their criteria should always be evaluated for its applicability to commercial enterprises.

some R&D staff exploit the information available in academia and national labs effectively; others do not even visit these remarkably cooperative institutions

The Customer Perspective

It is impossible to sell products that customers don't know they want. Customer interviews and focus groups are invaluable for comparing the company with the competition and obtaining ideas on how the company's products could serve customers better. The R&D organization should hear customers' responses firsthand, because at least some of the requested improvements will come directly from R&D efforts.

The University and National Laboratory Perspective

Some R&D departments effectively exploit the information available in academia and national labs; others do not even visit these remarkably cooperative institutions. Those companies that do take advantage of this relatively inexpensive and accommodating information source reap handsome rewards.

Open discussions with scientists from these organizations can reveal not only the results of their most recent findings but the investigator's expectations, limitations, enthusiasm, skepticism, and depth of effort. They may also provide the most informed vision of the technological future.

The R&D Staff Perspective

The corporate R&D staff is probably the best single source of ideas concerning science and technology development for the company. Selected R&D staff should accompany the laboratory chief when visiting other sources to ensure that the right questions are asked and that presumptions are challenged, to offer information that makes the meetings valuable to the hosts, to discuss their own relevant work, and perhaps to encourage future collaboration. In addition, including staff members makes them a party to the plan they are to implement. They are basic stakeholders and must therefore play a principal role in the exploration and decision-making process.

Articles, Books, and Other Literature

In conducting this research, certain articles took on new significance as other information sources were interviewed. Many agencies and consulting companies have studied R&D trends and have published their findings freely; these documents were invaluable in improving the R&D investigators' perspectives, especially because several of those agencies might become future customers.

EVALUATING THE RESEARCH

From the beginning, a screening device was required to limit the vast data that threatened to overwhelm the whole examination process. Therefore, the R&D

investigators established the following criteria to ensure that the technology explored would be appropriate and eligible for development during the next 10 years:

- The technology must be embryonic, uncertain, and possibly controversial, at least in regard to particular applications.
- The technology must be sound scientifically (i.e., its basic scientific character must be understood and must point clearly to important future applications).
- The technology must not be preempted by the competition, at least in terms of the probable applications.
- The company must be able to develop, manufacture, and market products created with this technology, perhaps with partners.
- The technology must be useful in several of the company's potential products.
- The technology must be revolutionary rather than evolutionary—It must not be an extrapolation of existing technology.

The examination process took about a year and was conducted principally by the R&D laboratory manager, a colleague trusted for his perceptiveness and scientific fluency, and by a number of specialists, who were consulted as outside experts. Each discussion with members of corporate management lasted several hours; to clarify issues, some senior managers were interviewed more than once. Discussions at universities usually lasted several days; at some universities with extensive research programs, they lasted as long as a week. Over all, about 240 authorities were interviewed personally, including domestic and foreign university, agency, and industrial R&D officials. The intense dedication to technology development around the world indicated that technology would continue to advance rapidly and that our customers expected advanced applications to be introduced throughout the next decade.

Although the details of the findings of this particular study must remain proprietary, the essential elements can be described. These are discussed in the following sections.

The Corporate Charter

The results of the discussions with members of management—from the CEO and senior board members to senior technical managers at the company's several operating companies—can be summarized as follows. The company is a systems company; however, in some programs, the products are integrated vertically, and that vertical integration may increase. The company is particularly responsive to customer interests and delivers on the promises made. It wants to develop new technologies that enhance current and new systems, but will not overlook profitable uses of older technologies. Production efficiency, quality, and reliability will be emphasized as the technology development strategy and the long-range operating plan are formed.

The company will not diversify beyond the current product lines but, under certain conditions, would consider applying current products into new markets.

Corporate growth will arise from the company's existing strengths; management anticipates no major corporate changes, acquisitions, or mergers. By the year 2000, the company should be largely information-technology oriented—that is, its products will be characterized by their skillful, real-time acquisitions capabilities, by their effective application, and by integration of information.

in forecasting the development of embryonic technologies, surprises and disappointments will occur, and final applications will frequently differ from those expected

Marketing and Customer Data

Research revealed that customers' plans either are highly specific and rely only on evolutionary improvements in current technology or are overly general and call for improved performance without specifying the technical advance necessary to achieve it. Customers incorporate a new technology in their plans only after the technology is largely developed. Thus, advanced technology development and customer long-range programs, though ultimately interdependent, proceed separately. Nevertheless, customers demand and expect substantially improved performance, from whatever source. In the end, the new technology will dominate. The R&D investigators selected around 70 technical topics from 12 major customers for further study.

The information provided by customers and the marketing information gave the R&D investigators a view of what customers—and therefore management—were looking for. This general type of information was intended to determine the technology to support products three or four generations into the future.

The Technical Staff and the Scientific Community

The data from the corporate R&D staff was extensive; the findings were sometimes inaccurate, sometimes incisive and hard to take, but always revealing. This information was particularly sensitive and must remain confidential. It touched many nerves because it revealed laudable efforts as well as imperfections; to the R&D staff, the imperfections seemed to overshadow their achievements. An evaluation of management decisions was included in this analysis.

University and National Laboratory Feedback

By far the greatest part of the survey concentrated on the probable state of technology in the mid- to late 1990s. As noted, this information was gathered in face-to-face meetings, usually at the investigators' laboratories, during which the scientists displayed not only their current work but also their expectations for extensions of this work during the next decade. They compared their work to the state of the art and to the work of colleagues around the world. R&D investigators tried to ask only significant questions, to acquaint the scientists with related work they had examined elsewhere, and to apprise them of the R&D department's relevant work.

These meetings were usually informal and animated, though all were worthwhile. The differences in these scientists' visions usually reflected different levels of optimism and expectations of necessary effort. The interviews didn't create the almost impossible information overload that was anticipated because, by the time we started this phase of research, we had narrowed the focus of the discussion to those areas that seemed to match our interests. We then explored those areas in depth.

All in all, R&D investigators visited 11 employees of national laboratories, 36 of federal agencies, 43 of US universities, 67 of foreign universities, 15 of US corporations, and 28 of foreign corporations. In every case, those interviewed were gracious, thorough, and pleased to answer questions on the investigators' terms. Perhaps they enjoyed talking about their work because they have confidence that the technology is likely to become important, and they welcome willing converts.

Before these meetings with scientists, the investigators tried to learn about their work so that they could interact productively and not waste their time. In addition, the investigators represented a major corporation that could offer them potential support and collaboration; these possibilities were considered during many of the meetings.

CREATING A NEW IDEAS BANK

The report on the findings of this study recommends new programs in areas that were hardly explored in the corporate laboratories at the time. The report also includes some important observations that should improve the health of the laboratory's overall R&D program. The company's senior technical personnel recognized that the emerging technologies examined in the study (and inadequately covered in the R&D program) might support some of the systems and products recommended for development. Thus, the case for new R&D programs developed easily.

In forecasting the development of embryonic technologies, surprises and disappointments will occur, and final applications will frequently differ from those expected. The uncertainty lies in the timing and limitations of each technology. In planning a decade ahead for the company, the R&D department made the most reliable choices. These choices will be assessed continually as the programs proceed; the direction of each program depends on consistently positive and promising results.

The report on this study recommended technologies and suggested their intensive development in the appropriate divisions of the company. Other observations, unique to the company, were also contained in the report, as were much technical data and many suggestions for the near-term improvement of the R&D program. Thus, the study fulfilled its original purpose.

Even more was gained from this study, however, than was originally expected. Foremost was the cooperation, enthusiasm, and total professional dedication of the several hundred scientists the investigators had the great fortune of meeting. The investigators were also pleased by the many customer organizations that discussed their interests and expectations; this source of essential but frequently overlooked information is available to all serious investigators. Within the company, many colleagues were supportive, awaited the results eagerly, and volunteered information and assistance; the corporate managers were frank and responsive and provided confidential reports and insights.

if the technology it is exploring is real and important, a company must dedicate substantial effort to achieving early and continuous results

Overall, the information received was extensive, candid, and interesting. In addition, the study yielded thoroughly practicable material; its data and revelations even overturned several of my own convictions.

Although the recommendations are only as good as the data and the insights of the investigators, the essential uniqueness of the study's results springs from its structure—in particular, the investigative team's attempt to determine the appropriate blend of information about the nature of the company (as projected for the next 10 years by its senior executives), the character of the marketplace over the same period (as seen by the company's own marketing executives and its major customer management and technical personnel), and the emerging scientific and technological opportunities (as seen by the scientists and engineers already preparing the enabling technology). This process worked primarily because of the candor and cooperation of all concerned and because the necessary time and expense to go directly to the source for each element of information needed was properly allocated.

Every successful industrial laboratory manager knows that long-term R&D is only one leg of a total R&D program; the other leg is short-term R&D. By their nature, long-term programs experience delays, failures, and disappointments. If the technology it is exploring is real and important, a company must dedicate substantial effort to achieving early and continuous results. Current and frequent results constitute an important way for a laboratory to earn its daily bread and become widely respected in the corporation while struggling with the greater challenges of the future.

LEAVING NOTHING TO CHANCE

Creating a Strategic Technology Plan

The benefits of strategic technology planning are well known, and many technology managers are eager to work with a strategic plan. The planning process, however, is not a simple matter. It requires making choices and entering into a process of managed evolution. The authors have faced some of these choices, and the evolutionary steps they have taken are outlined in this chapter.

Adrian Timms and William D. Torregrossa

Why a company decides to adopt technology planning determines how it goes about it. In many cases, the adoption of strategic planning systems is motivated by a crisis; for example, a CEO is unfamiliar with the technological basis of the company's business and hence frustrated because the technology development process isn't working. In such cases, organizations are often willing to pay high consulting fees to progress rapidly up the strategic-planning learning curve.

In our case, readings in R&D management led to an interest in technology planning. There was no crisis, just the desire for excellence and the recognition that technology was becoming an increasingly important means to gain competitive advantage. We felt that the future role of technology in the relatively low-tech food products industry was too important to be left to chance. We opted to use internal resources to increase the staff's technology-planning skills.

TAKING THE FIRST STEPS

The approach we took to implementing strategic planning emphasizes home-grown learning, which invariably entails making mistakes and having to live with certain inefficiencies during the early stages. For example, it took a detailed analysis of the technologies required to produce existing projects to drive home the importance of structured planning for decision making and of finding the appropriate level of planning detail. This investment of analytical sweat brought to light things that should have been obvious, including crossovers in technical functions, new opportunities, and wasteful practices.

ESTABLISHING LEADERSHIP AND SUPPORT

Although the push for strategic technology planning does not have to come from the top, having the support of senior executives is vital. In many cases, the CEO

ADRIAN TIMMS is the associate manager at the Hershey Foods Corp Technical Center in Hershey PA.

WILLIAM D. TORREGROSSA is the director of R&D planning and analysis at the Hershey Foods Corp Technical Center.

is a driving force in implementing strategic technology planning. In our case, the impetus for planning did not originate at the executive level, but we did have senior-level executive support. Senior management was receptive to our efforts, but the burden was on us to sell the planning process and enlist the support of all levels of management.

The company's current CEO encourages investment in R&D, which allows R&D management to focus on determining the right mix of R&D activities rather than continually justifying R&D projects. Knowing that support for technology planning is maintained only when executives feel that there is value in the level of investment, we instituted systems to evaluate R&D investments in a business manner. As part of this process, financial managers work with the development team, which helps to sell the R&D efforts to senior executives. An additional benefit of these managers' involvement is that they provide a much needed alternative perspective on strategic technology planning, which is particularly critical during the early stages of the process.

Selling Concepts Internally

We gave one manager, who reported to an R&D director, full-time responsibility to determine what was needed in the program, to experiment with alternative approaches, and to implement one that suited the company's needs. Having one person devoted full-time to this job ensured that the planning would take place.

From the outset, we took a customer-oriented approach to technology planning (the customers being the internal business managers). When any manager is brought into these activities, individual needs and motives must be addressed. We selected key players to participate in the programs and to champion the system in their areas. These individuals each have their own reasons for backing the program, and their needs are addressed accordingly.

Hurdling Obstacles

Several obstacles were encountered in the form of attitudes and questions. An R&D manager looking for support for such a program should be prepared to address these types of issues. Some typical questions were:

- If we are already a successful company, why do we need strategic technology planning?
- Shouldn't any competent manager know instinctively what an operating unit's technology needs are?
- Isn't technology planning just for the high-tech guys? Our industry is low tech.

One way to address these questions is through seminars and training workshops presented by technology planning consultants—an effective use of outside consultants. To some extent, however, the key players need to be supportive of the program beforehand so that they can deal with objections from their staffs early in the process. We have found that a combination of one-on-one meetings with key managers and general awareness-building techniques is helpful here. One such teaching device is the use of the language of strategic technology management in meetings and correspondence.

strategic technology planning should be a routine management activity that is intrinsic to an organization's culture

Planning from the Top Down or Bottom Up

The majority of technology planning practitioners are consultants. Perhaps because of the source of their incomes or because their experience has taught them that it is effective, consultants usually address senior management directly and put the burden on these executives to provide functional decision makers with the motivation and commitment to perform the technology planning activities. There is no doubt that this approach is expedient, especially in the context of a relatively brief consultancy.

Much of the literature on technology planning says that the process must be top down—that is, that the impetus must come from senior management. We subscribe to this view insofar as it is essential for the motivation for technology planning to be shared at the top. Our view of strategic technology planning, however, is that it should be a routine management activity that is intrinsic to an organization's culture, not a one-time process prompted by senior management. For continuous strategic planning that obtains input from the practicing engineers, we aim to strike a balance between top-down and bottom-up develoment of planning, and we pay deliberate attention to selling concepts at the engineers' level.

Focusing Resources at the Shop Level

Organizing personnel and activities in groups associated with chosen technology areas not only helps to focus resources on strategic thrusts but facilitates grassroots, bottom-up planning. Individuals with the commitment to lead the group should be encouraged to make the plan happen. Two methods for accomplishing this are discussed in the following sections.

Centers of Excellence. One of the aims of technology planning is to identify key areas for strategic thrust. Some of these are core areas of the company's business, and in such cases we have not waited for technology planning to start setting them up. Instead, we use centers of excellence as vehicles for planning and coordinating development related to a specific core technology.

A center of excellence is not a building; it is an organizational construct, a multidisciplinary congregation of people who share the same strategic objectives for that technology. The mission of a center is twofold: to maintain strong links between its technical and strategic business goals and define its long-term technical goals, and to continually pursue world leadership in that technology.

Even in an organization delineated by functional departments (e.g., by academic discipline), a task force approach can be used to bring collective focus to a particular technology. The center is not a task force because it is permanent until such time as that technology is divested; it can even be treated as a budgetary cost center. Because the technologies chosen are of primary importance to the corporation, divestment is unlikely. We make it incumbent on each center, however, to be on the lookout for potential substitute technologies and to plan accordingly.

Centers of excellence give us opportunities to bridge departmental, rank, and divisional boundaries among participants. These bridges not only lead to better strategic vision but disseminate technology information and planning techniques to a wider audience. Our current stage of evolution involves only technical representatives in the centers, but we plan for them to ultimately include nontechnical business managers.

Through the centers, we are able to perform strategic technology planning even before the adoption of formal planning processes. The centers will continue to function effectively and to become an increasingly integral part of the overall strategic planning system.

Technology Champions. Technology champions actively take on leadership roles in a technology development group, playing a role analogous to that of a brand manager. A brand manager is charged with developing and deriving maximum benefit from a brand asset; a technology champion does the same for the technology asset. The champion must maintain development impetus, forecast and strive for the future state of the art, promote effective use of the latest technology, and use all available channels to sell management on the need for continued investment in that technology.

MANAGING AN EVOLUTIONARY PROCESS

Implementing a technology planning program is an evolutionary process, and it should be managed as such. We drew up strategic objectives for implementing technology planning as if the planning process were a major development program in itself. This plan should evolve as the technology manager developing it learns more about the plan and process; however, in its early form, it will help the managers prepare for the successive stages. We have emphasized the evolutionary nature of planning activities and regard the planning workshops as a laboratory for developing the appropriate planning techniques.

One of our initial actions was to put down on paper what we had learned from literature and seminars as part of a grand design for an overall strategic technology planning process that was tailored to our organization. At first, we intended to implement this process by starting with one small business unit and introducing the process to all business units sequentially; however, on examining the potential culture shock that would ensue, we opted for a phased, evolutionary approach. We felt that this would permit the development of sound methods built on basic principles and enable us to avoid working with untried techniques straight from the planning department.

Only as we became more experienced with the techniques did we find it appropriate to involve non-R&D managers; if the multidisciplinary team concept is to be realized, their eventual involvement is imperative. During the formative stages, we analyzed and drafted the plans and then reviewed them with the R&D staff. After staff members had acquired additional formal technology planning skills, they began to perform some of the planning activities themselves, and R&D management facilitated the overall process.

two factors will hinder efforts to institute technology planning: silence, and activities that are inconclusive or not useful

Structuring the Program

Technologies readily lend themselves to hierarchical structure and thus many levels of detail. For purposes of scanning technology and acquiring intellectual property, detailed distinctions among technologies may be desirable. We found fairly quickly that the level of detail appropriate to strategic allocation of resources can be (to the technologist) embarrassingly simple.

Much the same is true for projects. A laboratory does not need to be unusually large to find itself handling more than 200 projects. If these projects have been independently instigated by a wide variety of customers, the technology manager will probably need to group hundreds of projects into a manageable number of programs. As our strategic planning of R&D becomes more big picture–oriented, we expect the definition of programs to lead to the definition of projects.

From an analytical viewpoint, there is a distinction between technologies and projects as they relate to business goals. A one-to-one relationship among products, technologies, and R&D programs rarely exists; the concept is a simplification frequently used to describe technology planning techniques. In assessing the needs of these development programs, the technology manager is bound to encounter analytical complexity before learning to differentiate these entities and establish their various interrelationships.

Evaluation techniques should be appropriate to each strategic unit. Technologies can be regarded as capabilities—as a set of strategic assets rather like tools that could be put to use through projects. Most R&D projects apply technological capabilities to producing measurable commercial results rather than to developing the technology itself; other R&D projects are dedicated to furthering technological capabilities.

After projects are divided into those that develop and those that merely apply technology, measures of value, risk, and reward need to be consistent with the types of projects. For example, the concept of return on investment for a developing technology becomes strained because return on investment should apply only to specific applied technology projects; however, individual technological capabilities can be evaluated for their relevance to future business needs and opportunities.

The Triad. The triad is a grouping of business goals, technologies, and R&D projects. In the center of the triad are various planning techniques that emphasize the links among the three strategic entities. The techniques used to evaluate projects and technologies are at opposite ends of a spectrum of technology planning methods. Technology management consultants typically have specialized expertise associated with different points along this spectrum. For example, some consultants are particularly adept at applying portfolio management techniques to R&D projects.

Design Principles for Technology Planning

We decided at the outset that the following ground rules would govern any planning systems we adopted:

- The planning process and tools must be simple and easy to operate—The logic of the planning system should be straightforward and useful; meaningful decisions should result.
- There must be multifunctional, multilevel corporate involvement in the planning process.
- The role of technology planners must be to catalyze and facilitate the process of planning, not to make strategic decisions.
- All technology planning must be oriented to product attributes deemed significant by the consumer; that is, attributes that drive the consumer's willingness to purchase the product must be the focus of the planning efforts.

The Need for Visible Output

Results of the technology planning process should be reviewed at regular intervals. These results should be shared with the development team, especially if they help to communicate technology planning principles.

While the technology manager is developing planning methods, the internal customers should be involved in and kept abreast of progress made. Two factors will hinder efforts to institute technology planning: silence (which slows momentum) and activities that are inconclusive or not useful. Especially with R&D colleagues, we feel it is necessary to keep technology planning in their minds through frequent involvement in useful and productive activities. A planned evolutionary process will help to accomplish this.

Technology-monitoring activities are useful in this regard. We systematically scan patent data bases and copy relevant items to interested parties, provide reports to those charged with gathering competitive intelligence, and make presentations of company and technology profiles in selected areas. We plan to publish regularly an awareness newsletter for technical and nontechnical personnel to highlight trends and emerging technologies. All these efforts provide necessary input into technology planning and serve as useful, visible evidence of planning activities.

The Technology Planning Toolbox

We found that setting up a technology planning system is expedited when certain tools and skills are available, particularly a data base, trained facilitators with effective interactive skills, expert systems software, and graphics analysis packages. These tools and skills are discussed in the following sections.

Data Base Accessibility. R&D management will need to draw on a store of historical and planned project data to make sound decisions on products and processes. A data base is the ideal medium for storing and experimenting with different treatments of the data.

training in such techniques as group decision making and focus groups is invaluable to technology planning activities

In this case, we already had an ongoing data base on which we planned and tracked the costs of R&D projects. The data base had been established primarily to justify budgetary resource requirements for that year rather than to perform strategic planning. Nevertheless, it served as a launching pad for quantitative analysis of our current project portfolio. As our capabilities and needs in strategic technology planning grow, so do the systems and structure of our project data base. This interdependence is reinforced by involving the data base administrator in the development of planning systems.

If technology managers do not currently maintain such a system to track project costs in their organizations, developing one to meet both strategic and operational needs will be both a challenge and an opportunity. The challenge is to design the system with enough flexibility to support experimental data use; the opportunity is to make planning easy and effective by creating such a system.

For example, one data base application in strategic technology planning is resource allocation, which entails coding projects for reporting purposes; such data as cost estimates are totaled and formatted to produce a graphic output. Much of this task is iterative and well suited to a data base. It is recommended that someone with data processing experience and skills be included on the technology planning team.

Interactive Skills. Planning is a creative activity that occurs in a workshop-like process, so we trained ourselves in group facilitation. Training in such techniques as group decision making and focus groups was invaluable when applied to technology planning activities. Without these techniques, it would have been extremely difficult to achieve the desired results.

Consultants are often used for this facilitating role; however, with proper training, in-house facilitators can be developed and used to lead technology planning as well as other problem-solving groups.

Expert Systems Software and Graphics Analysis Packages. In addition to a data base, expert systems software can help a planning team get through evaluation without the need for a full-time domain expert. Such software can assist managers in analyzing their projects, defining objectives, and focusing resources better. For example, an excellent system for project portfolio analysis is the one developed by Strategic Decisions Group (Menlo Park CA). In addition, it is helpful to present information graphically. Group discussion is greatly facilitated by diagrams and visual representations.

EXPLORING THE CONSULTANCY OPTION

In developing a strategic technology planning process, immense value can be gained from using technology management consultants. The extent to which consultants are used is dictated by the budget resources available and the management team's willingness to learn from its mistakes.

A variety of high-quality consulting companies specialize in technology management. Because of the strategic nature of technology management decisions, these consulting services are priced accordingly. If the six-figure fees that are standard for help in making R&D management decisions constitute a minor part of a department's budget, this route is an excellent way to rapidly gain knowledge of a variety of techniques. If handled correctly, the use of consultants also resolves some of the political issues that arise when multiple functions are involved in the planning activities.

We have corresponded with many companies in other industries concerning the implementation of planning systems. If their experience is typical, technology managers should be prepared to engage several consultants before finding a good fit with their own needs. As in any other field, consultants vary in their areas of expertise and style.

When such factors as gaining support and establishing new responsibilities and relationships are added to those of developing the actual planning processes, strategic technology planning becomes a full-time activity. To accomplish these tasks using only outside consultants, it is necessary to engage the consultants on long-term retainers because the selling process is gradual and requires a continued presence.

However, consultants have a role to play even if a manager chooses to proceed with the planning process using internal resources. As we learn more about the distinctions between different technology planning approaches, we have been able to define more specifically the particular hurdle that we need a consultant to help us overcome. In our experience, condensing the projects list to the appropriate level of detail—that is, from many small projects into a few strategic programs—would not have been effective use of a consultant; however, getting an external audit on our portfolio would have been. The external audit would have helped us maintain a realistic view of our competitive position and the true value of our mix of R&D programs.

CONCLUSION

Getting an organization to realize the benefits of strategic technology planning requires the technology manager to oversee the process of evolving its planning systems. This applies whether the company is implementing such planning where none exists or improving existing technology planning systems. We hope technology managers can use the points we have covered to help formulate strategies for achieving this goal.

BACK TO THE FUTURE

Conducting a Corporate R&D Retrenchment

During the past 10 years, Weyerhaeuser Co rapidly expanded and diversified its business. Then, in quick succession, it was forced to downsize, refocus on core areas, and retrench, in agreement with corporate strategy. This sudden change in corporate policy required realigning corporate R&D and technology commercialization quickly and cost-effectively.

Mark T. Hehnen
Norman E. Johnson
Edward L. Soule

Since April 1989, Weyerhaeuser Co has been changing its direction. To increase shareholder value, it is currently seeking to regain the leadership position in the forest products industries. During the previous five years, the company had shifted its focus from developing new products to seeking manufacturing excellence in the production of high-volume, commodity goods. Although the rationale for this shift is familiar to industries throughout the country, an understanding of how the switch affected Weyerhaeuser's corporate research and technology commercialization efforts, which had been strongly supported from 1985 to early 1989, may prove valuable.

A less-than-desirable performance during the first half of the 1980s prompted Weyerhaeuser to decentralize in 1985. It became a quasi holding company, shifting attention toward customers and the marketplace and away from large, raw-material resources. Soon afterward, research, development, and engineering were realigned to satisfy this new corporate structure. Whereas the focus of research had been on process, it was now on developing new products to meet customer needs.

What was once a centralized research, development, and engineering organization was split into four components; three were to serve the three main sectors of the newly reorganized company, and the fourth was formed as a holding-company research, engineering, and commercial development unit. That unit can be characterized as follows:

- Its mission was to help Weyerhaeuser gain strategic and competitive advantage through the use of technology.
- Its goal was to develop, within three years, many new business and product opportunities in all states of commercialization for various Weyerhaeuser businesses.

MARK T. HEHNEN is director of technology commercialization at Weyerhaeuser Co, Tacoma WA.

NORMAN E. JOHNSON is vice-president of corporate research, engineering, and technology commercialization at Weyerhaeuser.

EDWARD L. SOULE is vice-president of corporate research and engineering at Weyerhaeuser.

- Its job was to:
  - Provide technology support services to the operating companies.
  - Keep abreast of key technologies that could threaten the company.
  - Evaluate, manage, and develop technologies that would offer business opportunities and competitive advantage.
  - Develop commercial opportunities from technologies that had a strategic fit with the company.

The commercial development function of the unit was responsible for:

- Developing new products and processes beyond the scope of other corporate units engaged in new product development.
- Staying abreast of key technologies that could threaten or benefit the corporation and ensuring that technological information was included in corporate strategies.
- Maximizing the income from developed technologies through selective sales and licensing agreements.

From the outset, the mission of the corporate group was to create commercial and strategic advantage through technology. To accomplish this, the company decided to leverage existing and developing technologies in several core areas:

- Strategic biology—To support the company's plant-growing activities.
- Materials science—To improve the price/performance ratio of materials, particularly composites or laminates that contain wood and other raw materials.
- Environmental sciences—To better use the technology and skills the R&D staff had developed over many years.
- Process technology—To further the development of sensors.

Lacking an express corporate strategy, R&D management selected these four areas because they represented significant commercial opportunities supported by demographic, economic, and technology trends as well as technologies important to Weyerhaeuser. In addition, these were areas in which the R&D staff had acquired strong skills during the last decade. As a result of this effort, many new materials, products, and business opportunites were developed; in fact, some major commercialization efforts are currently under way.

The primary strategy was to build around a technology base that the R&D group had largely mastered. In addition, the areas R&D management chose to pursue were supported by significant market trends. It was important not to venture too far from familiar markets, and when Weyerhaeuser did, it tried to establish marketing alliances with companies that had strengths in those markets.

R&D management's philosophy was to think big but start small, and its approach was to make existing businesses more successful and to give them first choice to implement a new technology. In addition, R&D management set up boards of advisers to help improve its business decisions; R&D management also established quarterly meetings with the president and chief executive officer to verify their support and obtain their advice.

Weyerhaeuser's aim was to make existing businesses more successful and give them the first choice to implement a new technology

A CHANGE IN DIRECTION

By the end of 1988, however, the bloom was off the rose for new business and product development. Wall Street clearly told Weyerhaeuser that shareholder value must be increased. Corporate finances indicated that the company was involved in too many areas and that it needed to focus on those areas that had once made Weyerhaeuser the leader in its industry. At the shareholders meeting in April 1989, senior management announced the beginning of a major refocusing effort. The financial situation was inarguable; in part, the culprit was diversification, which stretched scarce managerial resources, talent, and capital over too many businesses.

Weyerhaeuser consequently decided to cut back its diversification efforts and focus on core strengths. For example, a major diversified business group that had led the company into such fields as nursery products, skin care, diapers, and salmon ranching was discontinued; other such businesses were sold or closed. In 1987, there were five new business development groups spread throughout the company; by 1989, the only commercial development group left was corporate R&D.

Weyerhaeuser was not the only company forced to undertake such retrenchment. Over the years, Weyerhaeuser had expanded and diversified as rapidly as such companies as Campbell Soup, Exxon, and Kodak; now these companies were also reassessing their new businesses.

This turn of events strongly suggested that R&D should reexamine its commercialization program. Ideally, this should be done in the light of a new corporate strategy, because corporate R&D and technology commercialization management had constantly stressed the need to match technology strategies to corporate strategies. However, the new corporate strategy was not to be determined until the results of ongoing, business-by-business strategic reviews were in. In the meantime, the corporate R&D and technology commercialization function tried to understand its current situation and what its options might be.

A RATIONALE FOR TECHNOLOGY COMMERCIALIZATION

As R&D management examined its situation and considered the lessons learned, it reflected on how the departmental rationale and implementation approach had evolved over the preceding five years. Certain core technologies are critically important to the future of Weyerhaeuser, whatever form the company may assume. R&D management determined which technologies fulfilled this role by assessing which ones were sources of competitive or strategic advantage for the business units. Every core business activity has essential technologies that are important for gaining competitive advantage. For example, if growing trees is fundamental to a company, various forestry technologies are essential because they maintain strategic positions in pulp making, paper making, or sawmilling.

In addition, some technologies cut across business lines; they form core competencies.[1] In 1985, when R&D management identified four potentially key technologies, the order of priority was materials science, plant biotechnology, environmental sciences and equipment, and process technologies. Although Weyerhaeuser still considers these areas essential, the order of priorities has changed with the emerging corporate strategy—environmental sciences and process technologies now lead the list.

Applying Core Technologies to Support Strategic Excellent Positions

Strategic excellent positions are key business areas that are developed by the allocation of resources.[2] Appropriate resources must therefore be committed to those core technologies that support strong strategic positions across the corporation. Once strategic excellent position has been developed, it can be maintained only through consistent, long-term corporate support through the ups and downs of business cycles. Developing strategic excellent positions is a medium- to long-term activity that can be easily disrupted when the length of the cycle between reorganizations is shorter than the time required for commercializing a new product.

At this juncture, Weyerhaeuser faced the critical strategic choice of whether or not to invest at the corporate level in developing core competence technology assets to support its strategic excellent positions. To make the correct decision, several questions had to be addressed:

- Should the company's individual businesses be responsible for their own technologies?
- Should Weyerhaeuser stay among the leaders in the industry in supporting new technology at a time when competitors are retrenching to increase shareholder value?
- How can the company use its technological base to greater advantage?

Underlying these questions is the principle that if a company decides to invest in an R&D program, it must commit resources to seek maximum return on that investment. It must find the people and create the organization to support this effort. Maximizing return means not only seeking the benefit that accrues to the main businesses from having strong core technology competence but seeking value through all legitimate avenues.

THE CURRENT APPROACH TO TECHNOLOGY COMMERCIALIZATION

In Weyerhaeuser's experience, creating new products or businesses that fit easily into its existing business units was limiting and difficult for the R&D group to accomplish because of natural impediments in the internal system. Consequently, R&D management now considers all the following commercialization options (none of which are mutually exclusive) as early in the technology commercialization process as possible:

- A new product for an existing business.
- A new Weyerhaeuser business.
- A joint venture.
- A spin-off venture.
- Licensing or cross-licensing.
- Trading technology for strategic position.

some technologies cut across business lines, forming core competencies—these were essential to the business

To maximize R&D resources, studies were conducted to determine how other companies commercialize technologies. Through conferences, visits to exemplary companies, conversations with those who had experience hiring consultants, and training courses, much was learned about the options available. The commercialization approach Weyerhaeuser R&D finally adopted emphasizes corporate strategy and exploring all options before making an investment.

R&D now seeks to align research programs with business and corporate strategies. If these strategies are not written down, R&D management must push, tug, and cajole senior management to explain them; R&D management then writes down these strategies and tests and retests them.

In addition, R&D management follows and assesses important demographic, social, economic, political, and technical trends that can benefit or threaten the company. Technologies that support the primary business and corporate strategies are pursued relentlessly, and an aggressive technology commercialization effort helps uncover opportunities outside the company.

Weyerhaeuser R&D now evaluates and sponsors commercialization projects by establishing rigorous business plans, seeking early market involvement, and developing cross-functional teams that include members from the marketing, engineering, research, manufacturing, and finance divisions. It is always easier to start a project than to stop one; this approach tends to discourage premature project termination. Every successful project has its dark days, and many managers understandably are reluctant to kill projects in their early, difficult stages. The commercialization process must be sufficiently rigorous at each stage to keep projects from going on forever yet not so tough as to stop innovation.

R&D management now aggressively seeks to extract value from promising ideas; the organizational environment is designed to be conducive to innovation and commercialization, so these ideas are heartily encouraged. The importance of a separate organizational home for projects not yet ready to be picked up by operational units is emphasized. These temporary homes act as greenhouses to nurture young projects until permanent organizational homes can be found.

Ideally, once they are developed, a product or an emerging business can be sold to a buyer—at the right time. The best time to approach a prospective buyer is after the product's value is apparent but before its commercial value is threatened by outside forces. Weyerhaeuser has tended toward gaining the early involvement of the businesses in order to gain their support sooner.

DON'T MISS THE BOAT

The typical corporation has immense resources to prevent errors of commission but almost never has the equivalent resources to prevent errors of omission. A technology commercialization process such as Weyerhaeuser's is at least one attempt to gain shareholder value by not missing new opportunities.

Notes

1. C.K. Prahalad and G. Hamel, "The Core Competence of the Corporation," *Harvard Business Review* (May–June 1990), pp 79–91.
2. Strategic excellent positions are defined and discussed by C. Pumpin in *The Essence of Corporate Strategy* (Brookfield VT: Gower, 1987).

SHIFTING SECTORS

The Evolution of R&D for Services

R&D efforts in the service sector are often structured according to frameworks that evolved in the manufacturing sector. Although R&D for services is equally determined by developing technology, a different approach—one that focuses on the business strategy for developing services—is required. At GTE Laboratories, this shift to service-oriented R&D provides one model for a service-oriented R&D strategy.

Graham Mitchell

The service sector in the US has grown to the point at which it now accounts for more than 70 percent of the gross national product (GNP) and provides employment for 75 percent of the work force. Despite this growth, the character and importance of service industries are widely misunderstood, and the growing technical sophistication of service industries is often underestimated.

More than a decade ago, Theodore Levitt pointed out that new technology—particularly information technology—had advanced to the point at which highly distributed and even mobile service organizations were becoming industrialized in much the same way as had occurred in manufacturing.[1] Although a few case histories of technical innovation in service industries have recently been published, the most familiar examples of successful industrial research are drawn from the manufacturing and process industries. Inventors are best remembered for creating products and possibly processes; few are famous for inventing services. Similarly, the models that are used to characterize the innovation process in industry often reflect this weight of experience in manufacturing and disregard precedents in the service industries.

This point is significant because the generic strategic priorities and the resultant technical needs of service enterprises often differ from those of manufacturing. For example, in a manufacturing corporation, it is usually apparent to both business and technical management that new and improved materials, designs, and manufacturing processes can lead to higher revenues and lower costs. As a result, corporate R&D budgets in manufacturing industries usually allocate significant resources to materials, product designs, and manufacturing processes.

For service businesses, however, the same overall technical focus is not so clearly appropriate because strategic advantage in service companies may depend on different underlying business factors. Although it is critical for service companies to un-

---

GRAHAM MITCHELL, PhD, is the director of planning at GTE Laboratories in Waltham MA, where his responsibilities have included the development of the strategic planning process for the central laboratories and technology planning for the corporation.

derstand R&D for services, their understanding must extend beyond the service sector. Many technical advances have broad relevance to US industry in general.

In addition, many traditionally nonservice companies have grown in recent years by expanding or acquiring service operations. For example, expertise in finance, engineering, repair, or customer service may have been developed in autonomous and profitable businesses that subsequently became coupled with related businesses under a larger corporate umbrella. These services are provided in response to opportunities the corporation perceived in the service sector. Such a diversification of businesses usually presents R&D management with the problem of determining how to support new service operations.

The shift to services has been particularly dramatic at GTE Laboratories, which was originally established during the 1940s to support manufacturing businesses in lighting, materials, and consumer electronics. In the latter half of the 1970s, manufactured products were still the primary focus of the company's R&D efforts, and the laboratories supported many of the corporate manufacturing businesses and telecommunications product lines. During the last decade, telecommunications service businesses have become more important to the company. In addition, as the telecommunications industry has become increasingly deregulated, developing technology has become the means for corporations to diversify and to achieve competitive advantage. These trends have combined to raise the priority of technical programs for services to the point at which they now constitute a large part of the R&D budget of the GTE corporate laboratories.

ESTABLISHING STRATEGIC FRAMEWORKS FOR SERVICES

In developing a technology strategy for services, defining how services differ from products is the first issue that must be addressed. The most familiar strategic planning approaches and frameworks managers typically use to determine near- and long-range priorities provide little guidance in understanding the differences between service and manufacturing industries. Existing frameworks typically analyze the business strategy from the perspective of the individual firm and, to a large extent, build on structural relationships that are common to most corporations. Within these frameworks, all companies must address similar issues and options when dealing with generic customers, suppliers, and competitors. Because these frameworks are insensitive to the differences between service and manufacturing as classes of business, their use tends to further confuse the needs of R&D and strategic support for services rather than to clarify them.

Defining Service Industry Needs

Service and manufacturing industries are inextricably intertwined; economies and advances in one area usually provide improvements and opportunities in the other. To examine the differences between manufacturing and service industries, the perspective must go beyond that of the individual firm; segmentation by industry sector must be considered. Different industrial segments can be arrayed along a value-chain sequence to illustrate the value added at each stage of product manufacture and support, as shown in Exhibit 1.

Exhibit 1
The Value Chain

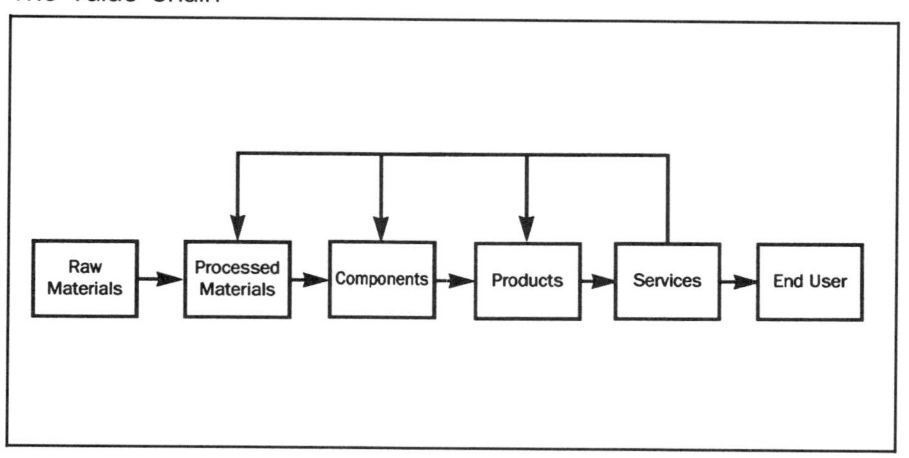

Successful companies exist at all points in the value-chain progression and operate with widely varying degrees of vertical integration. Each company develops a view of the world, with a corresponding understanding of technology, that is centered on its position in the chain. Although some of the output of the services, which occurs toward the end of the chain, is fed back to manufacturing or processing stages, many services are supplied directly to the general public because the market for much of the service economy is the individual consumer.

Exhibit 1 identifies the service sector as the major customer of manufacturing, which implies that many service industries have a basic need to specify or to control and influence in other ways the design of the products they buy. The specification and procurement activities of government, transportation, public utilities, communications, and financial service industries provide visible examples of this need. The second generic need in the service industries is to assemble, build, or otherwise integrate these products into a system or organization. The third is to operate the system so that it delivers appropriate services. Thus, the need to build, integrate, and operate systems as well as to understand and interact with the general public drives many of the programs that are set up in service companies, including those that involve major expenditures of technical resources.

DEFINING TECHNICAL SKILLS AND APPLICATIONS AREAS TO CLARIFY STRATEGY

The language used to describe industrial R&D plays a crucial role in defining the goals and strategic position of the R&D effort. When asked to define technology, the technical community usually thinks in terms of skills or disciplines—the input to the process. On the other hand, business management usually thinks of technology in terms of products of systems—the output of the process. Both points of view must be integrated to determine the strategic importance of specific technical areas in a particular industrial context, and in practice, technologies are often formally defined or recognized as having a skills component and an applications component.

Scientific knowledge and technical skills may be applied effectively during all stages of the aforementioned value chain; until these skills and expertise are coupled with applications that reflect the priorities of a particular business area, however, the overall strategy will remain unfocused. Although goods-producing companies are usually familiar with the technical skills needed in the service sector, they are often unfamiliar with the way service companies really work and are therefore unable to establish priorities for service industry applications or technology and R&D support for services.

How the Process Works at GTE Laboratories

In the R&D department at GTE Laboratories, the program balance has shifted from products to services in less than 10 years, a relatively short period. Although many of the same scientific and engineering skills have been needed on both sides of the business, the way these skills are applied and the criteria used to focus R&D efforts have changed significantly. Because this transition occurred in a single laboratory, the results are not colored by cultural and other factors that may be present in a study of two separate laboratories in the service and manufacturing sectors.

The evolution of GTE Laboratories has made it possible to observe firsthand the different attitudes of service companies and manufacturing companies concerning technological advances. For example, when a manufacturer of transmission equipment is presented with a breakthrough in fiber-optic transmission capability (e.g., higher-frequency, lower-cost sources or detectors), the immediate interest is in how these advances can be incorporated into the next generation of equipment design. Other technical issues—including access to components, patents and other forms of knowledge protection, and technology and product obsolescence—are among the initial concerns common to many product manufacturers.

For a telecommunications service company (e.g., a telecommunications network operator), the implications of the same research advance may differ significantly. One point of interest would be that bandwidth may be both more plentiful and less expensive in the near future. The most critical questions for the provider of such services are What will customers do with this new bandwidth? and What new video, data, or voice services will need to be created and marketed?

Because GTE Laboratories serves both product- and service-oriented businesses, its researchers must have an in-depth understanding of both perspectives. These points of view are used to formally link disciplines, techniques, and skills to the applications and market areas for each research area. A typical definition that is used to focus R&D efforts on telecommunications equipment manufacturing would state how the company expects to transform skills in fundamental physical sciences into products. For example, the following definition applies to integrated-circuit fabrication:

> The principal skills include lithographic techniques (e.g., photo, X-ray, and electron beam) for defining fine geometrics on semiconductors as well as high-temperature, solid-state chemistry and thin-film processing. These are applied to the fabrication of semiconductor integrated circuits, which are used in a wide range of switching and transmission products throughout the telecommunications industry.

the models used to characterize the innovation process in industry often reflect a weight of experience in manufacturing and disregard precedents in the service industries

For the service provider, applications must extend well beyond the individual product. For example, a typical description of an area that addresses network planning is:

> Major mathematical disciplines include decision, queuing, and game theory; simulation, regression analysis, and linear programming are required as well as algorithmic techniques applicable to the solution of computationally large network problems. These are used to develop network and engineering design tools and systems for public telecommunications networks.

However, the multidisciplinary nature of R&D for network services must reach beyond the physical data stream and address consumer needs directly. In contrast to the definition for R&D for network planning, the definition of R&D for network services would be:

> Major disciplines include statistics, market science, signal processing, and human factors as well as hardware, software, and communications systems engineering. These will be applied to develop techniques for the creation of interactive services for specific customer applications on the public network.

These definitions illustrate a progressive shift from manufacturing to services that parallels the transition the corporation has undergone in its business development. Exhibit 2 correlates the company's evolution through the stages of the value chain with its primary products and services that correspond to each stage. The services are defined as strategic technical areas because they establish the priorities and focus for R&D efforts.

When the majority of R&D programs in GTE's laboratories was directed at products, significant portions of the telecommunications R&D program were classified into such distinct areas as electronics, photonics, software, and telecommunications systems, reflecting the well-understood and intuitively recognized link between the skills required in these areas and their application in the telecommunications product line. As more of the program addressed services, however, increasing dissatisfaction was expressed with this classification system as it became clear that electronics, photonics, and software were ubiquitous in telecommunications services and were not useful concepts around which to build strategic priorities and to communicate with business management. Ultimately,

Exhibit 2
Strategic Technical Areas in the Value Chain for Products and Services

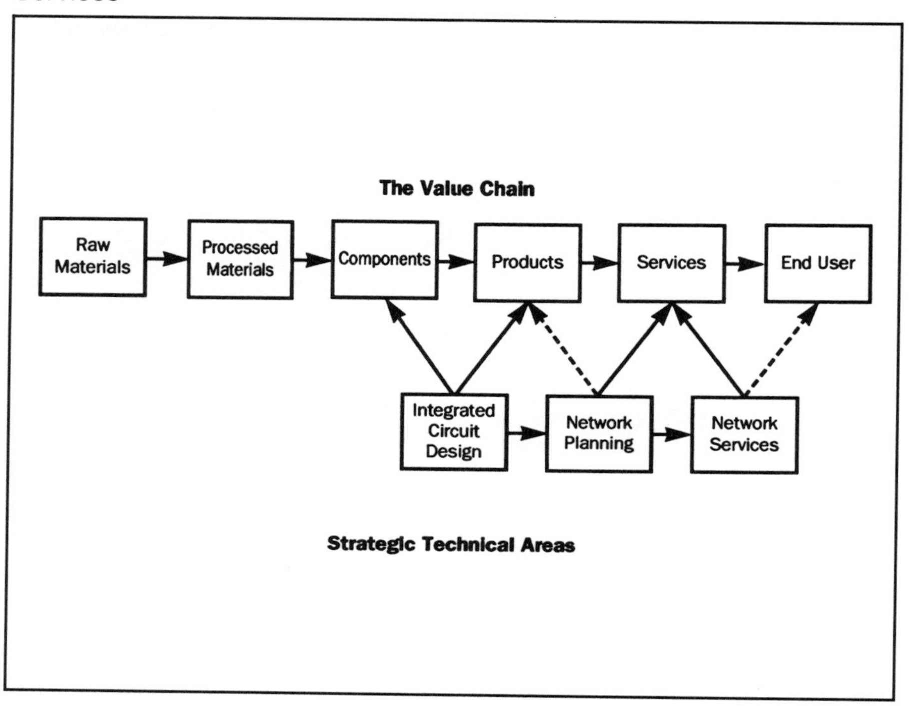

a new set of categories—namely, networks and systems, operation of networks, and applications or services—emerged as more closely portraying the purpose of the R&D program for services and as more useful terminology for setting strategic priorities. This change in language communicates the new R&D point of view and the relevance of the work more directly to the telephone operations management; more important, it is the outward expression of significant rethinking by the R&D staff about how to redirect and refocus the corporate R&D programs.

EVOLVING SERVICE-ORIENTED TECHNOLOGY DEVELOPMENT

The broad categories of R&D that have evolved naturally in manufacturing industries (e.g., materials, product design, and manufacturing processes) arose from the position of manufacturing in the value chain. The equivalent categories for service industries have not fully evolved yet. The foregoing experience at GTE Laboratories may suggest that R&D efforts for communications, transportation, banks and financial services, and even some wholesale and retail businesses are often focused around a few generic application areas: the physical system, the operation of the system, and the services provided (i.e., the applications or products of the system).

manufacturing companies are often unfamiliar with the way service companies really work and are therefore unable to establish priorities for service industry applications or technology

The Physical System

The physical system is usually built using equipment supplied by manufacturers and is usually assembled by the service company. Examples of such systems include:

- The network in a telecommunications system.
- The fleet and communications systems for airlines.
- Computer networks and information systems for banks and financial services.
- The generation, transmission, and distribution system for an electric utility.
- The network of wholesale and retail outlets, together with their formation systems, for trade.

Not only do service companies develop significant technical expertise in building these physical systems, but many are becoming increasingly sophisticated in managing equipment suppliers. This is true for giant utilities and government agencies as well as for newer entrepreneurial services. For example, in developing package-tracking systems, Federal Express soon discovered that developing specialized hardware in partnership with a vendor was a skill in itself.[2]

The Operation of the System

The second generic area of service-oriented R&D is operations, including the development of efficient maintenance and information systems and procedures. Often, these systems become a source of strategic advantage for service companies and may even provide a separate source of revenue (e.g., reservation systems for airlines, mutual funds management systems for banks, and directories for telephone companies). A wide range of mathematical and computational skills is often required, and operations research has provided numerous examples of added value in industries as diverse as government services, railroads and transportation, banking, and electrical power. Not surprisingly, senior management in service companies often exhibit considerable sophistication in their knowledge of information technology, probably because information remains for many service industries a means to achieve competitive advantage.

The Services Provided by the System

The third generic area for service-oriented R&D activities is in developing the applications or products of the system—the actual services. The technical skills required should help to ensure a close coupling of the services to the customers' wishes and needs. Many of these skills are not adequately represented in more traditional industrial research laboratories, where physical science and engineering disciplines may dominate. The overall need for such customer-oriented research is strongly influenced by the range, the variety, and the technical sophistication of the service-oriented products.

In the telecommunications, airline, and banking and financial service industries, deregulation is leading to an explosion in the variety and number of new service products. Understanding this and the role of technology in providing imaginative responses to new business opportunities in the service sector are among the most critical strategic issues facing many service industries. Innovative technology can often provide solutions to long-standing consumer needs, as it did when Citicorp recognized that ATMs would give a clear competitive advantage to Citibank by meeting known customer desires.[3]

AN ECONOMIC IMPERATIVE

The US economy can be viewed in terms of four interlocking and vital business sectors: agriculture, mining and material resources, manufacturing, and services. The largest sector—services—is in many ways the least understood, and the roles of technology and R&D in that sector are only slowly being recognized.

GTE Laboratories has managed its corporate R&D budget through a transition that occurred in less than 10 years and that witnessed a shift in primary business focus from products to services. Because this evolution took place in a single laboratory that had a stable environment, management staff, and research staff, the GTE experience offers more direct insight into the inherent differences between manufacturing- and service-oriented R&D than is usually available from comparisons across industrial sectors.

The GTE experience signifies that a better understanding of research and development processes for services is critical not only for the service sector but also for manufacturing companies because many leading technical contributions are pioneered by the service sector, as illustrated by the telecommunications industry example. During the past 20 years, many manufacturing companies have grown by adding service businesses. Technical executives in these companies who are challenged to support new service operations may find the three generic areas of service technology identified here helpful in providing strategic focus for R&D and for easing the transition from manufacturing- to service-oriented R&D.

Notes

1. T. Levitt, "Industrialization of Service," *Harvard Business Review* 54 (September–October 1976), p 63.
2. C. Nehls, "Custodial Package Tracking at Federal Express," *Managing Innovation: Cases from the Service Industries*, eds. R. Guile and J.B. Quinn (Washington DC: National Academy Press, 1988), p 57.
3. P.F. Glaser, "Using Technology for Competitive Advantage: The ATM Experience at Citicorp," *Managing Innovation*, Guile and Quinn, p 108.

Recommended Reading

Bitran, G.R., and Hoech, J. "The Humanization of Service: Respect at the Moment of Truth." *Sloan Management Review* 31 (Winter 1990), p 89.

Guile, B.R., and Quinn, J.B. *Technology in Services: Policies for Growth, Trade, and Employment.* Washington DC: National Academy Press, 1988.

Hackett, G.P. "Investment in Technology—The Service Sector Sinkhole?" *Sloan Management Review* 31 (Winter 1990), p 97.

Mitchell, G.R. "Research and Development for Services." *Research Technology Management* 32 (November–December 1989), p 37.

SECTION

2

MANAGEMENT OF R&D

This section mainly addresses the steps that R&D managers themselves can take to make the most of their departments' efforts. Four aspects of R&D management are covered: managing, evaluating, and training scientists and engineers; improving the quality of R&D; managing computer and information resources for R&D; and coordinating R&D with the rest of the company.

Stephen Fraenkel provides practical guidelines for managing the technical and nontechnical aspects of R&D personnel's work. He also elaborates on the measures R&D managers should take to maintain the respect of R&D personnel.

Donald Hoeg focuses on one aspect of managing scientists and engineers: evaluating them. In addition to addressing many facets of their evaluation, Hoeg considers the system of performance evaluation itself, emphasizing the importance of preserving the integrity of the system.

Ron Bhada discusses the need to provide university engineering students with elementary management training. According to Bhada, a lack of training in this area hinders many engineers in their careers. Bhada also reports that New Mexico State University has provided this training by introducing management concepts to engineering students in their junior years and enabling them to put these concepts into practice.

In the chapters on improving R&D quality, Robert Szakonyi provides examples of what an R&D department can do to improve the quality of its work, such as quality-related training for technical personnel. Szakonyi then discusses methods the R&D department and other company functions can jointly use to improve companywide quality. Finally, he discusses the use of computer-aided design and computer-aided management to improve quality.

Addressing the management of computer and information resources for R&D, Myra Williams relates Merck & Co, Inc's experiences in planning in this area. Williams covers the phases of the planning process, key issues that were addressed in the planning stage (e.g., management centralization versus decentralization), and Merck R&D's future use of computer and information resources.

Regarding the coordination of R&D with the rest of the company, Stephen Hellman outlines the methods that his company followed to integrate R&D and marketing operations. Hellman describes the company's former process of initiating R&D projects, depicting the problems that arose from poor interdepartmental coordination. He then explains the new project management system his company uses to select R&D projects more effectively. He also covers what will be needed to maintain this system of project management in the following years.

Robert Szakonyi discusses the issue of coordinating R&D and finance. He recommends taking three steps to achieve this: providing an evaluation and a rationalization of the R&D effort to the finance department, managing finance methods for improving coordination between them (e.g., creating a protocol for managing new product developments).

Finally, Szakonyi discusses the issue of coordinating R&D and finance. He recommends taking three steps to achieve this: providing an evaluation and a rationalization of the R&D effort to the finance department, managing finance operations to make financial figures more useful to R&D, and implementing mechanisms to satisfy the needs of both R&D and finance.

WALKING THE MANAGERIAL TIGHTROPE

Guidelines for Motivating Scientists and Engineers

R&D personnel have different management requirements than other company employees, particularly in their general allegiance to their technical professions and in their need to integrate their highly specialized technical work into a general business environment. R&D managers who effectively address these issues foster productive, profitable technology development.

Stephen J. Fraenkel

Scientists and engineers have many attributes that make managing them different from managing production, sales, or marketing people. For example, scientists and engineers are often more committed to their professions than they are to their employers. They also have more specialized skills than do most other company members.

Because scientists and engineers differ professionally from others in a company, R&D managers should take into account two basic issues in managing them. First, they should consider how R&D employees' professionalism influences their work and attitudes. Second, R&D managers should help their personnel integrate their technical work into their company's business. The 17 tips presented address both issues comprehensively.

DON'T ALLOW YOUR R&D STAFF TO BECOME OVERSPECIALIZED

R&D personnel need to maintain a balance between specialization in a field (e.g., electrical engineering or polymer chemistry) and intellectual flexibility. Unfortunately, technical personnel too often err on the side of overspecialization in one field. An R&D manager should prevent R&D staff from becoming too narrowly focused.

One way to achieve this balance is to recruit technical personnel who do not show a tendency to overspecialize. Moreover, a person who is a fast learner in a related technical area (e.g., an electrical engineer who can deal with questions related to physics) can often help a development laboratory more than a person who knows a great deal about only one technical area.

Furthermore, highly specialized technical personnel present two problems. First, it is difficult to keep such individuals motivated if their long-term professional aspirations are not fulfilled. Second, it can become difficult to use these people's technical skills if the needs of the company change.

As an R&D manager, I have seen many situations in which a technical person who was a fast learner did an excellent job. For example, I once had the task

STEPHEN J. FRAENKEL, PhD, is founder of Technology Services Inc, a Northfield IL consulting firm specializing in R&D management and materials engineering. Before founding Technology Services, Fraenkel had a distinguished career in industrial R&D management.

of overseeing the development of a process for rapidly coating a metallic web. The problems posed by the challenge of developing this process involved many technical areas, including heat transfer, mechanical metallurgy, process stability control, and chemistry. Unfortunately, in our laboratory at that time we could not assign—or even find—all of the experts in the various areas who were needed for this project. In addition, we were concerned that a committee of experts would not be able to solve the problems. Instead of forming a committee, we appointed a physicist to be in charge of the whole endeavor. This individual was a quick learner and knew just enough about all the aspects of the problems to ask the right questions and find the right technical assistance; therefore, he was able to solve the problems and develop the process we needed.

RECRUIT R&D PERSONNEL WHO KNOW WHEN TO STOP TRYING TO IMPROVE A NEW PRODUCT

In recruiting R&D personnel, the company's goal is to hire scientists and engineers who are interested in and capable of improving its products. On the other hand, it is important to ensure that they know when to stop improving a new product. Many scientists and engineers are perpetually dissatisfied with the performance of their innovations and downplay issues related to delivering a new product to the marketplace as quickly as possible. Although scientists and engineers are not the only ones in a company who should recognize trade-offs between technical performance, cost, and time, they still should take them into account in their technical work.

IMPRESS UPON THE STAFF THE NEED FOR R&D TO SERVE THE COMPANY'S BUSINESSES

Technical personnel often consider issues related to cost, schedule, safety, and producibility to be peripheral to their primary functions. From the perspective of a business manager, however, the results of an R&D project are purely academic if these issues are not considered. If an R&D engineer indicates tolerances that are not producible, specifies materials that are not procurable, or does not propose methods to dispose of the by-products of a manufacturing process, a business manager has no use for that person's work.

Consequently, the technical staff must be oriented toward serving the company's business needs. First, technical staff members should be evaluated to see whether they operate well in the business world, not just in the laboratory; that is, they should understand the company's business in its entirety and understand how such social concerns as safety and pollution control issues affect the company's businesses. Naturally, some are not interested in this broader perspective and consider the world of science and technology the only worthwhile arena. If these individuals cannot be shaken from this opinion, they may be better off working in either a university or a very large R&D organization that has a substantial organizational cushion between R&D and the management of a company's businesses.

Second, R&D managers should instill in technical staff a sense of urgency. Although not all R&D projects should be crash programs, even the most relaxed endeavor has to be undertaken as quickly as possible. There is never time in industrial R&D for a leisurely approach. Moreover, most business managers will accept approximate answers and near-optimal solutions if the situation is urgent.

R&D personnel should not be allowed to think of the process of discovery as an end in itself

Third, technical staff members should understand that they must produce results. R&D personnel should not be allowed to think of the process of discovery as an end in itself. Business managers do not really care about the intellectual elegance of a technical approach; they want useful answers to their problems.

LOOK FOR TANGIBLE EVIDENCE OF CREATIVITY

Every R&D manager would like to have a very creative technical staff. To build such a staff, however, R&D managers must do more than wish for good luck. In recruiting candidates, an R&D manager should seek tangible evidence of creativity. For example, does a candidate understand and use known techniques to stimulate creativity? These techniques include synectics, which is largely based on analogies; brainstorming; morphological analysis, in which a search for the structure of a problem is conducted; and value engineering. Evidence of creativity can also be found in published papers and patents.

INSIST THAT R&D PERSONNEL REALIZE THE LIMITS TO THEIR TECHNICAL COMPETENCY

No technical person can be expected to know everything. In most cases, R&D staff members need to seek advice or guidance or to discuss their technical problems with their peers. To facilitate this process, R&D managers should insist that technical personnel recognize the limits of their technical competency and work cooperatively with other specialized members of the department.

DON'T NEGLECT ISSUES RELATED TO EMOTIONAL STABILITY

R&D managers should pay close attention to the emotional stability of their staff members. The most important elements of an R&D project are the scientists and engineers involved. Most crises that arise in R&D projects stem from interpersonal conflicts rather than malfunctioning equipment, late deliveries, or unreasonable customers. For example, in Harvard Business School's collection of business case studies, those related to R&D show a striking number of problems that stem from technical personnel who are emotionally unstable, react defensively to criticism, or do not tolerate dissent. The emotional well-being and interpersonal relationships of R&D staff members can be crucial to the outcome of the project.

EXAMINE PERSONAL QUALITIES CLOSELY WHEN EVALUATING R&D STAFF MEMBERS FOR PROMOTION

When evaluating technical staff members for promotion, personal qualities as well as technical capabilities should be carefully noted. For example, how much supervision does the person require? A properly motivated technical person should not need—and indeed should resist—constant supervision. Conversely, how does that person react to supervision when the situation requires it?

The intellectual curiosity of staff members should also be taken into account. One evaluation technique is to ask questions that stray from the narrow confines of a person's technical specialty. The R&D manager should note whether that person has a record of continuous learning and whether the person can accomplish more in a greater variety of areas than, say, five years ago.

BE AN EFFECTIVE MANAGER IN ADDITION TO BEING TECHNICALLY ADEPT

Effective managers of other technical personnel must be able to manage their own activities effectively. For example, the ability to set sensible objectives is important. One of the most important of these is setting high standards for personal performance. In such intellectual endeavors as R&D, it is difficult to develop quantitative measures to evaluate how well such performance standards are met. Nevertheless, the R&D manager should insist that the staff be intellectually honest and, moreover, be perceived as such.

In addition, realistic work objectives must be established; those objectives should factor in the resources that are available. It is also important to set deadlines and hold the R&D staff accountable for meeting them. For example, if a project review is scheduled on the third Wednesday of every month, that review should be held as scheduled even if it takes only five minutes. If laxity is allowed there, laxity will also creep into other matters that are more important than a project review meeting.

Finally, the R&D manager should delegate authority to the lowest possible management level. In this way, the whole staff participates in activities and has a stake in making those activities successful.

HAVE THE R&D STAFF FIND OUT WHAT THE COMPANY'S CUSTOMERS NEED AND WANT

Depending on the nature of the company's businesses, it may be necessary to reach beyond the company's customers to formulate a market-driven R&D program. For example, the R&D staff I managed at a packaging-products company found that the input of our customers' customers was necessary to make technical work useful. What ultimately sold a package was not the opinion of the company that bought and filled the package but the opinion of the person who opened it and used its contents. In addition to having my R&D personnel visit such consumer-product companies as Proctor & Gamble, Lever Brothers, and Stroh Breweries, I had the staff arrange its own consumer focus groups to evaluate new package designs before the packages ever reached our company's customers.

At our packaging-products company, we also developed many new products for industrial applications, including containers for hazardous substances. In these instances, my R&D staff dealt with the users of those containers (i.e., chemical companies) as well as with the producers of the hazardous substances. The staff found for the most part that the practices and preferences of users were the key considerations in developing a new container.

If efforts to put the R&D staff in contact with the customers' customers cause conflicts with the marketing or sales department (which happened to us occasionally), these conflicts should be resolved rather than avoided. The R&D staff—not the marketing or sales departments—will pay the price if the new products are unsuccessful.

if efforts to put the R&D staff in contact with the company's customers cause conflicts, these conflicts should be resolved rather than avoided

ACCOUNT FOR ALL FACTORS WHEN ASSIGNING R&D PERSONNEL TO SPECIFIC JOBS

When considering an R&D staff memer for a job, the R&D manager should take all factors into account. For example, not only the nature of the technical job should be considered but also such secondary aspects of the job as the amount of communication required, the types of resources to be managed, and the commercial issues to be addressed. Ensuring that a technical person can deal effectively with the nontechnical aspects of a job can be as important as ensuring that the technical requirements are met.

As R&D manager at a paper-producing company, I observed a case involving a glaring mismatch between a technical person's abilities and the needs of the situation. Even though the person assigned was an expert in metallurgy—one of the basic technical requirements—he lacked communication skills. Therefore, when he was assigned to monitor the paper mills for corrosion, he worked alone rather than in coordination with personnel from the mills. First, he wandered around the mills on his own, taking photographs, then withdrew to his office and developed an action plan for the mills without consulting anyone. As it turned out, the mills rejected his advice and his whole effort failed.

DON'T SET GOALS WITHOUT THE STAFF'S INPUT

The technical staff should be involved in setting goals. Few activities increase staff commitment to effective R&D more than actually setting departmental goals. By allowing staff members to participate in this important activity, the R&D manager bolsters their professional status and their commitment to success.

When the technical staff is involved in setting goals, such financial goals as overhead charges and R&D budgets should be included along with R&D project goals. Staff members must be made to understand budget requirements and constraints—although they may not like these restrictions, they cannot then complain that the restrictions were imposed without explanation.

When I was the director of a laboratory, I always met with section and department heads to discuss their budgets for the following year. I asked them to make the first draft of their own budgets and encouraged them to explain their budgets to their staffs. This method of developing a total R&D budget worked extremely well. All members of the laboratory were kept informed of the budgetary constraints within which they would have to work.

DON'T MAKE ALL OF THE PRESENTATIONS AT MEETINGS YOURSELF

The R&D manager should not be the exclusive spokesperson for the R&D group; if that is the case, the professional status of the R&D staff is demeaned, albeit unintentionally. Therefore, technical staff members should be given a chance to speak publicly, particularly about ideas they originate. Assuming this role provides many benefits, including:

- Acquainting business managers with the staff.
- Enhancing the morale of the staff.
- Minimizing misunderstandings that arise when technical ideas are explained by someone who is not intimately familiar with them.
- Providing the staff with an incentive to learn how to communicate with business managers.

ENCOURAGE R&D AND NON-R&D STAFF TO ROTATE JOBS

The more the R&D staff and non-R&D personnel in the rest of the company understand one another's capabilities and needs, the better the R&D projects will serve the needs of the company's businesses. One way to hasten such mutual understanding is to implement job rotation among members of the R&D staff and non-R&D personnel.

Unfortunately, in most companies, R&D staff members are seldom assigned to the marketing or finance departments. It is even rarer for a marketing or finance person to be assigned to an R&D department. There is no reason, however, that R&D personnel cannot function effectively in planning, marketing, and finance departments. These individuals are skilled in analyzing phenomena in quantitative terms—a talent that can benefit those departments as well. Temporary transfers of non-R&D staff to an R&D department are more difficult to arrange but not impossible.

As R&D manager at the packaging-products company, which was also an operating division of a large oil company, I had the opportunity to compare the two scenarios because the parent company rotated jobs and mine did not. In the parent organization, engineers worked in the planning and finance departments and finance personnel worked in the R&D department, particularly in R&D planning. This experience led me to conclude that no disadvantages stemmed from rotating R&D and non-R&D personnel; on the contrary, there were many advantages, especially in terms of integrating R&D into the company's businesses.

BREAK DOWN BARRIERS BETWEEN R&D AND TECHNICAL SERVICE PERSONNEL

In laboratories, barriers can develop between R&D and technical service personnel. Sometimes, these rifts occur because R&D personnel regard their technical services colleagues as operating at a lower technical level and having fewer capabilities. The R&D manager should try to prevent such an attitude. This view of technical services personnel's capabilities is not necessarily accurate and can destroy any chance of fostering the much-needed cooperation between the R&D and the technical services staff.

the underlying challenge is to keep the R&D staff intellectually flexible rather than to prevent technical obsolescence

USE A DUAL LADDER FOR ADVANCEMENT OF R&D PERSONNEL

In many companies, R&D personnel who want to advance are forced to become managers, even if they are not suited to the position. Using a dual ladder to advance R&D personnel solves that problem by allowing such individuals to climb a technical ladder if they are not suited for or are not inclined to pursue a management career path.

For a system of dual career ladders to operate effectively, qualifications for positions in both paths need to be spelled out as specifically as possible. This is usually easier to do for positions on the management ladder than it is for positions on the technical ladder.

Problems with a dual-ladder system generally arise for two reasons. First, such a system is sometimes misused by R&D managers who use the technical ladder as a place to put R&D personnel who have failed as managers. Second, a dual ladder system does not work well if no genuine equity exists between corresponding positions in the technical and management ladders.

HELP YOUR R&D STAFF COMBAT TECHNICAL OBSOLESCENCE

Although technical obsolescence can be a problem in any company, it presents particular difficulties for companies whose technologies change rapidly. In many organizations, however, the main technical problems involve applying existing knowledge rather than making cutting-edge technical breakthroughs. Nevertheless, the problems of applying existing technology can still be complex and difficult. The underlying challenge is more often to keep the R&D staff intellectually flexible rather than to prevent the staff from becoming technically obsolescent.

An R&D manager can ask several questions to evaluate the intellectual flexibility of the R&D staff, including:

- Can the person reason by analogy from a biological system to a mechanical one?
- Does the individual have and apply the knowledge that the same differential equation describes both current flow and mechanical vibrations?
- Can a wear problem in a wood-based product be dealt with by using experimental results from mechanical metallurgy?

The R&D manager can decrease technical obsolescence and increase intellectual flexibility by helping the R&D staff gain skills with broad applicability. One such skill is the use of statistics to analyze and interpret research findings. It is amazing how many highly trained R&D professionals grind out reams of data but have little knowledge and make no use of statistical techniques, probability theory, and inference methods.

Statistical techniques are applicable to all branches of science and engineering, and they are increasingly useful in business as well. In addition, statistical techniques provide insights into the meaning of data that cannot otherwise be gained. As manager of an R&D department, I required everyone on my technical staff, including technicians, to take a training course in statistics, and R&D project leaders were required to take a course in experimental design.

TEACH THE R&D STAFF HOW TO COMMUNICATE

R&D managers should encourage their staffs to write clearly. The litmus test of clear writing is whether it can be understood by the most senior managers to whom it is addressed. Because senior managers are often business managers with minimal technical education, technical jargon should be avoided.

Furthermore, R&D managers should encourage R&D staff to be more effective in their oral communication. One way to do this is to have a capable instructor conduct a short course on making oral presentations. As an R&D manager, I regularly had instructors teach my staff how to give speeches and make presentations at meetings. These short courses on public speaking helped my staff members develop their stage presence, self-confidence, and ability to think on their feet.

ACHIEVING DEDICATION AND EXCELLENCE

The points included in this chapter are based on my personal experience as an R&D manager. They worked in my situation, and I believe that they are effective rules of thumb that can be applied in almost any technical development environment. Because all R&D staff members are individuals, as are their managers, the generalizations made here will not apply in all cases; however, for the sake of offering some useful advice, a few generalizations can be made. The basic premise is not to characterize R&D personnel as lacking in fundamental areas but to show that encourging commitment, pride in work, team spirit, and respect for oneself and one's peers encourages dedication and technical excellence.

BRINGING OUT THE BEST IN R&D STAFF

How to Manage Scientists and Engineers Effectively

The R&D manager treads a narrow path between responsibilities. On one hand is the technology under development; on the other, the management of the R&D staff. By developing the staff's capablities, the successful manager empowers R&D staff members to become more effective researchers and colleagues in their own right.

Donald F. Hoeg

Although R&D managers must understand the technology they are responsible for developing, their success is determined more by the R&D staff's performance than by the manager's knowledge of technology. Therefore, successful managers place a high priority on developing the capabilities of their R&D staff.

To develop staff capabilities, a manager first must understand what motivates R&D professionals. R&D staff members want to be productive and responsible; they also want to be treated as productive and responsible. Consequently, when evaluating the performance of R&D staff—and when providing them with training—the manager must keep in mind that these people deserve respect for their abilities. This article presents tips on how managers can effectively develop the capabilities of R&D personnel.

EMPHASIZE LEADERSHIP

The quality most needed for successful R&D is leadership. When evaluating and training staff, the manager must emphasize the development of leadership. Leadership is important in R&D management because technological innovations threaten the established order. Consequently, R&D professionals inevitably become agents for change.

In practice, the R&D manager should not become immersed in the details of a personnel evaluation system (i.e., in such details as how performance reviews are conducted, how job ratings are described, and how performance measures are distributed). Defining the details of a training curriculum for technical staff is not the R&D manager's primary responsibility. The most important issue to keep in mind in evaluating and training technical staff is the development of leadership.

DONALD F. HOEG's career includes three decades of intensive involvement in R&D, technology management, and new venture business development. The former general manager of Borg-Warner Corp's research center in Des Plaines IL, he has served in numerous professional organizations. An adjunct professor at the Illinois Institute of Technology and a management consultant at the IITRI Center on Technology Management, Hoeg works as an independent management consultant and teacher in research and technology management and new technology development. Based in Mt Prospect IL, he is a frequent author and speaker on technology management. He holds several patents.

ENCOURAGE RISK TAKING

For R&D professionals to become leaders, they must be allowed to take risks. With risk, however, comes occasional failure.

R&D staff members can take risks in different ways. An independent researcher can take risks in challenging a current theory to test a new concept. An R&D manager can take risks in trying a novel approach to introducing a new technology into a business. Whatever the case, departing from convention to achieve worthwhile objectives at times results in failure. Even if the R&D professional's decision seems right, an innovative effort will not necessarily succeed.

Besides encouraging R&D staff members to take risks, the R&D manager should teach them how to make sound decisions regarding their innovative efforts. The more they are helped to make better decisions, the more likely these efforts will pay off.

PROMOTE PEOPLE FOR THE RIGHT REASONS

Many technical staff members are promoted for the wrong reasons. This happens most often when they are promoted to management positions. In this case, typically the most creative and productive researcher is selected to be an R&D manager. R&D managers who promote researchers in this way are guilty of evaluating a person's technical capabilities while neglecting to consider whether that person has the required management skills.

PROVIDE FIELD EXPERIENCE

Rather than thinking first of the training courses that R&D staff could use, managers should think first of how and when staff members can be sent on special assignments to the company's manufacturing plants and to customers' facilities. At these sites, the R&D staff will learn far more about the problems and needs of the users of their technical work than can be learned in a classroom.

PROVIDE MANAGEMENT TRAINING

After promoting a researcher to the position of R&D manager, the supervising manager should make sure that the person in question is provided with necessary management training. This person typically has invested years in advanced technical study and most likely will benefit from management training to deal effectively with both technical and managerial aspects of the new position.

REINFORCE LESSONS LEARNED

Companies often allow poor integration between the training that R&D staff members are given to improve their work and the reinforcement of that training in their work environment. This often occurs with training programs run by the company. For example, a group of technical personnel might be trained in computer-based program management, the statistical design of experiments, or the use of microcomputers in the laboratory. After this training is completed, these technical personnel typically are sent back to the same work environment—where everything is still done the same way. R&D managers who train their technical staff in this way fail in their training efforts by not planning how to reinforce that training. If they want the training they provide their staff actually to be translated into everyday use, R&D managers must also commit to changing the work environment in which those technical personnel operate.

to preserve harmony, R&D managers are often tempted to avoid evaluating a technical staff member's performance honestly

PLAN FOR CONTINUING EDUCATION

Another area of career development often subject to poor planning concerns company-subsidized continuing education programs that technical personnel often are encouraged to participate in. Technical personnel often gain only limited benefits from these programs because no one has focused clearly on the specific goals for any technical staff member's continuing education and the value of that education for that person and for the company. An R&D manager must help R&D staff members plan their continuing education to help their career and their company.

Many technical personnel have been disappointed after taking educational courses at night or on the weekend. Although initially encouraged by their manager to take these courses, they found after completing their studies that there were no new positions to which they could advance. Their manager had encouraged them to take courses without paying attention to where this new education could be used. Unfortunately, what usually happens in this situation is that the technical staff member leaves the company and the manager loses a very capable technical performer to poor planning.

MAINTAIN THE INTEGRITY OF PERFORMANCE EVALUATION SYSTEMS

Many R&D managers consider conducting performance evaluations of their staff to be the second most unpleasant managerial task—just behind firing an employee. Performance evaluations almost always involve some criticism of a technical staff member's work. To preserve harmony, R&D managers are often tempted to avoid evaluating a technical performance honestly. The following is an actual case that illustrates what can happen when an R&D manager sacrifices the integrity of a performance evaluation system for the purpose of maintaining harmony.

An R&D manager who had just been appointed the head of a department of advanced materials technology at a large industrial R&D center ambitiously set a goal of building his department into a first-class group of technical performers. One of his first steps in attaining that goal was to evaluate each member of his staff. After preliminary evaluations, the manager's assessment was that his entire staff was technically very strong and he had inherited an effective system of performance evaluation. Unfortunately, his initial appraisal was wrong.

One of the manager's senior technical performers, a senior staff research engineer in the process research section, had worked at the R&D center for 15 years. Just after the R&D center had been established, he joined the staff with a group of friends who had all worked together at another company. The senior engineer was a friendly, gregarious person who seemed to get along well with everyone.

Although the R&D manager had been acquainted with the senior engineer before becoming head of the advanced materials department, he had only superficial knowledge about his employee's technical work. After becoming familiar with the employee's work on a regular basis, however, he became concerned about the senior engineer's performance and—more important—about his technical abilities.

At the R&D center, senior research staff were expected to be independent R&D professionals who knew how to conduct technical projects and to solve the technical problems that arose in these projects. It was in these areas that the senior engineer's skills seemed particularly weak. For example, at meetings at which the senior research staff reported on their project's progress, the senior engineer always reported data, but never any research results. When asked about the principles involved in his technical projects, he gave vague answers. He was unable to define and conduct a critical final experiment for his projects. Finally, he never had a new technical idea or a new project proposal.

Because he was concerned that there might be a performance problem, the manager reviewed all of the senior engineer's records. These records included previous performance reviews, fitness reports, promotion recommendations, and other records related to the employee's work over the past 15 years. Amazingly, these records showed that his work had been excellent all along.

Because the records contradicted what the manager had observed, the manager started speaking about his employee to others in the R&D center. The senior engineer's immediate supervisor agreed that the senior engineer was not performing at the expected level. He also admitted that he had overstated the senior engineer's performance in the last performance evaluation, but, as he explained, "He has been here a long time and seems to be a good worker."

The R&D manager then talked with the people to whom the senior engineer had reported previously. He asked them whether there had been problems with his work in the past or had something happened to make his work deteriorate. Each of the people said essentially the same thing: the employee had always had performance problems, but his managers would not mention these problems in a written performance report. To do so would have jeopardized the senior engineer's chances for salary increases. The R&D manager also found out during these conversations that the senior engineer had earlier failed as a manager and that one of his managers had then recommended him for the technical ladder.

After these conversations with his employee's former managers, the R&D manager finally understood what he had inherited. All of the previous managers had committed a fraud. What was probably most damaging was that everyone in the advanced materials department also knew of the fraud. The system of performance evaluation had lost credibility with the whole department. All of the members of the department were aware that the senior engineer's rank and salary were inconsistent with his performance.

The R&D manager also worried about another problem. Because the records of the senior engineer's past work were not truthful, he questioned how truthful the work records of the other members in the department were. Were the senior engineer's records just the tip of an iceberg?

managers should always coach the R&D staff, providing them with guidance whenever appropriate

Faced with this potentially massive problem, the R&D manager decided to revamp the system of performance evaluation and to change the whole department's attitude toward performance evaluations. He established clear expectations about the kind of performance required for promotion. He also made it clear to everyone in the department that the technical ladder was not to be used as a dumping ground for management failures.

Even though he had a problem with his senior engineer, the R&D manager also recognized that his current situation offered him the opportunity to demonstrate to his department how strongly he supported the philosophy behind his new system of performance evaluation. He did not intend to fire the senior engineer abruptly. After all, who had actually committed the fraud? It was not the employee, but the people to whom he had reported. The senior engineer had worked loyally for the company for 15 years.

Nonetheless, it was clear that the senior engineer was in the wrong job and had to be removed. For the sake of the health and credibility of the advanced materials department, he could not be allowed to stay.

The R&D manager then had a candid discussion with his senior engineer. The manager's candid comments, however, did not surprise him because he knew that he had not been performing well. The manager gave him the choice of being transferred to a less demanding job in the company or taking a special assignment while the manager found a new job for him at another company. The senior engineer chose to take the special assignment and then a job at another company. The R&D manager eventually found a job at another company that was appropriate to the senior engineer's interests and talents.

When the members of the department realized how the R&D manager was dealing with the employee, his reputation was greatly enhanced. The members of the department recognized that the manager was taking appropriate actions in two areas. First, he was improving the system of performance evaluation. Second, he was handling the employee fairly and with dignity.

As this case shows, when the integrity of a performance evaluation is sacrificed, many people are poorly served: the ones who are promoted but should not be, their co-workers, and their managers. In addition, the performance of the group itself is affected adversely.

OFFER TIMELY FEEDBACK

Developing the capabilities of technical personnel has to be a continual process. An R&D manager should not wait until a formal evaluation to indicate problems to technical personnel or to praise them. Instead, a manager should always coach the R&D staff, providing guidance whenever appropriate.

Early in my career, I came very close to firing a technician whom I had personally hired. Had I fired him, the company would have lost a future outstanding worker. When I hired this person, he had solid credentials and seemed to be very bright and ambitious. During the first few months as a technician, however, he kept making minor errors in his work, as though his mind were someplace else. In addition, he made two major errors on critical experiments that I had prepared.

Rather than wait until a formal evaluation to address his shortcomings, I had a frank discussion with him a few months after he had begun work. I told him very clearly what I expected from him in the laboratory. I also told him that I would give him six months to improve. At first, he made only minor improvements. Little by little, however, his confidence grew. By the end of the six months he was excelling. Furthermore, it was evident that he was capable of being more than a superior technician. He went back to college in the evenings to finish his courses for a bachelor of science degree and then went on to receive a master of science. Since his early months in the company, he has advanced from one position to another; currently he is a senior executive.

In looking back over this person's progress, I do not think that a change in his abilities was the key factor. Rather, the key factor was his change in attitude. Once his priorities became clear, he performed well.

This example shows the effect an informal discussion with a staff member at the appropriate time can have. If I had waited until the formal evaluation to criticize this technician, it may have been too late to help him change direction. Moreover, one of the great benefits of having an informal discussion at the appropriate time is that the person being criticized is more likely to really listen and make changes. It is likely that if I had made similar comments to him during a formal evaluation, he would not have listened as well.

EVALUATE THE PERFORMANCE, NOT THE PERSON

Another common error is focusing criticism on the person, instead of on that person's R&D performance. R&D managers should evaluate employees' activities and the results of their work, not their personal characteristics.

Furthermore, it is possible to establish a method of performance evaluation such that employees can also evaluate their own performance. If performance goals are delineated effectively beforehand and clear ways to measure performance are established, employees can include their own assessments during a formal performance evaluation.

EVALUATE PERSONNEL AND PROJECTS QUANTITATIVELY

Many methods for measuring R&D performance have been developed during the last few decades; some of them are quite elegant and thorough. Unfortunately, most of these methods have limitations, usually because they are difficult to use.

An effective method of evaluation was introduced by a senior R&D manager in one company. For more than a decade, this method was used in the laboratory to evaluate every R&D project. The results from these evaluations were then used as part of personnel performance evaluations.

This method focuses on evaluating how well R&D projects have been carried out. A project that fails to meet its goals receives a score of zero. If a project meets most of its goals, it receives a score of 1. A project is assigned a score of 2 if its results meet the goals established beforehand. If the results of a project exceed the goals, the project receives a score of 3. Examples of results that have exceeded a project's goals are a breakthrough discovery or completion of a project with half the allotted time or money. The scores for projects also should be weighted in terms of the importance of the project or the amount of money spent on the project.

if R&D managers rank everyone as excellent performers, they do not adequately recognize those doing outstanding work

Many people in the laboratory, including individual contributors, played a role in evaluating R&D projects. In addition, managers in the operating divisions of the company used this method to evaluate R&D projects.

This method of evaluation was useful because it concentrated on measuring only what was under the control of the R&D staff—that is, the technical progress of a project in relation to its goals. In using this method, managers did not try to measure how many new products or processes were introduced or what the economic impact of an R&D project was. Measuring these items would have involved consideration of the activities of many functions other than the R&D department.

This method of evaluating projects also served as an excellent tool for communicating with the operating divisions, especially in situations in which there was a large discrepancy between the scores given an R&D project by an operating division and those given it by an individual researcher. Also, by evaluating R&D projects with this method, managers in the laboratory clearly demonstrated to the operating divisions that they were strengthening R&D performance.

Finally, R&D managers used the results from these project evaluations, along with other information, to conduct personnel performance evaluations of R&D staff. Instead of judging the technical personnel's contributions to the laboratory by getting an overall impression of their miscellaneous technical activities, managers were able to use information about each person's work on specific R&D projects in making an evaluation.

CONTROL STAFFING LEVELS

The morale of an R&D department is weakened greatly if the department managers are forced to cut many R&D people from the staff. Frequently, after a boom period, such cuts can occur because the R&D managers during the boom period were too optimistic about how much they should increase their staff. Therefore, R&D managers should resist the temptation to add unnecessary staff during boom periods.

When I was the manager of a laboratory, I would not hire someone unless I was certain there would be a lasting relationship between the laboratory and this new researcher. I simply would not hire someone for the short term.

For many years, until my laboratory was restructured, the other R&D managers and I never had to cut many staff members at once. We used many strategies to help avoid doing this. For example, we tried to keep the number of staff members constant during difficult periods and boom periods. We used outside consultants, especially retired researchers, when we were overloaded. We hired flexible researchers who were capable of changing jobs as our research needs changed. By pursuing all of these strategies, we were able to keep most of our R&D staff for years.

COUNTER CREEPING INFLATION IN PERFORMANCE RANKINGS

Many laboratories suffer from a creeping inflation in performance rankings. Because R&D managers usually consider almost all of their staff to do very good jobs, they are reluctant to rank anyone as an average performer. However, if R&D managers rank almost everyone as excellent, they do not give adequate recognition to individual researchers doing truly outstanding work. Consequently, R&D managers must find ways to counter this type of inflation in performance rankings.

One method involves the use of a middle-range ranking called competent performance, for technical staff members who are doing very good work. By giving someone this ranking, a manager indicates that the person has fully met all criteria set for someone at that level. We considered this competent performance ranking to indicate that someone had met 100% of the requirements regarding quality. Because few people object to receiving a 100% ranking, few researchers object to being ranked as competent performers.

AID THE BEST AND THE WORST PERFORMERS

In many laboratories, R&D managers evaluate their staff, identify the best, worst, and average performers, but then do nothing on the basis of these performance evaluations—until the next formal evaluation. This is a mistake. R&D managers should take special steps to aid the best and worst R&D performers.

An R&D manager specifically should develop action plans to fit both cases. For the superior performer, an R&D manager should develop a plan that helps that person become at least a competent performer in a job at a higher level. For a poor performer, the R&D manager should develop a plan to help that person change jobs, either within or outside the company.

RECOGNIZE TECHNICAL EXCELLENCE

Most researchers, like most workers of any kind, want to be recognized for excellent performance. Therefore, R&D managers should find any opportunity to congratulate researchers for their excellent technical performance. To thank publicly someone who deserves it is one of the most effective ways of motivating researchers. In addition, the manager should look for ways to show this appreciation for excellence in the presence of the staff, the family of the individual performer, and the public.

R&D managers should take special steps to aid the best and worst performers

EVALUATE SCIENTISTS AND ENGINEERS WITH SIMILAR METHODS—WHEN THEY PLAY SIMILAR ROLES

Scientists and engineers are different from one another. This is partly because they have been educated differently: scientists are taught to challenge established theories, and engineers are taught to follow certain rules and to abide by specific standards to achieve certain results. Nonetheless, scientists and engineers frequently play similar roles in a laboratory. Engineers are often required to find imaginative new solutions to technical problems, not just to follow standard engineeering practices. Scientists are often required to apply their research findings, not just produce them. When scientists and engineers play similar roles, managers should use similar methods in evaluating them. When evaluating scientists and engineers, managers need to emphasize performance, not a person's position or profession.

CONDUCT A SELF-EVALUATION OF PERFORMANCE AS AN R&D MANAGER

In addition to evaluating the R&D staff's performance, R&D managers should also evaluate their own performance. R&D managers can check their performance by considering information gained from the exit interviews of researchers leaving the company. In addition, staff members can be asked to assess laboratory procedures and practices, and ideas about improvement can be solicited from employee committees (e.g., quality circles).

BROADEN NONTECHNICAL KNOWLEDGE

R&D managers should always broaden their knowledge about nontechnical areas. Four of the most important nontechnical areas are:

- Information about the company—A primary element of effective R&D management is an intimate knowledge of the company and how it works. Also essential is a working knowledge of the company's competitors. Efforts to acquire more knowledge about the company and how it works can expand the manager's communication channels with many people in company functions outside R&D.
- Finance—Just as mathematics is the language of science and engineering, finance is the language of business. To foster communication with non-R&D managers in a language they understand, the R&D manager must become literate in finance, through becoming very comfortable reading the balance sheet of a profit-and-loss statement and learning to understand the company's flow of funds. Some financial concepts can be learned in a business course, and a focused course in finance also may be helpful. In addition, several books on finance are specifically aimed at the technical manager. Short courses, such as a course on finance for the nonfinancial executive, are also effective.
- Sales skills—R&D staff members sell one of the most fragile and perishable of

commodities: ideas. To succeed in carrying out innovation requires skills in selling and communication. Innovative ideas do not speak for themselves; they need champions. Many short courses can teach R&D managers how to sell their ideas effectively to business managers and customers.

- Marketing—A marketing function can be a strong ally of an R&D department. Indeed, in some cases, the marketing department can be the champion of technological innovation in a company.

Consequently, the R&D manager should try to learn as much as possible about marketing strategy to work effectively with the marketing department. One way to do this is to take courses in marketing, particularly in market development.

CONCLUSION

Although a company's long-term growth and success depend on development of new technology, successful R&D managers understand that their primary focus must be on people rather than technology. Technical expertise continues to be important; however, effective R&D managers concentrate on providing their staff with leadership, motivation, opportunities for training and education, and an environment that encourages innovation.

THE MISSING RUNG

Teaching Management Concepts to Engineers

Engineers who are capable of managing are made, not born. Managerial skills, like engineering skills, require training and experience. Few engineering programs, however, offer training in technology and project management, which places the burden for basic management training on the commercial sector.

Ron K. Bhada

More than 20 years ago, after graduating from the University of Michigan, I joined a major industrial organization. I served in many positions at that company during the next 28 years, starting as a junior engineer and working my way up to manage a department, reporting to the division head. During this period, I also taught outside courses to novice engineers.

In the course of this experience, I observed in my company and in other organizations that excellent engineers are frequently limited in their careers because of their deficient management skills. These deficiencies usually occur in such areas as interpersonal relations, group dynamics, planning and controlling projects, and general business acumen—all aspects of management principles that can be learned at universities.

It is difficult to see how more nontechnical training could be added to a four- or five-year curriculum that is already bursting with necessary scientific and engineering training; indeed, adding management courses is not the solution because new courses entail giving up some necessary engineering training. This chapter presents a few select cases indicative of specific deficiencies that could have been resolved by early attention at the academic level. In addition, an alternative approach that one university tried and found successful is presented.

NOT JUST A PERSONAL LOSS

The scenarios detailed in the following sections illustrate how a lack of management training hurts both engineers and companies that employ them. These problems, far from being exceptional, are quite common; in conversations with managers around the country, I hear the same stories. In these three cases, the root of the problem seems to be a lack of basic knowledge of the principles of management and interpersonal relations.

Engineers as Supervisors

Industrial organizations often appoint engineers as technology supervisors; typically, they are excellent engineers but lack management training. Such supervisors

RON K. BHADA, PhD, is currently department head and professor in chemical engineering at New Mexico State University. He is also director of the New Mexico Waste Management Education & Research Consortium. Prior to this, he worked for more than 28 years at the Babcock & Wilcox Co; his most recent position there was manager of advanced products and strategic planning.

have the ability to identify potential problem areas in any new process and provide corrective solutions before the problems become a reality, an ability that is invaluable to senior management of the company. When it comes time for promotions or advancement, however, these individuals are usually bypassed because they are totally unacquainted with such basic management principles as motivation, interpersonal relations, and group dynamics. As a result, they often unintentionally alienate their superiors, subordinates, and peers.

These engineers are not insensitive human beings; they simply were not introduced to the basic concepts of management early enough in their careers to be able to practice them in their work environments. Many of these individuals could have gone much further in their careers, profiting both themselves and their companies, if management skills had been added to their excellent technical knowledge.

Engineers as Project Leaders

Another example that points to this problem is that of young engineers who are appointed project leaders for major projects. These engineers possess excellent technical abilities and yet often fail because they lack the training in strategic and operational planning and control that is essential to completing large multimillion-dollar programs successfully.

Misdirected Technology Development

In addition, I have observed many cases of companies that have realized only a minuscule return after investing several million dollars in developing advanced technologies and new products. At the same time, the companies' R&D divisions issue many reports that point to successfully developed products. The products they mention are technical successes but commercial failures.

Although the reasons behind these failures are manifold, they all share a common denominator: the developments are executed without a business plan. Typically, the researchers identify what they think is a commercial need, identify a technical approach that could lead to a product that satisfies this need, test the technical feasibility of the approach, develop a prototype, and declare the product a success. Essentially, they follow the procedures for a well-executed laboratory experiment.

What is lacking is a business approach to new product development. This approach should involve marketing, manufacturing, and business planning in addition to These market-oriented functions are vital to commercial success, yet they are not participants in the formative stages. Nor is there any planning or development for the market, manufacturing, or commercialization phases. In general, the R&D staff members are not avoiding such a team approach; they are unaware of the need.

MORE MANAGEMENT TRAINING

During the last couple of decades, many universities have taken to heart the advice that industry gave them regarding improving students' communication skills.

In schools across the country, in every curriculum, oral and written communication is stressed in classrooms as a part of regular course work, and graduates' communication skills have improved noticeably.

Now, similar training in management skills is needed. This is even more important today because technical professionals are expected to play more of a business role in industry. These individuals should not necessarily be aiming at a management career. Management expects every technical person to have certain basic skills without which professional staff will be relegated to the roles of technicians performing routine functions.

management expects technical professionals to have certain basic management skills without which professional staff will be relegated to performing routine technician-level functions

TWO-PRONGED APPROACH

At New Mexico State University, the department of chemical engineering has a two-pronged approach. The curriculum includes the introduction of basic management concepts to engineering students in their junior year, and the practice of these concepts as part of upper-level technical courses and laboratory coursework.

In spring 1989, the department introduced basic management concepts to technical students in their junior years or early in their senior years in a one-credit, one-hour-per-week seminar course on professionalism. The department did not have the luxury of conducting full 3- or 4-credit courses on this subject; given the limited time available, this course aims to familiarize students with the basic concepts. The application of these concepts is up to the individual students.

The department did not expect a large number of students to register for the first semester of the course. It was gratifying that at least several students registered, and as the course progressed, others heard that it taught students the practical aspects of business life. Several students who were not registered in the course started attending the seminars (without receiving credit); about 25 students participated eventually. The course covered many areas, including the ones discussed in the following sections.

Planning and Control

This subject area includes the basic concepts of strategic and operational planning, which were previously unknown to these students. The subject was introduced with a study of the methods of evaluating the background environment, setting goals and objectives, developing strategies and strategic directions, developing tactical and implementation plans, and developing techniques for tracking and controlling the plans. Industrial case studies were used, and students also developed their own plans.

One plan that was studied in the course was a five-year business plan for the university's chemical engineering department. The plan identified the mission, the environment, the goals, the objectives, and the implementation actions in the following areas vital to the business of a university department:

- The student body.
- The research aims.
- The teaching objectives and curriculum.
- The faculty.
- The facilities.
- The department's external relationships.

On the personal side, each student developed a plan for life beyond graduation, and the students as a group analyzed each of these plans. The students learned how to set their own long-range goals and then set an action plan to achieve these goals within the limits of each student's expected resources. Both the departmental plan and the individual students' plans are maintained in the course file to monitor how they compare with the achievements of the specified goals during the next 5 to 10 years.

Business Concepts

Technical personnel cannot work in a technical vacuum, even if they are not managers. They must be aware of the interactions of technology with the marketing and manufacturing functions and other business needs. As Bruce Merrifield has pointed out, "The crucial understanding is that industrial technology is not a technical decision, but rather a business decision."[1]

To deepen their interest in general business concepts, the students studied cases in which technology alone did not solve business problems. For example, in the environmental area, the federal government has worked hard to develop the technology for the burial of transuranic nuclear wastes. However, technology alone does not satisfy this issue. The government is struggling to develop information relative to economics, risk analysis, legal implications, public policy, and marketing concepts. Without this information, the plans for burial of the nuclear waste will never become a reality.

Other examples involved the development of new products. The students analyzed cases in which a new product that was developed by a company never had commercial success because of a lack of coordination of the R&D function with the manufacturing and marketing functions during the development phase.

Motivation Concepts

Although many technical graduates believe that what motivates them motivates every other preson in the world, not everyone is motivated by the potential for a technical breakthrough, particularly if it may not make money. This often leads to interpersonal problems for engineers in a professional atmosphere, and the engineer is not even aware of the problem's root cause: technical myopia.

I have observed that interpersonal relationships can be greatly improved if engineers are familiar with motivational research and theory. To address this need, the New Mexico State University course discussed basic concepts of the hierarchy of needs.[2] The understanding that human needs start with basic needs (e.g., physical safety and love), which when satisfied lead to higher needs (e.g., esteem and self-actualization), was a revelation to many of the students.

managerial concepts are put into practice as part of existing technical courses and laboratory coursework

The course then focused on how achievement can be learned. The students read further to understand in finer detail what makes people perform better to achieve the goals of the organization and also satisfy their own personal needs. Several case studies were discussed that brought out the application of the theories.

Engineers traditionally think analytically; these motivational theories emphasize that motivation is often provided by feelings, which are nonanalytical. By studying motivational theories and cases, the engineering students learned to understand what makes people behave the way they do. The case studies further illustrated how careers of skilled engineers have been limited because of their lack of understanding in this area.

Group Dynamics

In today's world, every engineer must learn to work as part of a team. The students learned the factors that affect organizations and team performance and the importance of informal leadership on the functions of a team. It was the intention of the course instructor to follow up with fictitious case studies; however, this purpose was served by students who introduced cases of group dynamics from their past social or classroom experiences. For the engineering students, it was a revelation to understand why a particular group situation in the past had developed in a certain way. The student groups analyzed their past experiences, identifying how the situation might have been used to advantage if the group dynamics had been understood. Certainly this training will assist in future group situations.

After the Course

These concepts are put into practice as part of existing technical and laboratory coursework. This is similar to what has been done in communications training, in which students give oral and written reports as part of their technical coursework. For example, planning and control is practiced in laboratory courses. The students set objectives and goals, develop action plans, set schedules, and perform other related tasks and then measure their progress as they work on particular projects.

Group dynamics techniques are practiced in group projects. The group projects can be a learning experience if some time is devoted at the end of a project to analyze how the group functioned—that is, answering such questions as, Who were the formal and informal leaders? How did communication proceed? and Was there consensus or compromise?

Business concepts are naturally practiced as part of design courses and engineering economics courses. The students perform market evaluation, manufacturing plan development, commercialization plan development, and pro forma balance sheets and income statements.

GETTING A HEAD START

An introduction to management concepts is essential for all engineers. This can be accomplished without overloading or extensively revising the technical curriculum. A professionalism seminar would be an asset to any curriculum, and even a small degree of interpersonal development practice would provide a head start for the engineers as they enter the industrial world.

Notes

1. B. Merrifield, "Selecting R&D Projects for Commercial Success," *Research Management* (January 1981).
2. A. Maslow, *Motivation and Personality* (New York: Harper & Row, 1954).

MAKING A TEAM EFFORT

Integrating R&D with Quality Improvement Efforts

In recent years, companies have grappled in various ways with the issue of quality improvement. Traditionally considered a manufacturing and marketing issue, quality improvement is affected equally strongly by engineering and R&D. Now, companies are discovering the challenge of redirecting their engineering and R&D forces toward overall company quality.

Robert Szakonyi

Integrating R&D with efforts to improve quality has become an important issue for US companies. During the 1980s, many companies began making great efforts to train their employees to improve the quality of their products and processes. Until then, US companies, on the whole, had neglected the issue of quality. During the 1970s and 1980s, however, strong competition from foreign companies, especially Japanese companies, revealed many weaknesses in the quality of US companies' products and processes. Many US companies recognized and began to combat these weaknesses.

Most US companies typically have reacted to these perceived weaknesses by requiring dozens—or even hundreds—of their employees to attend training courses on improving quality. These employees usually would come from all of a company's functions (e.g., manufacturing, purchasing, marketing, R&D, engineering, and quality assurance). Everyone who attended these courses learned the same general principles of quality improvement. After participating in these training courses, however, the managers of each function had to adapt these general principles to the specific conditions of their function.

For some of these functional managers, the task of adapting these general principles of quality improvement has been quite straightforward. Traditionally, issues related to manufacturing, purchasing, and quality assurance have constituted the bulk of the issues related to quality improvement. However, issues related to R&D, engineering, and marketing have not been acknowledged as being relevant to quality improvement. This is because quality traditionally has been limited to those aspects that involve the inspection of products after they have been produced. When quality began to be defined more broadly as related to the whole company's operations—which is what the employees in US companies were taught in those training courses—R&D, engineering, and marketing also were recognized to be involved in company efforts to improve quality.

For managers of R&D, engineering, and marketing, who now find themselves

ROBERT SZAKONYI, PhD, is the director of the Center on Technology Management at IIT Research Institute, Chicago. He has performed consulting work for many companies in a variety of industries and has written two books and more than 40 articles on technology management.

responsible for quality improvement, the new challenge is to adapt general principles of quality improvement to their own function's specific operations. They have had to rethink how their function should operate with regard to the issues traditionally associated with quality (i.e., issues related to inspection). More importantly, these managers have had to consider how their function should operate in light of new quality-improvement ideas that encompass a whole company's operations.

Although marketing managers face challenges in integrating their function's operations with company efforts to improve quality, this chapter does not deal with those challenges—except in relation to the issues that R&D or engineering managers confront. This chapter focuses only on how R&D or engineering managers can integrate their function's operations with company efforts to improve quality.

The starting points for considering what R&D or engineering managers can do in this area are the general principles of quality imrovement. The following section discusses these general principles. Because an R&D or engineering department's efforts to improve quality can be understood fully only within the context of a whole company's improvement efforts, this chapter also examines company efforts to improve quality. Six cases that this writer knows from personal experience illustrate the many aspects of improving quality within a company.

This chapter continues by describing how the R&D engineering department can participate in its company's efforts and adapt general quality-improvement principles to its own operations. Finally, a special case is described in which an engineering department and a manufacturing department can work together to improve quality. This case involves computer-aided design and computer-aided manufacturing (CAD/CAM).

GENERAL PRINCIPLES OF QUALITY IMPROVEMENT

The general principles of quality improvement are quite simple. The most important tenet is: Make products right the first time, rather than spend a great deal of time and effort inspecting for all of the defective products.

This first tenet provides the basis for examining a company's entire operations in terms of quality. Once a company commits itself to producing products right the first time, it is forced to prevent mistakes rather than to correct them.

To accomplish the goal of initially producing products right, a company should concentrate as much on improving the processes as on the products themselves. Consequently, implementing statistical process controls in manufacturing plants is as important as inspecting the products afterward.

Just as important as statistical process controls are the procedures through which managers manage their activities and the attitudes of the employees. Such methods as statistical process controls will not yield better results unless the managers and employees who manage and use them are willing and able to do a thorough job.

Significant improvements in the way products are manufactured will not improve quality much if the company's other functions do not concurrently improve their own operations with regard to quality. Marketing must improve its definition of product requirements. R&D and design engineering must improve the producibility of their designs. Manufacturing engineering and quality assurance must improve the methods and tools of statistical process controls. Purchasing must improve the quality of materials used in production. Unless all of these functions work together, the company cannot attain major improvements in the quality of its products.

the costs of making defective products outweigh the resources needed to improve quality

It is the customers, not the company, who are the ultimate judges of product quality. Customers judge quality in terms of how well a product conforms to their needs. It is from the customers' perspective that the first tenet really takes on importance. The problems that arise from delivering defective products to customers can last for years. Consequently, a company must know about the deteriorating quality of its products before its customers do so that corrective action can be taken.

For any company to take all necessary actions to improve quality, it will have to shift resources from other activities. The obvious question is, Will improvements in quality pay off? Ideally, quality improvements pay off in two ways: they allow a company to cut quality costs, and they help a company become more competitive.

The term *quality costs*, contrary to its apparent meaning, concerns all of the costs that stem from making defective products. These costs include expenses from conducting inspections and tests, expenses from reworking or replacing defective products in manufacturing, and expenses from reworking or replacing defective products at customers' facilities. In addition, there are many other costs, such as those stemming from product recalls or from liability litigation.

In general, the costs of making defective products outweigh the resources needed to implement company efforts to improve quality. They also make a company less profitable. As John M. Groocock wrote in *The Chain of Quality* (New York: John Wiley & Sons, 1986), quality costs are unnecessary. Whereas many of the costs involved in developing, producing, selling, and distributing products can be cut only somewhat (e.g., through shrewder negotiations for raw materials), quality costs are wasted resources. Quality costs can be cut considerably—they can be almost eliminated. Therefore, improvements in quality help cut quality costs.

Improvements in quality also make a company more competitive. Groocock pointed out that, according to studies conducted by the Strategic Planning Institute, the quality of a company's products has about the same weight as a company's market share in determining the profitability of a company. The studies are consistent with much anecdotal evidence showing that Japanese companies' business successes have been due in large part to the quality of their products. Consequently, by improving the quality of their products, companies become more competitive.

SIX CASES OF COMPANY EFFORTS TO IMPROVE QUALITY

Before considering the R&D or engineering department's integration with its company's efforts to improve quality, it is useful to examine existing companies' experiences. Six cases of such company efforts are described. These companies come from six different industries—chemical, electrical products, food processing, adhesive products, aerospace, and automotive parts and electronic components.

These companies achieved different degrees of success in their quality-improvement efforts. One of the companies failed, four succeeded in varying degrees, and one had tremendous success. The critical factor in the companies' success was the degree to which a general manager in a company led efforts to improve quality. Together, these cases illustrate the various kinds of company quality-improvement efforts that an R&D or engineering department should consider in its own efforts.

Case 1: Failure to Take Off

A medium-sized chemical company faced major problems in remaining competitive. Its products were mostly mature commodities. In addition, the company's costs in producing its products were much too high.

Within the company, two corporate staff groups tried to get their company to improve the quality of its products and processes: a management information services group and a manufacturing systems group.

The management information services (MIS) group tried for many years to persuade manufacturing managers to use statistical process controls; they never succeeded. The group also tried persuading managers of other company functions, including the R&D managers, to exploit computers more thoroughly in their operations. Most company managers, including the R&D managers, resisted the MIS group's efforts and continued to use computers mostly for accounting purposes.

At first, it appeared that the manufacturing systems group would have more success than the MIS group in getting the company to improve quality. Supported by the head of manufacturing, the manager of the manufacturing systems group pressured manufacturing plant managers to accept many new manufacturing systems (e.g., materials requirements planning, CAD/CAM, management information systems, and group technologies). The manufacturing managers and the workers in the plants, however, were not really eager to implement these systems.

Shortly after these new systems were established in the plants, the head of manufacturing left the company. From then on, the plant workers ignored the new manufacturing systems. Thus, the manufacturing systems group had succeeded in getting its new systems established, but failed to get them used.

Case 2: Limited Progress

At an operating division of a large electrical-products company, the group that pressed for improving quality was a manufacturing technology group. This group's mission was to provide support to the largest manufacturing plant in the operating division.

The manufacturing technology group was able to establish some activities related to improving quality within its division. For example, it developed courses to train design engineers in designing for manufacturability. It was able to establish in-process quality controls in a few of the manufacturing lines of the plant.

the manufacturing technology group had reached a plateau in its efforts to improve quality

Nevertheless, the manufacturing technology group was able to make only limited progress. Because the quality of the operating division's products had been reasonably high for 20 years, the manufacturing technology group had continual difficulty in getting senior manufacturing and senior R&D managers to pay attention to quality problems. For example, in the manufacturing plant, some machines were operating only 40% of the time because the operators of the machines did not understand how to run the machines efficiently. In addition, because the R&D people did not always test their new product designs thoroughly, the manufacturing plant still ran across too many problems in producing new products.

In sum, the manufacturing technology group of this operating division had reached a plateau in its efforts to improve quality. On its own, it was able to make some progress. Because it did not have the support of the general manager, however, this group was limited in what it could persuade senior manufacturing and R&D managers to do.

Case 3: An Important Corner

At a large food-processing company, it was the corporate R&D managers who pressed for improvements in quality. Manufacturing managers were able for years to resist the R&D managers' efforts to give higher priority to quality. One reason for the manufacturing managers' successful resistance was that the senior corporate managers did not appreciate the significance of poor quality and thus paid no attention to the R&D managers' ideas.

From the R&D managers' point of view, the two main issues to be addressed in the company were replacing the antiquated equipment in the manufacturing plants and overcoming the short-sightedness of the manufacturing managers in planning new capital investments. The antiquated equipment made it very difficult to produce high-quality products. In addition, the manufacturing managers preferred simply to patch up this antiquated equipment, rather than invest in newer equipment.

The corporate R&D managers made little progress in their efforts to improve quality until a new general manager was appointed in the largest operating division. What made this appointment significant was that he had different ideas on the management of this operating division's businesses. Because he had managed the small company that had been acquired by this food-processing company, he had experience in managing operations as well as in marketing.

Within several months of his appointment, this general manager realized that his operating division would not meet his business goals unless the manufacturing plants were modernized. He then concentrated on completely reexamining the operations of the manufacturing plants. He also arranged to have the corporate engineering department of the company placed under his authority. For years, manufacturing managers and R&D managers had argued about what the responsibilities of this engineering department were. Now, that problem was solved.

In this company, an important corner was finally turned with regard to addressing issues related to quality. Even though the R&D managers had tried for years to foster improvements in quality, nothing happened until a general manager took the lead.

Case 4: Perseverance

A small adhesive-products company manufactured its products extremely inefficiently. With some products, the company had to scrap 70% of the output from its plants because the products did not conform to product specifications. Given the state of the plants, this was not surprising. Pumps leaked, thermometers were missing, and some of the equipment was broken. Furthermore, many of the people in the plants did not know how to run the equipment they were operating. Finally, no one in the plants kept records. For years, no one in the plants even knew how much of the plants' output was off specification.

Two new managers in this company led the campaign to improve quality—a head of manufacturing and a manager of a process engineering group, which was located in the R&D department. After taking charge, the new head of manufacturing developed a five-step, long-term strategy for the plants' improvement. These steps are:

1. Reeducating the manufacturing managers who reported to him.
2. Repairing all of the equipment in the plants.
3. Redirecting efficiently the flow of materials through the plants (e.g., changing the flow of materials so that a chemical reactor would be filled in two or three hours instead of the eight hours that it was taking formerly).
4. Using electronic equipment in the plants to monitor the manufacturing processes.
5. Having the plants run with the help of computers.

It took the head of manufacturing a few years to accomplish the first three steps in his long-term strategy. Because reeducating manufacturing managers, getting the equipment repaired, and changing the flow of materials were activities that he had the authority to get done, he could proceed on his own in these areas. In integrating electronics controls and computers into the plants' operations, however, he had to proceed more cautiously, because these activities had to be approved by the president of the company.

While the head of manufacturing was improving the overall operations of the plants, a newly-appointed manager of process engineering was tackling the specific problems in the plants, one by one. First, he worked on assuring that existing products were manufactured in conformance with their specifications. During the first 18 months, he helped the plants improve their operations to reduce the 50% to 70% of products off specification to only 5%. Then, he developed a system of documentation to enable the people in the plants to keep records of their operations and equipment problems.

The perseverance of these two managers eventually paid off. Within a few years, the manufacturing plants were operating at a much higher level of performance. Although the head of manufacturing was not able to accomplish his goal of completely automating the plants, he was able to persuade the president of the company to approve two significant items.

the company was told that if it didn't improve the quality of its products, prospects for future contracts would be hurt

First, the president agreed to purchase $3.2 million of automation equipment. The head of manufacturing convinced the president that this equipment would quickly pay for itself by correcting a major manufacturing problem. Until then, the plants had been producing one bad batch of adhesive materials per week, which had to be discarded. The head of manufacturing demonstrated that the company needed to generate $100,000 of new business a week to compensate for the losses from one bad batch. Second, the head of manufacturing persuaded the president to allow every employee in the company to attend a one-week training course on improving quality.

In this company, perseverance paid off in the end. The head of manufacturing and the manager of process engineering were able to improve operations in the plants with the president's support.

Case 5: Recognition of Quality Problems

An operating division of an aerospace company was informed by its primary customer in the US Department of Defense that its products were defective. The company was also told that, if it did not improve the quality of its products, its prospects for future contracts with the Department of Defense would be hurt. Because the general manager of this operating division participated in all discussions with the customer, he was forced to recognize how serious the problems were.

This general manager led new company efforts to improve quality, and the operating division nearly set records for the speed with which it changed its operations. Two of the general manager's most important actions involved sending the 250 senior-most managers of the division to outside training courses on quality improvement and taking the quality assurance function away from the manufacturing organization and having it report directly to him.

With support from senior management, the quality assurance group played a vital role in helping the operating division correct its problems. One of its main tasks was to change the way the management information system was run. Although the management information system collected an enormous amount of data, it did not then provide the data in a useful format. The quality assurance group's goal was to change the format of the information system so that its data alerted company managers to problems—particularly to those concerning quality. To achieve this goal, the quality assurance group enlisted the help of 20 primary users of the management information system within the operating division. These persons served on a task force to redesign the format of the infomation system to serve all users.

After the task force completed its mission, this renovated information system provided many advantages. Design engineers received better information about design-related problems that appeared after products left the plant. Manufacturing engineers and manufacturing foremen were able to spot problems more quickly. Purchasing managers had better information for identifying patterns in defects in purchased raw materials.

The combination of a more useful management information system and managers throughout the company educated in quality improvement enabled this company to make great progress. When a general manager takes the lead in recognizing and dealing with quality problems, as in this case, great progress can be made quickly.

Case 6: Sponsorship from Senior Corporate Management

The last case concerns a large automotive parts and electronic components company with many operating divisions. In this company, the senior corporate managers sponsored improvements in quality without first experiencing customer pressure to recognize quality problems. These managers gave wholehearted support to a corporate quality assurance group, which was charged with helping change corporationwide management practices related to quality.

One of the new management practices that the quality assurance group helped put in place related to planning. Each operating division was required by the corporate senior managers to develop an annual plan delineating its concrete goals for improving quality, which was then reviewed by the senior corporate managers.

This corporate quality assurance group was responsible for a wide range of activities. For example, it established a company training college to improve quality, which hundreds of employees from the divisions have attended each year. It also publishes a monthly newsletter for educating all company employees about key issues related to quality improvement.

One of the corporate quality assurance group's most important methods to help the divisions incorporate quality-improvement ideas into their daily operations involves the technique of competitive quality evaluation. The corporate quality assurance group trains managers in the divisions to compare systematically their own products with those of competitors, in terms of various criteria (e.g., how well the product matches requirements and how it performs after delivery).

The great support that the senior managers of the corporation have given the corporate quality assurance group has allowed this group to develop a philosophy of quality for the whole company, which the senior corporate managers have endorsed. This philosophy serves as a basis for gaining total commitment by all company employees to produce superior products—products, that is, that can enable the company to gain a dominant share in each of its markets.

As demonstrated in these six cases, when general managers are more involved in company quality improvement efforts, these efforts are more successful. Also, these six cases show the many aspects of improving quality within a company. To varying degrees, all functions within a company must play a role in improving quality. Finally, these cases illustrate the kinds of company efforts within which R&D or engineering must work to integrate its own efforts to improve quality.

IMPROVING QUALITY WITHIN THE R&D LABORATORY

each operating division was required to develop an annual plan for its goals

When integrating its operations with company efforts, an R&D department first must define its own goals concerning what it can do within the laboratory and with the other functions in the company. This section addresses what it can do within the laboratory. The following section deals with what it can do with the other functions.

To define quality improvement–related goals that are also meaningful to scientists and engineers is a challenging task. For example, Lester C. Krogh, who was vice-president of R&D at 3M, wrote that his R&D department did not find "manufacturing's quality program readily applicable to the laboratory environment."[1] In a similar vein, Leon Starr, who was president of Celanese Research Co, wrote that "force-fitting the Crosby Quality College program across the corporation was just not going to work." Therefore, Starr recommended that general principles of quality improvement be restructured "to connect with the research values we wanted to promulgate."[2]

An R&D department can guide its efforts within the laboratory with the following five actions for improving quality, which are also relevant to an R&D or engineering environment:

- Strengthening a laboratory's technical capabilities.
- Improving the management of R&D.
- Increasing technical personnel's training related to improving quality.
- Improving the work environment within a laboratory.
- Strengthening analytical, information, and computer capabilities within R&D.

Strengthening a Laboratory's Technical Capabilities

Almost all laboratory managers work to strengthen their laboratory's technical capabilities. An effort to improve quality can also help broaden the perspectives of technical personnel regarding technologies outside their discipline. The author visited laboratories (e.g., in the health-care industry) in which scientists, although qualified, resisted developing new technologies outside their own technical disciplines. The author also visited an instrument company in which the engineers, most of whom were mechanical engineers, for years ignored advances taking place in electronics. Eventually, the only way for this company to remain competitive was to base its products on electronic, not mechanical, devices. Unfortunately, most of its engineers were unprepared to make the transition.

R&D managers can use various tactics to keep technical personnel's minds open to new technical opportunities outside their discipline. The R&D managers of a materials company established a technical forum in which a diverse group of company technical personnel presented recent R&D results. The R&D managers of a pharmaceutical company assigned a small group within the laboratory to monitor many new technologies and to keep the technical staff abreast of them. The R&D managers of a computer company made sure they recruited at least a few new college graduates each year from a variety of technical disciplines so that the laboratory as a whole could be kept abreast of the latest technological discoveries outside the company.

Improving the Management of R&D

Another aspect of improving the quality of operations within R&D involves improving the management of R&D. One way to do this is to track R&D activities more effectively. Three R&D activities that R&D managers should particularly observe are:

- R&D projects—Many R&D departments do not have formal reviews of ongoing projects. These reviews should be conducted by technical personnel who are not involved in the projects.
- Design changes—Many changes in design that occur during the latter phases of a new product development are called design improvements by engineers. According to Groocock, however, "80% or more of the drawing changes are for corrections of design defects." Therefore, design changes as a whole should be tracked to identify the major reasons for defects in design.
- New product developments—The author visited companies in which marketing or quality assurance departments had to develop a multiple-step protocol for managing the various phases of new product developments. This responsibility fell on these other departments because the R&D department would not define the necessary milestones to develop a new product.

Other aspects of a laboratory's operations do not lend themselves to being tracked. R&D managers can analyze the technical skills of a laboratory in terms of the company's needs. They can examine the costs of running a laboratory to discover whether those costs are consistent with the laboratory's priorities. They can evaluate the procedures for approving the laboratory's technical reports to ensure that these procedures are not too bureaucratic.[3] And they can audit failed R&D projects to discover the reasons for their failure.

Increasing the Training of Technical Personnel Related to Improving Quality

R&D managers can also improve the quality of the work by increasing the training of technical personnel. Six areas in which technical personnel can be trained in this regard are: design for manufacturability, experimental design, problem solving, value engineering, statistical process controls, and quality function deployment.

Design for Manufacturability. This training emphasizes coordinating product design and manufacturing process planning. Some of the basic rules of design for manufacturability are:

- Keep the design as simple as possible.
- Select materials that are not difficult to clean.

- Minimize the use of fasteners and adhesives.
- Conduct a complete tolerance analysis of components.
- Design for the most economical production methods.
- Provide enough room for the assembly of parts.[4]

an effort to improve quality can help broaden the perspectives of technical personnel outside their discipline

Experimental Design. This training involves combining engineering and statistical methods to develop a design that could be manufactured with high quality and at the lowest possible cost. This training helps engineers avoid problems that are created by considering one factor at a time when developing a design. Experimental design allows engineers to examine the interactions of a wide range of factors in experimental settings to maximize the robustness of a design under a variety of operating conditions.[5]

One particular approach to experimental design is the Taguchi method developed by a Japanese engineer, Genichi Taguchi. This method allows engineers to identify the factors with the strongest functional relationship to a product's requirements. In addition, engineers can use this method to isolate the effects of various manufacturing conditions on a product. Finally, engineers can determine the effect of uncontrolled variation in the manufacturing process on the quality of a product.[6]

Flow Charting. Flow charting is a method for examining the process through which work is done. To develop a flow chart, one must take into account the people involved in the work, the procedures used, the machines employed, and the purposes of the work. For example, an R&D or engineering department can examine how requests for laboratory equipment are handled, from the initial recognition that equipment is needed to the final decision about a request. Flow charting can help R&D managers understand all that is involved in an activity. It also can help them identify areas in which they need to take corrective action.[7]

Value Engineering. Value engineering is an approach for identifying unnecessary costs in the manufacturing of a product. For example, ways of reducing production or material costs are considered; alternative designs are also considered. This approach provides engineers with tools to optimize the manufacturability and quality of a product.

Statistical Process Controls. One of the most important ways of improving manufacturing is through statistical process controls. All process engineers should use this method whenever possible, and design engineers should at least be knowledgeable about it. By using ths method, engineers can understand the variability of a manufacturing process. The heart of this method consists of analyzing statistical variations in the production of a product at different times. By doing this, engineers determine how much variation occurs naturally in a manufacturing process and whether any special problem agents must be eliminated to control a manufacturing process. After analyzing a manufacturing process, engineers can establish statistical process controls that enable the process to consistently yield high-quality products.

Quality Function Deployment. In contrast with the other approaches, quality function deployment involves information about what a company's customers want. This information is then used in developing a product design. The most important tool of this approach is a chart that examines a set of market quality requirements in relation to a set of product quality features. By analyzing systematically how certain product features could affect—positively or negatively—the product's ability to meet specific market requirements, engineers are able to develop designs that more effectively address customers' needs.

Improving the Work Environment Within a Laboratory

To improve the work environment within a laboratory, R&D managers first must evaluate the laboratory climate. The questions they can ask themselves include:

- Are members of the technical staff encouraged to ask why they are supposed to carry out specific tasks?
- Do technical people understand the goals of the laboratory and the goals of the company?
- Are people in the laboratory encouraged to make improvements in quality?
- Are they rewarded for making improvements in quality?
- Are there recognized channels through which new technical ideas or innovations can be passed for evaluation?
- Do members of the technical staff have sufficient opportunities to attend technical conferences outside the company?
- Do scientists and engineers have some time (as much as 10% of their total work hours) to investigate technical ideas not necessarily connected to project work?
- Are efforts made to prevent barriers from arising among different technical groups in a laboratory?
- Are technicians in the laboratory given adequate opportunities to develop as well as due recognition for their accomplishments?

Strengthening the Analytical, Information, and Computer Capabilities Within R&D

Some of the important groups of an R&D or engineering department are those that provide services to the line groups. In a large R&D or engineering department, these groups include analytical services, information services, and computer services. In a small R&D or engineering department, one group (or even one person) may provide all these services. Another possibility in a small R&D department is that company groups outside of R&D provide these services to the R&D or engineering department. Whatever the case, the importance of these groups' services often are not fully appreciated within the R&D or engineering department.

quality function deployment takes into account the needs of the company's customers

For example, analytical tests can play a critical role in determining an R&D project's success. Consequently, there are times when a technical person from an analytical services group should be included as a member of an R&D project team.

Similarly, specialists in information services can contribute a great deal to technical efforts. The author visited a laboratory in a materials company that used its information specialists to uncover many new technical opportunities. These information specialists investigated abstracts of chemical research, abstracts of scientific research conducted by the US National Aeronautic and Space Agency, and newspaper articles from the Reuters wire service. Information specialists can save substantial research time by studying the technical literature and patents related to an R&D effort. R&D managers should also assess how well their laboratory uses information services.

Finally, computers—and any group that provides computer services—can help an R&D or engineering department in many ways. Computers can improve the quality of information that a laboratory uses. Computers can provide insight into phenomena that could not otherwise be gained when physical experiments might not be feasible. Computers can make members of a technical staff more efficient and productive through modeling, laboratory automation, statistics, and data base handling. R&D managers, therefore, should look for ways to strengthen their computer capabilities and to tap the potential of computers.

INTEGRATING R&D EFFORTS WITH OTHER COMPANY FUNCTIONS

When integrating its operations with company efforts to improve quality, an R&D department should also work with other company functions. The two functions with which R&D needs to work most closely are marketing and manufacturing. In addition, R&D should work with the purchasing department and with company groups responsible for dealing with safety and environmental issues.

Working with Marketing

An R&D department should work most closely with a marketing department in specifying the requirements of a new or modified product. How the requirements of a product are specified has a tremendous impact on all subsequent efforts to develop and produce a product. If quality-related issues are not dealt with effectively when the product requirements are defined, then later actions in the product development process—for example, designing for manufacturability and using statistical process controls—will not significantly improve the quality of the product.

Within companies, there are two major problems related to specifying product requirements. The first is that issues linking customers' needs to product requirements are not nearly as well understood as the issues that are dealt with later in product design and production. The second is that marketing people, on the whole, have not satisfactorily identified what their responsibilities are in the question of quality.

Issues of linking customers' needs to product requirements are not understood well because: companies often have not been able to—or have not tried to—define customers' needs clearly; and, even if companies have defined customers' needs clearly, it is still very difficult to judge whether certain product requirements really meet the customers' needs. An important aspect of this problem is that it is hard to evaluate how much quality is gained or sacrificed when making value judgments about the trade-offs among various product features or the trade-offs among certain product features and specific prices.

With regard to the second problem, many marketing people lack an understanding of their responsibilities concerning quality. In *The Chain of Quality*, Groocock wrote:

> Most marketing managers have little appreciation of their key responsibilities for quality. Typically, they believe it is the responsibility of the company's quality people to maintain a "market" level of quality (i.e., more or less as good as the competitors') and that sales growth and profit will then result from the practice of marketing skills—pricing, negotiation with customers, advertising, and so on. Many marketing managers are unfamiliar with the PIMS results (concerning quality having about the same weight as market share in determining profitability), which indicate that pricing changes are usually rapidly matched by competitors, but *superiority* in quality cannot be quickly matched and gives a much better basis for growth of market share.

When visiting companies, the author found the same weakness among marketing managers. At a chemical company, marketing managers were preoccupied with the pricing of products and with customer relations, neglecting questions about the product's value to customers. At a food-processing company, marketing managers continually did an inadequate job when filling out a two-page document concerning the specifications of the product they wanted to sell. At a health-care products company, marketing managers would not even fill out a one-page document on product specifications; instead, the R&D managers in the company completed the document and attempted to get the marketing managers to approve it.

Given these problems, there are three things that R&D managers and marketing managers can work on together to improve the quality of their company's products: identifying customers' needs, conducting evaluations of the quality of competitors' products, and participating in technical service calls at customers' facilities.

Identifying Customers' Needs. R&D and marketing managers can use various approaches to work together in identifying customers' needs more effectively. One approach consists of organizing a conference that includes customers. For example, at a computer company the marketing, engineering, and quality assurance departments organized a conference examining the quality of their products. Representatives of the customers were invited to participate and to air their opinions about the quality of the products.

A second approach for better collaboration involves having R&D people instruct salespeople concerning the performance characteristics of their products. The author visited an electrical components company at which the sales people knew very little about their products and could not talk intelligently with customers about product performance. The author also met salespeople (e.g., at a food-processing company) who desired to become more informed about their company's products in order to avoid embarrassing situations when talking with customers. The salespeople in this food-processing company, however, could not find the right mechanism within the R&D department to learn about their product's performance.

if quality issues are not effectively handled in the early stages, later actions will not significantly improve product quality

A third approach for improving collaboration involves persuading R&D managers to help marketing managers develop quality-directed market research. Such market research is aimed at discovering customers' perceptions of the quality of both the company's and competitors' products. This market research could cover such issues as the consistency of the company's and competitors' products, these products' ability to perform under various operating conditions, and the caliber of the technical assistance provided by the company and by its competitors.

Conducting Evaluations of Competitors' Products. Besides asking customers to evaluate available products, a company can conduct its own evaluations of the quality of each product. As mentioned earlier, a company the author visited conducts such evaluations regularly. Some of the rules this company has used in conducting these evaluations are (adapted from *The Chain of Quality*):

- In comparing competitors' products with one's own products, consider no more than three competitive products, to avoid making the comparison too complicated. Group any remaining competitive products into a category called other and include this category of products in the comparison.
- Compare only products that satisfy the same customer need at about the same price.
- Use four rankings when comparing all of these products:
  - — Best (no product has comparable quality).
  - — Joint best (no product has better quality).
  - — Average (product quality is average).
  - — Worse (product quality is worse than average).
- Consider all of these products in terms of several areas of performance, such as:
  - — Match between the product requirement specifications and design.
  - — Conformance of the product to requirements when delivered.
  - — Performance of the product after delivery.
  - — Overall product quality.
- Apply the less extreme ranking when there is doubt about a ranking (e.g., joint best instead of best, average instead of worse).

These evaluations of a company's products in relation to competitors' products can give R&D managers and marketing managers many ideas about how to improve product quality.

Participating in Technical Service Calls at Customers' Facilities. Technical people can work with marketing or salespeople by participating in technical service calls at customers' facilities. The author visited several companies that gained great competitive advantages by having their technical people participate in calls to customers. The mission of these technical people usually was to ensure that the products were performing as required (or were being used as required). For example, the technical people at a steel company worked closely with customers at the customers' facilities, teaching them how to design and manufacture parts made with the company's steel.

At an adhesive-products company, a technical services group helped its company enormously with customers because it had an excellent reputation for responding quickly to problems at customer facilities. In addition, this technical services group played a major role in helping customers define the requirements of a product that they were considering buying.

At a computer company, multidisciplinary field service teams visited customers one month after their company's product was delivered. One or more engineers always served on these interdisciplinary teams. The teams were valuable in helping a customer deal with problems that had just cropped up or that had not yet become apparent to the customer. In other words, these teams helped ensure that their company's computers really performed at the level they were supposed to.

A final tactic an R&D or engineering department can use with a marketing department to improve quality is to audit the status of the technologies used in a company's products. The author visited two companies that were particularly effective in this area: one is a consumer-products company, the other, a materials company. The philosophy of both companies has always been that none of its product lines is a cash cow (i.e., a product line near the end of its life cycle). Consequently, neither company lets its products slip in terms of quality. The R&D departments of both companies are always active in trying to find ways to revitalize product lines that other companies would neglect.

Working with Manufacturing

An R&D department should also work closely with a manufacturing department. There are two major issues that R&D (or engineering) managers and manufacturing managers should address together: improving the manufacturability of product designs and improving the manufacturing processes.

Improving the Manufacturability of Product Designs. The author visited many companies in which the product designs transferred to manufacturing were inadequate. At a building-materials company, the manufacturing plants had to scrap 30% to 40% of certain products for many years because the product designs were flawed. At a health-care products company, the manufacturing managers regularly had to conduct expensive tests in the plants to bring the working product designs up to standard.

R&D or engineering managers can take six steps to improve the quality of product designs in collaboration with either manufacturing managers or with managers of other functions in a company.

salespeople could not find the right mechanism within R&D to learn about their own products' performance

First, R&D managers can increase technical people's knowledge about manufacturing needs. Four kinds of training programs relevant here, which were mentioned earlier, are design for manufacturability, experimental design, value engineering, and statistical process controls. Some of the instructors in these training programs for technical people could be manufacturing engineers from the manufacturing department.

Second, members of a quality assurance group could be used to evaluate the quality of product designs. The author knows of three companies—a tire company, an aerospace company, and a consumer-products company—that all made great progress in this area by getting quality assurance people involved early in evaluating product designs. Two major issues quality assurance people can examine are: How well does the design match the product requirements? Are the tolerances specified in a design really necessary for high performance, as the design engineers believe?

Third, informal relations between R&D and manufacturing can be nurtured. R&D or engineering people often specify the tolerances within a design in an arbitrary way. As Groocock pointed out, they sometimes also indicate tighter tolerances than needed because they do not trust manufacturing's ability to fulfill requirements. If representatives from R&D and manufacturing can informally discuss their mutual concerns, the product designs transferred to manufacturing will be more manufacturable.

Fourth, design engineers and manufacturing engineers together can plan the development of a new product and the manufacturing processes needed to produce it. Studies show that collaboration between design engineering and manufacturing engineering has led at times to a "reduction in parts of nearly 73% and in material costs of nearly 40% for a given component."[8]

Fifth, formal reviews of product designs, in which representatives from R&D or engineering, manufacturing, quality assurance, marketing, and purchasing are involved, can improve the quality of the designs. The purpose of a design review is to identify inadequacies in a design that will make manufacturing difficult and that will impair the quality of a product.

Sixth, an R&D or engineering department and a quality assurance group together can conduct tests on a product when development is complete. The purpose of these tests is to ensure that the product design is correct and that it meets the requirements. In such an arrangement, R&D and engineering might carry out the tests and provide the resources for the tests, whereas quality assurance might audit the tests to make sure they are carried out properly, as Groocock has suggested.

Improving the Quality of the Manufacturing Processes. R&D and engineering can become involved in many ways to improve the manufacturing processes of a company. One way is to develop models of the manufacturing processes so that manufacturing people could better understand these processes. For example, the R&D department of a building-products company developed a model of a plant's processes that helped the plant save over $7 million. The R&D department's model highlighted all the inefficiencies in the manufacturing processes, enabling manufacturing managers to take corrective actions. In another example, the engineering department of a computer company developed a model of manufacturing processes that showed manufacturing managers they could carry out a certain procedure with just two machines, instead of the three they had planned to use.

Another way in which an R&D department can help improve the quality of manufacturing processes is to audit the technologies used in the plants. With this procedure, the R&D department of a chemical company helped the plants in its company in a variety of ways. By identifying those technologies, used in all the plants, that allowed the company to compete most effectively, the R&D department provided each plant with new standards to upgrade the particular technologies in which it was weak.

The R&D or engineering department can also help improve the quality of manufacturing processes by developing a data base containing the engineering and manufacturing practices involved in product manufacturing. For example, the engineering department of an aerospace company built a computer data base that served as the corporate memory for the way its airplanes were manufactured. Similarly, an engineering group in a process-controls company developed a data base concerning all of its plants' manufacturing technologies.

Working with Purchasing

Whereas an R&D or engineering department regularly deals with the marketing and manufacturing departments, it often does not work as closely as it should with the purchasing department. For example, the author interviewed purchasing managers at an automotive-parts company and at a food-processing company, who stated that they seldom talked with technical managers and were not aware of developments in R&D and engineering. Poor communication is costly. To illustrate this point further, in many industries 50% of the costs involved in producing a product stem from materials and components purchased outside the company. If R&D or engineering and purchasing are not communicating effectively, many of these materials and components may be bought at too high a cost or at a lower quality than needed. There are two main ways in which R&D (or engineering) and purchasing should work together.

First, the R&D department should specify with the purchasing department what items are to be purchased, when alternative items might be purchased instead, and what concessions might be made regarding these items. To be sure, getting R&D and purchasing to work together is not always easy. An R&D department and a purchasing department each has its own priorities. In general, R&D wants to get the best materials that it can, while purchasing wants to cut costs.

Nevertheless, as the author observed at a highway-equipment manufacturer, differences in priorities between R&D and purchasing can be worked out through continual negotiations. The R&D and purchasing departments at this company repeatedly debated the feasibility of relying on one source for materials. Relying on a single source allowed the purchasing department to buy at lower prices. The R&D department, however, was concerned about the potential risks of relying on one source. Eventually the R&D and purchasing departments agreed that the company should use a second source for a small percentage of the materials. This would ensure that the company would always have materials available when necessary.

the R&D department always works on finding ways to revitalize product lines that others might neglect

Second, the R&D department should work with the purchasing department to make sure that the materials and components received are of high quality. An R&D department can provide special help in evaluating any new materials to be used in production. In carrying out this part of its mission, the R&D department must often work closely with the suppliers of these materials by helping them produce the materials at the desired level of quality.

Addressing Safety and Environmental Issues

An R&D or engineering department should also help its company address safety and environmental issues. For example, it can help test a product's safety. It can help ensure that the technical content of instructions related to the company's products is correct. It can also help plan the safe reuse or disposal of the company's products and of scrap and waste from the company's manufacturing plants. Finally, it can make sure that the company considers safety and environmental issues when conducting research and development.

CAD/CAM'S POTENTIAL TO PROMOTE QUALITY

Engineering and manufacturing departments in companies have used various computer tools to improve the design and manufacture of products. Three of these tools are computer-aided design (CAD), computer-aided manufacturing (CAM), and integrated computer-aided design and manufacturing (CAD/CAM). In some companies it is only the engineering department that uses computer-aided tools; in others it is only the manufacturing department. Finally, in a relatively smaller number of companies, both the engineering and manufacturing departments use computer tools.

In this section, the potential of an integrated CAD/CAM system to improve quality is considered. First, a CAD/CAM system is described, particularly with regard to its potential benefits for improving the design and manufacture of products. Second, critical issues related to exploiting these benefits are examined.

Potential Benefits of CAD/CAM

Two great potential benefits of CAD/CAM are that it can reduce defects in product designs and it can improve the producibility of product designs.

CAD/CAM can help reduce defects in product designs by allowing an engineering department to eliminate many tasks involved in translating a rough design to a final version ready for manufacturing. Certain design rules in CAD eliminate the need for detailed drafting, and design changes can be put directly into the system. Also, standardized parts used in other designs can be incorporated automatically into any design. By using CAD, engineers are generally more likely to develop their design right the first time.

CAD/CAM also helps make product designs more producible because it contains rules that guide engineers on issues related to manufacturing costs and quality. In addition, CAD/CAM serves as an important communication tool for coordinating the efforts of design engineers and manufacturing engineers. For example, it allows them to work from the same data base. Furthermore, CAD/CAM facilitates a transfer of technical information from engineering to manufacturing.

Finally, CAD by itself can help engineers in a variety of ways. It can help engineers make their drawings more precise, which allows parts to be designed with tighter tolerances. For example, Harold Salzman pointed out in "Computer-Aided Design: Limitations in Automating Design and Drafting" (*IEEE Transactions on Engineering Management*, November 1989) that engineers in electronics companies can put more components and connections on the same size board that they could previously. CAD helps engineers save time. For example, engineers do not have to redraw a part after a design change. CAD also helps cut costs by decreasing the number of hours that engineers and technicians must spend developing a final design.

Critical Issues Related to CAD/CAM Use

There are five critical issues concerning companies' use of CAD/CAM that relate to the effective exploitation of CAD/CAM's potential. Because this chapter concentrates on the integration of R&D or engineering with a company's efforts to improve quality, the issues addressed in the following sections pertain to the use of computer-aided design more than of computer-aided manufacturing.

General Problems. Despite its potential, many companies have not used CAD/CAM as effectively as they could have. Ann Majchrzak and Harold Salzman wrote:

> Recent research has found that up to 75% of the advanced manufacturing technologies—of which CAD is a part—which have been implemented in US firms within the last five years have NOT achieved their intended benefits.[9]

After studying 13 US companies' use of CAD/CAM, Paul S. Adler came to similar conclusions. In "CAD/CAM: Managerial Challenges and Research Issues" (*IEEE Transactions on Engineering Management*, August 1989), Adler wrote that, in the nine electronics companies of the 13 companies that he studied, the productivity increases stemming from CAD/CAM had been modest, the process of developing new products had not been significantly helped, and the development of printed circuit boards of identical complexity had become more—not less—expensive. Adler also found that, of four aerospace companies of the 13 companies that he studied:

purchasing managers who seldom talked with technical managers were not aware of R&D and engineering developments

> None . . . had experienced any design-to-market time improvements; and, to take a revealing example, engineering change requests coming from assembly still took the traditional three to five months to review and implement.

Adler's and Majchrzak and Salzman's diagnosis of the source of these general problems is that managers in these companies using CAD/CAM have not adequately assessed issues of management and organizational change. Instead, company managers have focused almost entirely on the technical aspects of CAD or CAD/CAM. The author can confirm those conclusions on the basis of his personal experience with various companies. For example, the author visited an agricultural machinery company and a computer company in which efforts to implement CAD/CAM had stalled. In both cases, the sources of the problems were issues related to management, not technology.

Slow Progress in Getting Engineers to Use CAD Fully. Although CAD has much potential for helping engineers design better, many engineers resist using CAD to its full extent. Adler wrote that engineers at the 13 companies he studied resisted the tighter procedures inherent in CAD, which they believed restricted their traditional autonomy. Adler also noticed that none of the engineering departments in the nine electronics companies that he studied attempted to improve the quality of their work through the use of CAD. As he concluded:

> Engineering still thinks in terms of inspecting quality into their product—i.e., their designs—rather than building it into their processes.

The author has observed the same phenomenon. For example, the managers of a consumer electronics company had, for years, had difficulty getting engineers to use CAD.

Adler also pointed out what happens when engineers make a half-hearted effort to use CAD. At the 13 companies he studied, engineers took as long to work on engineering drawings in electronic form as they did to work on them in paper form. Also, the engineers used CAD/CAM in only a limited way, for example, to automate the exchange of information. They avoided analyzing how they actually passed on information and how quickly they passed this information.

Engineers' lack of interest in fully using computer tools can waste much of CAD/CAM's potential. At one computer company, engineers were uninterested in following CAD/CAM guidelines that could have helped make their designs more manufacturable. The engineers needed to expend only 5% more effort to check their product designs against these manufacturing guidelines; to have done this would have helped the manufacturing people in their company enormously. Nonetheless, the engineers in this company would not make the added effort because they would not accept the outside discipline in their own development of product designs.

Slow Progress in the Coordination Between Engineering and Manufacturing. The use of CAD/CAM has improved only somewhat the coordination between engineering and manufacturing. Adler wrote that design teams involving engineering and manufacturing are still rare and that the most common form of project coordination is still the sign-off. Being able to sign off means that the manufacturing department has the right to check documentation related to a design, but not to challenge the quality of the design. Similarly, Jan Forslin, Britt-Marie Thulestedt, and Sven Andersson wrote that after using CAD/CAM, engineers still did not work any more cooperatively with manufacturing.[10]

Adler provided examples of what still existed in the 13 companies that he studied despite their use of CAD/CAM:

- None of the companies used opportunities created by CAD/CAM to download data from the design data base to the manufacturing environment (i.e., translating the data into manufacturing terms).
- All of the companies still had design rules that were weak in terms of producibility guidelines.
- Most of the companies had only started developing two-way communication between engineering and manufacturing—Until then the manufacturing department in these companies could not send suggestions about design revisions back to the design data base.

The author found similar engineering-manufacturing coordination problems in companies that were using CAD/CAM. As an illustration, the engineering and manufacturing departments at one computer company continued to clash over technology transfer issues even though they used CAD/CAM. The essence of the dispute concerned the quality of the data that engineering would put into the data base. To preserve its flexibility during the developmental process, the engineering department wanted to avoid going into great detail in its designs. To make the production process as standardized as possible, manufacturing wanted all of the details. Both sides found it difficult to accept the other side's demands.

Methods for Improving CAD/CAM Use. A precondition to improving the use of CAD/CAM is getting engineers more committed to using CAD fully. In this regard, Harold Salzman wrote, "There was not a single case of 'successful' automation of design reported or observed among any of the designers [he] interviewed." Salzman's point is that CAD itself does not produce a good design. Creative and experienced engineers are always needed to develop a design.

What Salzman's conclusion means, in practice, is that engineering managers must persuade engineers that CAD is only a tool to help them work more efficiently.

Engineering managers must also convince engineers that an important part of being a productive, effective engineer involves developing a design that is manufacturable and of high quality. The training courses mentioned earlier—for example, concerning design for manufacturability and experimental design—can help broaden engineers' perspectives in this regard.

by using CAD, engineers are more likely to develop their design right the first time

Assuming a greater commitment by engineers to use CAD fully, one method for improving how effectively CAD/CAM is used in a company was outlined by Adler in "CAD/CAM: Managerial Challenges and Research Issues":

- Simplify engineering and manufacturing processes as much as possible, but also make sure that all relevant engineering processes are included within the CAD/CAM system.
- Build islands of automation—In doing this, concentrate more on making each island operate effectively than on developing a comprehensive scheme for all of the islands.
- Integrate these islands of automation as much as possible by developing a consistent set of policies that applies across the islands of automation.
- Modify the company's management procedures so that they reinforce the CAD/CAM system.

The actual implementation of these four steps takes great adeptness and management skill. The author visited an aerospace company at which major progress had recently been made in integrating CAD and CAM (e.g., by putting strong guidelines concerning producibility in the design rules). Strangely enough, the actual management steps that had occurred in the company to bring about this change were not clear even to engineers in the company. For example, engineers were not quite sure who had taken the initiative in getting strong producibility guidelines written into the design rules. They did believe, though, that their department had played a vital role in getting this done. As it turned out, the manufacturing department had orchestrated these changes. The manufacturing department had done all of this very subtly, so that the engineering department would think that their own department had brought about the changes. Then, because the engineers believed that their department had initiated these changes, they also accepted them.

Ways to Improve the Design of CAD. One way in which engineers could be persuaded to use CAD more effectively is by improving the design of CAD. The more CAD seems useful to engineers, the more it will be exploited.

M.A. Sinclair, C.E. Siemieniuch, and P.A. John have indicated a couple of specific areas in which CAD might be improved. One area involves the tools that a CAD user has for browsing through data bases. Another area involves visual representations, which turn out to be more important to engineers than the early designers of CAD had thought.[11] Salzman mentioned that CAD software should be improved so that engineers can consider alternative designs more effectively, as well as consider the viability of their current design.

CAD/CAM has great potential benefits, especially for reducing defects in product designs and for making product designs more producible. Nonetheless, several major problems prevent many companies from exploiting its full potential: resistance by engineers to examining their own engineering processes and poor coordination between engineering and manufacturing. Companies can overcome these problems. To do so, however, they must deal effectively with the management issues involved in introducing computer tools.

CONCLUSION

In many companies, R&D or engineering managers must integrate their function's operations with company efforts to improve quality. To do this, R&D or engineering managers must adapt general principles of quality improvement to their own function's operations.

There are two major ways in which R&D managers can work on improving quality: by improving quality within the laboratory and by working with other functions to improve company quality.

Within the laboratory, R&D managers can strengthen the laboratory's technical capabilities, improve the management of R&D, increase the quality-improvement training of technical personnel, improve the work environment within the laboratory, and strengthen analytical, information, and computer capabilities within R&D.

In addition, R&D managers can work jointly with the managers of other functions to improve quality. They can work with marketing managers to improve their method of specifying product requirements. They can work with manufacturing managers to improve the manufacturability of product designs and the quality of the manufacturing process. They can work with purchasing managers to improve the quality of raw materials and components.

Finally, R&D and engineering managers should work on exploiting the potential of CAD/CAM more effectively. CAD/CAM has the potential to help engineers work more efficiently and coordinate their technical efforts with manufacturing.

Notes

1. L.C. Krogh, "Measuring and Improving Laboratory Productivity/Quality," *Research Management* (November–December 1987), p 22.
2. L. Starr, " 'Celanizing' the Quality Program," *Research Management* (November–December 1987), p 31.
3. T.J. Murray, "Meeting the New Quality Challenge," *Research Management* (November–December 1987), p 28.
4. S.B. Billatos, "Guidelines for Productivity and Manufacturability," *Manufacturing Review* 1, no 3 (1988), p 166.
5. D.C. Montgomery, "Experiment Design and Product and Process Development," *Manufacturing Engineering* (September 1988), pp 57–63; Murray, p 27.
6. N.E. Ryan, "Tapping into Taguchi," *Manufacturing Engineering* (May 1987), pp 43–46.

7. M. Tribus, "Applying Quality Management Principles," *Research Management* (November–December), pp 14–18.
8. W.E. Scollard, "Tooling Up for Quality," *Manufacturing Engineering* (November 1989), p 43.
9. A. Majchrzak and H. Salzman, "Social and Organizational Dimensions of Computer-Aided Design," *IEEE Transactions on Engineeering Management* 36, no 3 (1989), p 174.
10. J. Forslin, B.-M. Thulestedt, and S. Anderson, "Computer-Aided Design: A Case of Strategy in Implementing a New Technology," *IEEE Transactions on Engineering Management* 36, no 3 (1989), p 199.
11. M.A. Sinclair, C.E. Siemieniuch, and P.A. John, "A User-Centered Approach to Define High-Level Requirements for Next Generation CAD Systems for Mechanical Engineering," *IEEE Transactions on Engineering Management* 36, no 4 (1989), pp 262–270.

THINKING STRATEGICALLY

Maximizing Computer and Information Resources Use In R&D

Many R&D departments do not take full advantage of the automation and information resources at their disposal. In part, this is because the computing requirements of an R&D department diverge markedly from those of the rest of the company and therefore require careful attention. Taking a strategic approach to building computer and information resources for the R&D staff can greatly enhance the overall effectiveness of the new product development process.

Myra Williams

The pharmaceutical industry has become increasingly competitive, as drugs experience shorter life cycles and new marketplace entrants (e.g., chemical and biotechnology companies) pose additional challenges to traditional companies. Such intense competition demands that organizations strive to improve their overall effectiveness. Three ingredients are required for success: hiring the most highly qualified, talented people, selecting the right projects, and providing the necessary resources, including automation and information resources. In fact, the effective management of information is a key to future success, and strategic planning can help build information resources.

To build and provide for the effective management of computer and information resources for its R&D staff, Merck & Co Inc took a strategic approach to this task to ensure that the company could continue to compete successfully in a highly competitive environment. Strategic planning for computer and information resources does not, in this case, mean performing an enterprise analysis or focusing on systems or data architecture. Although detailed database development plans and modeling tools can be useful, they are not required in the process described here. This case study discusses developing a strategic plan for information systems, not defining annual equipment and support requirements. The strategic plan at Merck was developed to provide a framework for guiding future decisions concerning information systems, automation, telecommunications, data security, end-user support, and training.

THE REQUIREMENTS FOR STRATEGIC PLANNING

In generating two such strategic plans at Merck, the company found that the following requirements must be met for a plan to be successful:

MYRA WILLIAMS, PhD, is the executive director of the Information Resources & Strategic Planning group at Merck & Company Inc, Rahway NJ.

- The planning process must have management support.
- The planning should be conducted by end users with assistance from computer and information experts.
- The leader of the planning activity should have broad knowledge of corporate functions and information systems.
- The plan must be closely integrated with the overall strategic plan for the R&D laboratory.
- The plan should reflect, but not be dominated by, the corporate plan for computer resources.

The following sections describe the approach to planning that was found useful in the Merck laboratories.

Establishing Management Support

The first requirement is that the head of the laboratory support the process. Unless there is a reasonable expectation that the plan will be implemented, which requires management support, it will be difficult to obtain the commitment of all participants. In this case, the president of the Merck Sharp & Dohme Research Laboratories (MSDRL) initiated the process because of his concern that the R&D laboratories were not making optimal use of the computers and information available.

Initially, meetings were held with the R&D area heads to discuss the objectives of the plan and to request that they select individuals to develop sections of the plan for their particular areas. Throughout the planning activities, the R&D area heads were kept informed of the plan's progress. Some of the area heads became personally involved, and others almost completely delegated their planning responsibilities during the early stages.

Establishing a Planning Process

The entire planning process is illustrated in Exhibit 1. The individual components of the process are described in the following sections.

Creating a Steering Committee. The R&D area representatives were members of a steering committee that was responsible for managing and planning computer and information resources at MSDRL. The leader of the steering committee was a scientist who had a broad understanding of the laboratory and who was reasonably knowledgeable about computer and information technology. The steering committee also included the divisional strategic planner and an information specialist as well as experts in telecommunications, laboratory automation, and systems and programming. Although several consultants were interviewed, it was decided to make only ad hoc use of consultants instead of retaining any individual for the complete project. This approach gave the committee access to the specialized expertise of the consultants and forced the committee members to assume more personal responsibility for the development of the plan.

Exhibit 1
The Planning Process

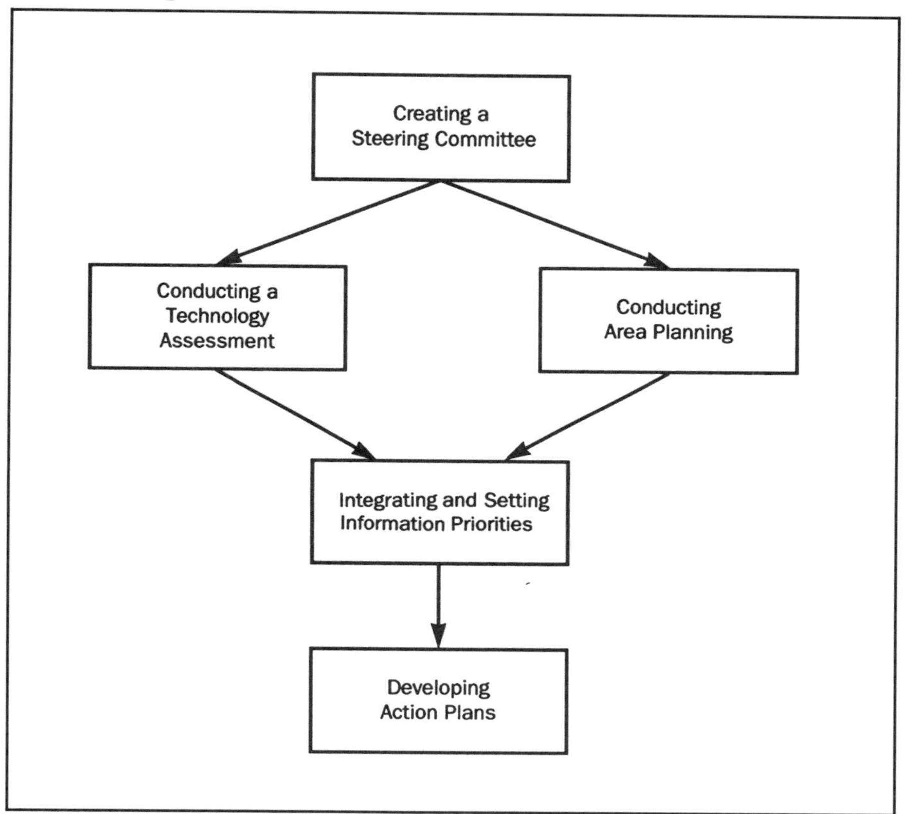

An important factor in the success of the process was that the steering committee not only developed the plan but established annual priorities for projects and support in coordination with senior management. Determining priorities in such a manner greatly facilitated the implementation of the plan.

Conducting a Technology Assessment. One of the first activities performed was a comprehensive assessment of the available information and computer technology. The purpose of this evaluation was to develop a shared sense among the steering committee members of what was feasible given the state of the technology. At Merck, numerous approaches were taken in performing this assessment, including an extensive analysis of published information, presentations to the steering committee by experts, workshops on topics of broad relevance, and task groups.

The technology assessments examined the possible future directions of software, hardware, and telecommunications. In addition, groups of laboratory experts (not necessarily members of the steering committee) were formed to make recommendations in particular areas. For example, a group of scientists and scientific programmers focused on numerically intensive computing requirements in such fields as molecular modeling, X-ray crystallography, and nuclear magnetic-resonance spectroscopy. In addition, a task force was organized to address optical storage technology and its potential for enhancing the way information is acquired, stored, and processed globally. Information security was another area closely scrutinized because it was deemed necessary to protect computer and information resources against inappropriate use without adding unnecessary buffers that would inhibit appropriate access.

Conducting Area Planning. While the technology was being assessed, the steering committee members began to develop plans for their own respective areas. Most committee members formed subcommittees that became actively involved in the process. The first assignment of these subgroups was to identify the critical issues facing their areas and to determine ways to address these using computer and information systems.

Integrating Information and Setting Priorities. The area subcommittees then completed an inventory of current systems, evaluated the area needs through interviews and questionnaires, and defined the opportunities and systems that would have the greatest positive effect on improving the areas' effectiveness.

Developing Action Plans. The subgroups then set priorities for these items, and area action plans were developed. The action plans differed from the strategic plan in that the action plans tended to be more tactical than strategic.

Defining the Driving Force

After area plans were endorsed by area management, they were presented and discussed at the steering committee meetings. During these sessions, the driving force of the laboratorywide strategic plan began to evolve and was explicitly identified. The driving force is the central focus of the laboratorywide plan and all subsequent recommendations as well. The driving force of a plan will vary markedly from company to company. Specific examples are discussed in a later section.

THE STRATEGIC PLANNING MEETING

When the area planning processes were completed, the steering committee met off site for two days to establish the general framework for the strategic plan for information and computer resources. Preparation material, which was distributed one week in advance of the meeting, consisted of the area-level action plans, summaries of technology assessments, and some thought-provoking literature. The action plans described the current environment in each area, identified critical issues and opportunities, and defined short- and long-term plans for computer and information systems. The literature covered such topics as system centralization versus system decentralization, the role of prototyping information systems, microcomputer-based applications development, and object-oriented programming.

The meeting began with a presentation and discussion of MSDRL's strategic plan by way of reviewing the issues, opportunities, and strategies of the laboratory as a whole. The attendees then split into smaller groups to address areas that were critical to the long-term plans. Although most of these subgroups focused on broad areas of technology (e.g., overall systems architecture), one group addressed the organizational issue of whether to centralize or decentralize systems, programming, and laboratory automation specialists. Each working group presented its analysis and recommendations to the committee, and key components of a strategy were adopted. Both an objective scoring mechanism and more subjective discussions were used to identify the highest-priority opportunities for the laboratories.

the leader of the steering committee was a scientist who had a broad understanding of the laboratory and was knowledgeable about computer and information technology

Developing the Five-Year Strategic Plan

Less than one week after the off-site meeting, each subcommittee completed written summaries of its analyses and included the conclusions of the steering committee. A small group of committee members used these reports along with all the other background material to draft an overall strategic plan for the development of laboratory information and computer resources. The entire committee then reviewed the draft in depth to ensure that in all aspects it reflected the recommendations of the steering committee.

The strategic plan highlighted key issues and included environmental assessments and action plans for each of the organizational entities represented by the steering committee. Such topics as system architecture, applications development, information integrity, and administration and support were addressed jointly for the laboratories. The plan clearly defined the limitations of current systems, the requirements for future systems, and the overall priorities of the laboratories. Policies concerning the management of information, hardware and software standards, information security, personnel training, and interdepartmental data flow were established. A summary of anticipated benefits and required resources was also included.

Issuing Two Consecutive Strategic Plans

MSDRL has now issued two strategic plans for information and computer resources, the first in 1984 and the second in 1988. When the first plan was issued, the laboratory was moving from a highly centralized computing environment to one that was at least partially decentralized and therefore more responsive to individual needs. Although the available telecommunications technology seemed to be changing rapidly, this area received a high priority in the first plan because the overall framework of the plan depended on having a reliable network in place. Another focus of the first plan was to increase the availability and the use of existing resources through education and the acquisition of relatively inexpensive devices (e.g., terminals).

In the area of laboratory automation, rapid prototyping and modular design were identified as approaches that would allow researchers to make better use of automation. Another goal was to provide improved connectivity between computer systems.

By 1987, it was clear that most of the objectives defined in the first plan had been accomplished and that computers were well integrated into the operations of the laboratory. Moreover, the available technology had changed sufficiently to make a new strategic plan both desirable and timely. Hence, the planning process was initiated again, and a revised plan was issued in 1988.

The second planning process made it apparent that information needed to be readily available regardless of its source or location. Therefore, the driving force of the second plan became integrated information management. High priority was given to the development of a standard user interface that would facilitate access to and integration of information across applications and minimize the need to learn computer jargon specific to each application.

In addition, the need for modular tools that operated across multiple platforms and operating systems was affirmed, as was the importance of an open architecture. High-speed networks were recommended to connect the geographically dispersed laboratories, and improved teleconferencing and videoconferencing were identified as highly desirable. A key issue in the plan was the proper balance between distribution and centralization in such areas as computer power, laboratory automation, applications development, systems management, training, and support. The decision was made to maintain the existing balance between centralization and decentralization. The plan also adopted functional standards rather than rigid approved lists; however, it was decided that preferred hardware and software would be identified and supported.

A CONTINUED FOCUS ON STRATEGIC INITIATIVES

Since the last strategic plan was written, the computing environment has changed sufficiently for MSDRL to develop a new five-year plan. Although the process is just beginning, such topics as client/server architecture, UNIX-based systems, object-oriented programming, numerically intensive computing on the desktop, and fiber-optic networks will undoubtedly be among the technologies explored. Connectivity among computers will probably evolve to seamless integration of computer systems. The focus of the plan is also likely to change from a user-friendly to a user-enabling environment—that is, to an environment that empowers the end user. The broad acceptance of the first two plans provides the foundation for strong management support as the planning process begins once again.

The primary benefit of the planning process is that it closely integrates the computer and information resources strategic plan with the strategic plan of the laboratorywide R&D program. The planning process also helps to establish a common commitment to laboratory priorities and provides a framework for future decision making. Because the steering committee annually reviews computer projects with senior management, the committee is accountable for the implementation of the plan and keeps abreast of changes in laboratory priorities and requirements.

As valuable as it may be, the strategic plan is not a blueprint; in fact, it is rarely referred to by members of the steering committee after it is issued. It is the process of planning that is most important, not the resulting document, which serves primarily as a vehicle for communication and organization. Because great effort is devoted to eliminating computer jargon, the plans are useful in communicating opportunities and issues to management.

although the available telecommunications technology was changing rapidly, this area received a high priority; the plan's framework depended on having a reliable network in place

Finally, because the steering committee is responsible for the preparation as well as the implementation of the plan, the priorities defined in the plan are reflected in actual advances that occur in the laboratories. For example, one direct result of the planning process is the Discovery Information System. Through a standard user interface, this system allows scientists to have access to integrated information across several data bases without having to master each individual software product. High priority was given to the development of such a laboratorywide system after integrated information management was identified as the driving force of the 1988 plan.

In retrospect, the fact that the plans were developed by scientists rather than by computer specialists or outside consultants has contributed to the successful focus on strategic initiatives rather than on building data bases and system architectures. As a result, significant progress has been made in the introduction of systems that have been crucial in improving the overall effectiveness of the corporate R&D effort at MSDRL.

R&D AND MARKETING TAKE THEIR VOWS

Improving the Project Planning Process

Not long ago, the relationship between the R&D and marketing departments of a certain consumer products company was a casual affair, with very little planning or commitment on either side and very little clear communication. Although this arrangement was convenient on a daily basis, the result was a lack of long-term product development planning and a dearth of new product introductions. So they decided to try talking to each other . . . .

Stephen Hellman

Marketing consumer products in the US is usually a fast-paced and aggressive process in which quarter-to-quarter changes in market share overshadow most longer-range considerations. Because R&D cannot be turned on and off with every quarterly performance report, coordinating the seemingly knee-jerk behavior of a company's consumer marketing department with an effective R&D operation presents an interesting mix of challenges.

These challenges are amplified by some general differences between the R&D and marketing departments. For example, because the marketing staff tends to turn over much faster than the R&D staff, marketing and R&D personnel assigned to a particular product sometimes have vastly different perspectives on the tasks at hand. Marketing product managers want to achieve results quickly—to increase market share during their tenure on a brand. The R&D view, on the other hand, is much longer because researchers will probably have responsibility for their product area for many years; they want their contributions to be realistic, meaningful, and well documented.

Another example is the difference between technical and marketing training. Technical training emphasizes the orderly collection of data, drawing conclusions only when sufficient data is available, with a heavy reliance on deductive reasoning. Marketing training, by contrast, tends to reinforce fast decision making in a data poor environment in which intuition and judgment are critical to effective performance. To work together effectively, marketing and R&D staff must understand and compensate for their differences in motivation, training, and temperament.

STEPHEN HELLMAN is the director of Technology Planning & Statistics in the Consumer Products R&D Division of a large multinational consumer products and pharmaceutical firm. He has worked for several Fortune 500 companies as well as an international management consulting firm that specializes in strategic business planning.

SWEETENING THE MIX

The following case history involves a large consumer products company that specializes in confectionery products for the US market. The company's annual sales of confections totaled approximately $1 billion, and unit sales were flat. A mature R&D organization with a staff of 75 supported this business and its international affiliates.

Because of a stalemate between R&D and marketing, virtually no new products or line extensions had been completed by R&D for several years, though marketing had made several requests for them. Finger-pointing had consumed a good deal of time and energy and had created an environment of suspicion and distrust.

A new department, called R&D planning, was established in R&D to identify the cause of this unfortunate situation and to implement corrective action. The first action of this new department was to organize a meeting between each business unit director in marketing and these directors' counterparts in the R&D department. The meetings had the following two purposes:

- To learn how marketing and R&D each decided which projects to work on and what each department's objectives were.
- To review the project list as perceived first by marketing and then by R&D to determine the extent of agreement on which projects were wanted and what their objectives and priorities were.

These meetings were quite lively and ultimately defined the agenda for the R&D planning department's first six months.

INITIATING R&D PROJECTS UNDER THE OLD SYSTEM

For many years, R&D projects had been initiated in the following manner. A marketing person would think of a new product and call a friendly colleague in R&D. After explaining what was wanted, the marketer would ask how long the project would take to complete. The R&D person, in the best spirit of cooperation and responsiveness, would come up with a time estimate. The satisfied marketing person would then issue a report that publicized this product delivery date. This simple and straightforward process was the reason that no new products had been delivered in many years. Some of the pitfalls associated with this process are discussed in the following sections.

The New Product Request Was Incomplete or Ambiguous

If market research showed that tart fruit flavors were a growing trend in confections, a marketing manager might ask an R&D manager to develop a new line of five tart fruit flavors. After R&D developed tart products in strawberry, lemon, lime, orange, and grapefruit, the marketing group may have, in the meantime, checked with its advertising agency and found that strawberry was ranked last in preference by consumers; could R&D please substitute cherry instead? Then

R&D might respond, Okay, but the change will delay the project and decrease the odds of an eventual market introduction. In such a case, had either marketing or R&D pursued the question of what specific flavors (or other relevant product attributes) were desired, this problem could have been avoided.

in the absence of professional R&D planning, what seemed to marketing like a detail in its request was in fact fatal to the project

The two departments might also have avoided the divisiveness that such miscommunication engenders. In this case (the story might continue some months later), the marketing people might say that they were sure that Joe, the former marketing assistant, had mentioned that strawberry should not be among the five flavors. But Joe left three months ago for another job, and there is no documentation of which flavors were discussed.

In an actual example of an incomplete request, R&D used natural raspberry flavor for a product-line extension. Although the new product tested exceptionally well in sensory tests, marketing canceled the project when it saw the cost-of-goods analysis that the plant worked up before starting production—the flavor was too expensive. R&D had not been given a cost-of-goods ceiling and hadn't asked for one. In the absence of professional R&D planning, what seemed to marketing like a detail in its request from R&D was in fact fatal to the project.

Other examples of how a marketing department's incomplete product specifications can derail a project include:

- The failure to describe the desired advertising claims (e.g., preferred two-to-one) in the initial request—Such claims must be designed into the product and proved in statistically valid tests for the media to accept advertisements that cite them.
- The failure to specify target markets and competitive standards, particularly when subsequent test markets or market research will use these standards to measure the new product's potential.
- The failure to identify special manufacturing constraints (e.g., the product must be produced on underused plant equipment at a specific location).

In the previous example, one obvious deficiency in the way the new flavors were developed was the failure to document what was wanted. Although oral agreements seem fine at the concept stage, they may later evolve into two different versions—one on each side of the functional boundary between R&D and marketing. Even informal documentation of an oral agreement helps to confirm everyone's understanding of the terms.

The R&D Project Plan Was Incomplete

Just as a marketing request can lack critical details regarding required product attributes, so too can the R&D response be incomplete. Critical R&D steps are often overlooked in companies that don't have R&D planning guidelines; such oversights often result in serious barriers to market introductions. For example, many corporate legal departments require clinical irritation testing before the introduction of many kinds of edible consumer products to avoid liability exposure. If such testing is not done in a timely fashion, a product launch might be aborted. Irritation testing is usually conducted on the early prototypes but not always on successive prototypes because of haste or the confusion resulting from poor documentation. In such a situation, the individuals responsible for documenting irritation testing may believe that requirements have been met because, being in a backroom support operation somewhat removed from the day-to-day changes in an R&D project, they may be unaware that product formulations have changed. Only when final documents are assembled to support a product launch may the deficiency surface.

The odds of overlooking a small but necessary portion of the R&D project become significant when the person responsible for developing the overall R&D plan has primary responsibilities other than planning R&D projects. Overlooking such necessary steps puts the entire project in jeopardy.

The Estimated Timing for Developing the Product Was Unrealistic

The process of developing a new product and transferring it to a plant involves many different R&D disciplines. A company's system for determining the time to market can be seriously flawed when the individuals estimating such timing do not adequately understand the responsibilities of R&D groups within the company other than their own.

To continue with the case of the confectionery company, product development chemists would, of course, consider the time they need and would probably also remember to account for the time consumed by analytical chemistry processes required for discovery and documentation. Perhaps they would also remember to account for the packaging stage; however, chances are they would forget the regulatory affairs procedures, in which ingredient copy is reviewed as well as the warnings required by the FDA (e.g., those regarding saccharin or aspartame). They would probably omit the time needed to analyze the statistics generated by the sensory tests; they might also forget that the product must be blind-packaged and computer-coded for randomness in all sensory tests.

When the timing estimates provided to the marketing department are unrealistic, the schedules for developing and producing advertising, test markets, plant scale-up schedules, and the business plans for the year are all thrown off. A delayed project's chances of reaching market are lower and lower each time another delay is encountered, and the relationships among R&D, marketing, and manufacturing are strained as well.

The Projects Lacked Management Commitment

When a member of the marketing department contacts an R&D person to initiate a project in the informal manner described, there is no guarantee that management supports the project. Therefore, even if the R&D and marketing departments successfully avoid all the pitfalls already noted, management may not authorize the start of production or an actual product launch. Indeed, management may never have been interested in the project initiated by the assistant product manager. If the project initiation process requires an assessment of management's commitment before the work starts, however, unsupported projects can be weeded out early and R&D resources can be saved.

the marketing department's third priority was reduced cost-of-goods chocolate bunnies—this product was not on the R&D department's list at all

ANALYZING THE PROJECTS PURSUED UNDER THE OLD SYSTEM

The initial meetings between the marketing and R&D departments to review the project portfolio were organized very simply. Each of the groups brought a list of the projects they considered active and ongoing, along with a brief description of these projects' objectives and rationales. Chaos and liveliness characterized these meetings; a summary follows.

Both groups agreed that a product, hypothetically called Power Peppermint Mints, was the top-priority project. Unfortunately, the marketing department's expectation was that Power Peppermint Mints would be a product-line extension to the existing peppermint line and would be developed in sizes similar to those of the parent line and with identical packaging. The R&D department's understanding was that the original idea was a line extension but that the concept had changed during the off-site meeting in Phoenix because of concerns over product cannibalization (i.e., new products eating up the market share of existing products to the extent that profitability suffered). Current R&D work was in fact focused on a larger product in single-format packaging, consistent with the power concept the ad agency had presented at the Phoenix meeting.

The marketing department's second priority was Mighty Mints. This product was number four on the R&D department's priority list—but at least it appeared on both groups' lists. Although product attributes for Mighty Mints as understood by both groups were compatible, the projected time to market differed because of the priority discrepancy.

The marketing department's third priority was reduced cost-of-goods chocolate bunnies. This project was not on the R&D department's list at all; R&D thought the product was not even under development because the chocolate bunny business was up for sale, and it had agreed in discussions with finance not to pursue that project. Meanwhile, the marketing department had promised their senior management a cost-reduction program and would risk missing their profit goals for the year if it didn't materialize.

This meeting was only the first of six, one with each business unit. During follow-up meetings, some projects were dropped and others were kept where they were. Most were carefully dissected and then put together again, piece by piece, until something emerged that clearly described what marketing needed and exactly how R&D would try to deliver it. One-page write-ups of objectives, rationales, the R&D plan, and estimated R&D feasibility, costs, and timing were assembled. The marketing directors used these plans in discussions with their management to determine which projects had management's support. Approval signatures were obtained—or not obtained. Project clutter and confusion began to thin.

IMPLEMENTING THE NEW R&D PROJECT MANAGEMENT SYSTEM

These meetings and the substantial amount of discussion and reworking of project objectives that followed served two purposes. First, the existing R&D project portfolio was pared down in a rational manner, and the surviving projects were streamlined and focused. The R&D project portfolio then comprised components of an overall R&D strategy that provided balanced support of the businesses. Second, the process heightened the awareness among both the R&D and marketing staffs that focused R&D project planning was a reasonable effort and could result in much more aggressive and realistic business plans.

Once the existing projects had been addressed, attention was turned toward new ones. A project request and planning system was needed to ensure a continuation of the process that had delivered improvements retroactively. The following system was eventually developed:

1. The marketing department initiates an R&D project by submitting to the R&D department a product blueprint that specifies all the information needed to service the request—If marketing wants a fully commercialized product (i.e., a major product launch), the blueprint must specify:
   - All the necessary product characteristics (e.g., weight, color, flavor, and ingredient restrictions).
   - The desired advertising claims.
   - A list of competing products and desired product performance relative to them.
   - The desired packaging.
   - A cost-of-goods target and approximate pricing.
   - The intended manufacturing site.
   - The desired shelf life.
2. If detailed product attributes are not yet established, the project is regarded as a concept-sample exercise and only the basic product attributes must be specified—In turn, R&D will not conduct process scale-up or major sensory or clinical testing of the prototypes.
3. All product blueprints submitted to the R&D department must include the signature of the vice-president of marketing to indicate management commitment to the project.
4. Within 10 working days of receipt of a complete blueprint, the R&D planning staff must assemble all R&D groups necessary to complete the project—A comprehensive yet broad-brush technical approach is developed, including estimates of worker hours, external costs, and timing. R&D management reviews and approves this plan.
5. The R&D planning staff then sends the complete project plan back to the originator in marketing, along with a nontechnical description of exactly what can and cannot be done—The feasibility of each requested advertising claim is

provided. Cost estimates are explained, and options for shortening timetables by increasing risk or expense are explained when appropriate. For unusually complicated projects, the R&D plan is presented in a face-to-face meeting.

6. If the entire package is acceptable to both sides (R&D and marketing), the originator obtains the required management signatures and returns the package to the R&D department.
7. R&D work on the project begins and is reviewed with marketing, both formally and informally, on an ongoing basis.

after several years under the system, many of the new marketing managers started to question the need for a structured system of R&D project management

MAINTAINING TIGHT R&D PROJECT MANAGEMENT

Approximately 18 months elapsed from the first review of the old R&D project planning process to the development, installation, and smooth functioning of the new one. All personnel affected by the changes involved were aware of the nature and reasons for what was happening. Senior corporate management supported the changes.

After several years, however, many of the marketing product managers and business directors were new to the organization. As a result, many started to question the need for a structured system of R&D project management. Even though a single piece of paper was all that was needed to initiate a project request, periodic accusations of bureaucracy and too much paperwork started to surface. Logic holds no sway in this kind of situation—if someone perceives a system as bureaucratic, then in the perceiver's eyes, it is. Still, it was hoped that if the rationale behind the system were understood, these objections would fade.

R&D decided to meet with each business unit separately to explain what the R&D-marketing relationship had been like before development of the established system of project management and why the system worked the way it did. This strategy was time-consuming for the R&D planning group but paid handsome dividends. Virtually all resistance faded, and the system continued to function well.

To guard against a recurrence of this situation, the R&D managers now meet for half a day with new marketing employees to orient them to R&D and explain to them how to access R&D resources most effectively. To date, this system works well. The time involved in such orientation is an investment in the continued smooth functioning of R&D and in the marketing-R&D relationship.

Consumer product R&D is a fast-paced and demanding environment. Because of the compressed time frame in which work is conducted and the rate of change in the marketplace (and often in the marketing organization as well), clear objectives and personal responsibility for R&D projects are critical. Although flexibility in project management is also essential, there are some basic pieces of information that every project must have before it is initiated. This project management system ensures that the needed information is obtained in a timely manner.

BRIDGING THE GAP

Improving the R&D and Finance Relationship

An effective relationship between the R&D and finance departments has been notoriously difficult to establish, primarily because of organizational distance and mutual lack of understanding. There are many ways to bridge these differences and to enhance the productivity and profitability of R&D in the process.

Robert Szakonyi

Of all the relationships an R&D department must build with other functions in a company, none is more difficult than its relationship with the finance department.[1] Although the relationships between R&D and such departments as marketing, manufacturing, and purchasing may be strained at times, the relationship between R&D and finance barely exists in most companies. In general, the R&D staff knows very little about finance, and finance personnel know very little about R&D. Frequently, R&D managers and finance managers have never even met.

The problems that arise because of poor communication between the R&D and finance functions can be viewed in two ways. First, most R&D departments have difficulty explaining and justifying their contributions to the company. Consequently, most R&D departments continually try to measure their productivity quantitatively to answer the questions finance and other non-R&D managers raise about whether investments in R&D are paying off.

Second, most managers, both non-R&D and R&D, have trouble answering the questions that arise when they are evaluating possible investments in technology, including:

- How large should the R&D budget be?
- What are the major factors contributing to the costs of new product development?
- Who should be responsible for making the financial projections for a new product?
- How much capital spending is required to develop a new business?
- How should the transfer of new products from R&D to production be paid for?
- What kinds of financial and nonfinancial justifications are relevant to decisions to invest in new manufacturing technology?
- How long should it take for a technological innovation to pay off?

When there is poor communication between R&D and finance, these questions are seldom answered satisfactorily.

ROBERT SZAKONYI, PhD, is the director of the Center on Technology Management at IIT Research Institute, Chicago. He has performed consulting work for many companies in a variety of industries and has written two books and more than 40 articles on technology management.

IMPROVING R&D AND FINANCE INTERACTION

Coordinating the efforts of R&D and finance can be accomplished by changing the way R&D results and financial data are communicated and by establishing new mechanisms for interaction. First, R&D efforts should be evaluated and explained in a way that allows a financial department to find out what it needs to know about the progress being made on R&D projects. Second, financial operations should be managed so that financial data is more useful to the R&D department. Third, mechanisms for satisfying the needs of both R&D and finance should be put in place. Each is discussed in detail in the following sections.

Changing the Way R&D Is Evaluated and Explained

R&D managers can improve their relationship with the finance department in two important ways. To start with, R&D managers should emphasize the tangible output of R&D projects in their evaluations so that finance managers can appreciate these results in terms that are understandable to them. Because the execution of R&D projects does not proceed in as orderly a fashion as the manufacture of a product does, however, the results of R&D efforts often cannot be accurately predicted. Therefore, R&D managers must also convey the broader message that the development of new technology cannot be viewed by finance managers simply in terms of the bottom line.

In their efforts to focus attention on their output, R&D departments in various companies have taken different approaches. For example, the R&D department of an aerospace company made all the documentation concerning its projects as precise as possible. It put particular emphasis on specifying the objectives and the technical milestones of its projects. This provided the finance managers with yardsticks to use in evaluating the R&D department's progress.

The R&D department of an electronics company went even further and developed a system of gates for tracking projects. Before an R&D project could continue beyond a gate, certain criteria had to be met—for example, experimental data had to be gained, or performance characteristics had to be obtained to proceed to the next stage.

Going a step further, the R&D department of an automotive products company established a system of report cards to measure its output. Both business managers and R&D personnel evaluated R&D projects in terms of how well these projects met their technical objectives, schedules, and cost projections. According to this system:

- Three points were given to a project that represented a technical breakthrough or met its schedule or cost expectations by half the desired amount.
- Two points were given to a project that met expectations on technical performance, schedule, and cost.
- One point was given to a project that met all but one of its expectations.
- No points were given to a project that failed to meet two of its expectations.

R&D departments have also followed different methods to clarify the uneven and disorderly nature of R&D progress. The R&D department of a natural resources company wrote case studies of the company's major technologies to illustrate this point. These case studies traced the development and successive applications of the company's technologies over several decades, thereby demonstrating the unpredictable and nonlinear course these technologies took and the profits gained through their development.

R&D efforts should be evaluated and explained so that finance can find out what it needs to know about a project's progress

The R&D department of a chemical company used contingency scenarios in its R&D project plans to highlight the many ways in which R&D progress could proceed. This R&D department also developed alternative budget projections to make financial managers sensitive to the many possible financial results of exploiting a technology.

Finally, the R&D department of a computer company dealt with differences in the ways R&D efforts could progress by establishing different criteria for evaluating various kinds of R&D work. For example, if the purpose of developing a technology was to maintain the competitiveness of a product, the R&D department used criteria related to sales and profits. On the other hand, if the purpose of developing a technology was to provide the scientific knowledge needed to eventually strengthen the technical capabilities of several company's products, the R&D department put greater weight on evaluating the results of R&D projects in relation to the state of the art in that technology.

Changing the Way Financial Data Is Presented

To improve the relationship between the R&D and finance functions, the finance department should present financial data so that it can be made more useful to the R&D department. For example, the finance department of a satellite communications company improved its system for tracking financial information so that the data that R&D managers received was only one and a half weeks old rather than six weeks old, as had been the case.

The finance department of an aerospace company improved the quality of the data provided by the company's management information systems, for which the department was responsible. The information had previously been presented in a format that helped only the finance department, and the system was improved so that the R&D department (and other departments) could use the information to diagnose and solve problems related to resources.

One way in which a finance department can learn how to make financial data more useful to R&D is to transfer a staff member from the finance department to the R&D department. The finance department of a food-processing company did just that, and this person subsequently became a vital bridge between the finance and R&D areas. The individual served on a committee that was set up to manage the growth of new businesses. Because of his financial background, he was called on to do all the financial projections for new businesses. As a valued member of the R&D department, he was also able to portray the vicissitudes of R&D project work accurately in his presentations to senior business managers.

The finance department of a consumer products company also transferred an individual from its staff to the R&D department. This person organized the laboratory's budget to reflect its ongoing expenditures more accurately. He also played a key role in analyzing the technical skills of various groups in the laboratory in terms of their contribution to the company's businesses.

Putting New Tools in Place

Aside from the individual actions that an R&D or finance department can take to make its own operations more compatible with the operations of the other, there are mechanisms that R&D and finance managers (in collaboration with the managers of other functions) can put in place to satisfy the needs of both departments. These are discussed in the following sections.

Cross-Functional Training. Training courses can be provided to the R&D staff on basic financial principles and to the financial staff on technology management. For example, a food-processing company offered its R&D managers seminars on accounting so that they would better understand the financial calculations business managers apply to assess the R&D area.

Cross-Functional Dedicated Personnel. A small staff of specialized financial analysts can be created in either the R&D or the finance department. The responsibility of this group would be to analyze the costs and benefits of technical efforts. The members of this group should have experience with R&D and some training or work experience in finance. Such a group was established at a chemical company, initially to take on straightforward assignments (e.g., analyzing the costs and benefits of commercializing a new product or implementing a new manufacturing technology). After its work was accepted, the group members became more involved in many other decisions that affected both the R&D and finance functions in their company.

This group's newest role in business decision making is to raise and help answer questions that fall between the R&D and finance managers' areas of knowledge. For example, the group raised questions regarding the best way to calculate the cost of materials needed to develop a new product. Depending on the situation, this group advised R&D managers to use standard costs, the incremental costs of getting more of those materials, or the market prices of such materials. The economic analysts at this company helped R&D managers in two other ways: by examining alternative ways of sourcing materials, and by identifying additional ways in which a technology might provide leverage to the company.

The economic analysts of a paper and pulp company tackled the assignment of translating the estimated potential of new technology into a format that was understandable to the company's business managers—that is, they showed how technology affected the company's income statement. By looking at the benefits of technology in terms of its effects on discounted cash flow, depreciation rates, capital projections, and corporate taxes, these economic analysts helped the business managers better appreciate the value of technology.

financial operations should be managed so that financial data is more useful to the R&D department

Joint Strategic Planning. R&D and finance can use the strategic planning process to build bridges between the departments. An R&D staff member can be assigned to the strategic planning staff with the mission of ensuring that technology plans are incorporated into the company's strategic plans. In this way, issues related to technical resources and the implementation of a company's financial goals will be addressed.

For example, the R&D department of a metals company assigned an experienced R&D manager to the company's planning staff and eventually found that he was able to contribute greatly toward making the company account for current R&D projects in its business plans. Initially, this R&D manager was told by other planners on the staff that technical considerations need not be included in the company's business plans. Through persistence, however, this manager was able to persuade the planning staff that the company's business plans should include the technical resources needed to meet the company's financial goals.

Dedicated Financial Resources. Special financial resources can be put in place to handle the development of new products or businesses. For example, the managers of an oil company established a financial system that served as a halfway house for nurturing new businesses that had significant commercial potential but that were not yet profitable. During the time in which they were in this halfway house, these new businesses were not burdened with as high an overhead as usual.

In another case, the financial managers of a steel company established a separate accounting fund to handle all costs involved in developing a new product after the initial research had been completed. This fund served as a supplemental system of accounting that allowed the manufcturing department to cover the costs involved in scaling up a new product to full production and also spared the marketing department the initial losses involved in commercializing a new product.

The goal set for those who managed this fund was to break even financially. Money was taken from the fund to compensate the manufacturing department for its cost in producing a new product. Profits from new products were given to the fund. In addition, the marketing department was able to separate its regular activities from those dedicated to the commercialization of new products by setting up a special group just to promote and sell new products. The R&D department benefited from this financial arrangement because it did not have to plead with the manufacturing and marketing departments for the financial resources needed to bring a new product to the marketplace.

WORKING WITHIN ONE SYSTEM

To facilitate interdepartmental communication and interaction, R&D managers should stress the results of their efforts whenever R&D projects are evaluated. In addition, they should help financial managers to understand the nature of R&D

work and to develop appropriate expectations regarding R&D activities. With this knowledge, financial managers can make financial data more useful to an R&D department. In both of these approaches, the key is to make an effort to find common ground and to make the work of each department accessible to the other. When companies take these actions, R&D managers will not have to continually explain their department's contributions to the company. Business managers will also have less difficulty making decisions about investments in technology.

Notes

1. Finance as defined here concerns such financial aspects of a company's technology development and commercialization as the economic analysis of R&D projects and R&D budgets. What is not included in this definition of finance are all other financial measures of a company's operations, including price/earnings ratios, debt leverage, and capital asset depreciation and its liquidity.

TO MARKET, TO MARKET . . .

Achieving Market-Oriented New Product Development

The two main reasons new products fail are that companies do not understand their customers' needs and that the products are not developed according to the known needs of customers. In short, there is no coordination between the R&D and marketing functions. This chapter outlines approaches for overcoming the barriers between R&D and marketing to achieve success with new products.

Robert Szakonyi

In 1972, the Science Policy Research Unit at the University of Sussex, UK, conducted a comprehensive study entitled "Success and Failure in Industrial Innovation."[1] By comparing pairs of innovations that were aimed at identical markets (in each pair, one succeeded and one failed), the Science Policy Research Unit was able to identify the reasons for success or failure. According to this study, two of the major reasons technological innovations succeed are that a given company understands customers' needs better and that the company pays more attention to marketing.

The results of this study are not surprising; still, most industrial companies neglect these two aspects of new product development. Many companies have problems integrating market information into decisions about technological innovation, and many managers have criticized their companies' operations on this account. For example, a technical manager at a chemical company admitted that his company did not know its markets well and was having problems in developing specialty chemicals. In another case, an engineering manager of an aerospace company, reflecting on the way his company lost an important contract, said that the company never really understood what the customer (the Department of Defense) wanted. Similarly, an engineering manager at an electronics company explained the failure of one of his company's most important new products as being caused by the engineering department's focus on performance advantages that turned out to be unimportant to customers.

Many such cases were identified in the Science Policy Research Unit study. In the 34 cases in which technological innovations failed:

- Six companies had conducted only brief customer research.
- Six companies were committed to preconceived designs, despite customer research.

ROBERT SZAKONYI, PhD, is the director of the Center on Technology Management at IIT Research Institute in Chicago. Szakonyi has performed consulting work for many companies in a variety of industries and has written two books and more than 40 articles on technology management.

- Four companies had conducted no customer research.
- Three companies failed to understand the working environment in which their product would be used.
- Two companies had ignored the results of their customer research.
- Two companies misinterpreted the results of their research.

These results show that approximately two-thirds of the companies that were introducing new products either did not find out what the market wanted or did not take it into account when this information was available.

Companies' neglect of marketing activities in technological innovation can also be seen from another perspective—that of the resources spent on marketing. One study indicated that companies spend between 0% and 3.5% of their new product development budgets on market research aimed at finding out what customers need.[2] From this remarkably insufficient expenditure, it is clear that R&D is often carried out with inadequate guidance concerning market needs.

IMPROVING COORDINATION BETWEEN R&D AND MARKETING

To increase the success rate new product development efforts, coordination between the R&D and marketing functions must be improved. Specifically, this means that marketing departments must provide R&D departments with more guidance concerning customers' needs. In addition, R&D departments must help marketing departments understand the company's technological capabilities so that the marketing department can do its job better. In other words, R&D and marketing must work as equal partners in developing new products.

Depending on the industry, either R&D or marketing may need to have somewhat more control than the other. In technology-driven industries (e.g., the pharmaceutical or aerospace industries), the R&D or engineering department usually should have more influence in this partnership. In market-driven industries (e.g., the consumer-products and food-processing industries), marketing should usually have greater control. In either case, a partnership is required.

Each task in new product development—generating a new product idea, identifying customer needs, setting objectives, establishing schedules and price/performance trade-offs, and receiving feedback from customers regarding product performance—requires collaboration between R&D and marketing. In such tasks as generating new product ideas, the R&D department should take the lead; in others, such as getting feedback from customers, the marketing department should take the lead.

To improve the coordination between the R&D and marketing functions, three types of weaknesses need to be corrected: weaknesses within the R&D department, weaknesses within the marketing department, and problems in the relationship between the R&D and marketing departments. These are discussed in the following sections.

STRENGTHENING R&D

to make an R&D staff more commercially oriented, the link between the R&D staff and the company's customers must be strengthened

Weaknesses in R&D are evident when the R&D department unilaterally selects R&D projects to pursue without asking for guidance from the marketing department or when the R&D department is dominated by the marketing department. R&D departments that ignore marketing data exist primarily in technology-driven companies. However, they can also be found in market-driven companies, which should be particularly attentive to market research. Two ways to correct this weakness are to establish more formal procedures in the R&D department and to make the R&D department more commercially oriented by strengthening the link between R&D staff and the company's customers.

Establishing More Formal Procedures in R&D

Before an R&D department tries to improve coordination with the marketing department, it should ensure that its own operations are managed effectively. A common weakness in R&D is inadequate discipline in managing projects and in establishing and maintaining priorities.

Five examples of inadequate discipline in R&D, specified by non-R&D managers in speaking about their companies' R&D departments, are discussed in this section. Each example is prefaced by the area of R&D management in which there should be greater discipline:

- Improving R&D project planning and control—A marketing manager of an appliance company advised his company's R&D department to plan and control its projects better. If the objectives and milestones of its projects had been more clearly defined, this R&D department could have had many more successful projects.
- Evaluating ongoing R&D projects—An economic analyst of a chemical company advised the company's R&D managers to conduct tougher evaluations of their projects earlier in the process. By the time this analyst was asked to evaluate the commercial potential of an R&D project, most of the resources of the project had already been spent and critical questions regarding its commercial potential were being raised. Most of these projects were eventually discontinued, but not until after many R&D resources had already been wasted.
- Maintaining priorities regarding R&D projects—A marketing manager of a consumer-products company advised the company's R&D department to stick to the priorities that had been agreed on for R&D projects. Because the priorities of important projects had been shifted by the R&D department, the company had missed or had almost missed several windows of opportunity in the marketplace.
- Optimizing engineering design tools and protocols—The president of an electronics company made its engineering department develop common design tools and design protocols so that each engineering project was not conducted in a vacuum. By agreeing on common design rules and procedures, the engineers

in this department were able to reduce the time needed to carry out their projects.

- Linking R&D projects to the business plan—A strategic planner of an aerospace company developed procedures for requiring the company's engineering department to link its R&D to the company's business plans. Traditionally, the engineering department had selected its projects without considering the business plan; consequently, many of the projects that were technically successful were never supported by business managers for commercial development. The procedures that this strategic planner established required the engineering department to clarify in business terms why it was allocating resources for particular purposes.

Establishing a More Commercial Orientation in R&D

For an R&D department to become more commercially oriented, the members of an R&D department need to change their attitudes and their perspective regarding their own work. For example, some R&D personnel are reluctant to become more commercially oriented because they fear that they will need to sacrifice the quality of their technical work. From their perspective, performing top-quality research and development and being market oriented are incompatible goals. R&D personnel need to be exposed to R&D departments that have successfully made the transition from being exclusively technically oriented to being both technically and commercially oriented.

In a study of 25 R&D departments that aimed to become more commercially oriented, it was found that greater awareness of market needs did not lead to less intellectual rigor in technical work. Instead, these departments continued to conduct the same amount of exploratory R&D in order to make their market-oriented efforts successful. To become more commercially oriented, R&D personnel must change their attitudes and expectations significantly.

For example, the typical attitudes and expectations of the R&D staff were expressed by an R&D manager in a consumer-products company who said, "The R&D personnel in our company do not even think of their work as developing new products. They see their jobs only as meeting the technical objectives of their projects." This viewpoint was supported by two R&D managers at different aerospace companies, who said, "It is easier to sell our customer in the Department of Defense on the idea of supporting an R&D project than it is to persuade a researcher in our company to carry it out. Our researchers want to work only on the projects that they select." In another case, an R&D manager of a household-products company said, "When I joined this company, I was shocked to find that most of the R&D people did not even understand many aspects of the current line of our company's products."

Five recommendations for making R&D personnel more commercially oriented are given in the following sections. Each tip is based on the experience of one or more companies.

Establish a Market Research Group. This group should be part of the R&D department. In a small company, one technical staff member can be designated to perform market research. The R&D department of a metals company set up such a group, known as the technical intelligence group. This group helped direct the laboratory's work toward new business opportunities. The R&D department of a food-processing company set up a similar group that was responsible for new product planning; this group provided estimates of the market potential of exploratory R&D.

from the R&D perspective, doing top-notch technical work and developing market-oriented products are incompatible goals

Evaluate Commercial Factors in R&D. The checklist used for evaluating R&D proposals should be expanded to include more items related to commercial development; these evaluations should be taken seriously. The R&D department of a power equipment company found that one way to broaden R&D personnel's perspective was to require that commercial issues be dealt with adequately before an R&D project was begun.

Solve Problems in Product Development. A large portion of an R&D department's exploratory work should be dedicated to anticipating and solving technical problems that arise—both unexpectedly and predictably—in carrying out development projects. The R&D department of a computer company found that solving these technical problems maximizes the impact of exploratory R&D. Although most development projects in this company initially seemed to involve predictable technical issues, many of the projects ran into unforeseen technical problems as they were being carried out. To solve these new technical problems, the R&D personnel were often required to gain a more fundamental understanding of a technology. Therefore, by directing almost all of its exploratory R&D toward solving unforeseen technical problems, this R&D department made sure that its technical work was commercially oriented.

Analyze the Competition. The R&D department should perform reverse engineering on competitors' products. The engineering department of an automotive company regularly did so—that is, it took the competitors' products apart to understand the technical principles on which they were based. By focusing on any competitive advantages that the competitors' products had, the engineering personnel of this company were better able to direct their own technical work toward market needs.

Perform Applications Development. Technical work should be partly directed toward helping customers use a company's products. The R&D department of a materials company found that technical work focusing on how the company's products could be used was as important as technical work devoted to developing new products or processes. By gaining a better understanding of how customers could use its company's products, this R&D department was able to select new product and process developments that had a greater commercial impact.

R&D Should Have Customer Contact

One aspect of making R&D personnel more commercially oriented has great significance and should be considered separately: the practice of R&D personnel contacting customers directly. The way this contact is established depends on the industry. For example, in the consumer-products industry, it is impossible for R&D personnel to meet the millions of potential customers; in an industry in which a company's products are sold to another company or to the government, however, R&D personnel typically visit customers personally.

Nevertheless, rare is the company whose R&D personnel visit customers as much as they should. Although the R&D personnel in consumer-products companies must confront greater difficulties in visiting customers, they still do not usually take advantage of certain limited customer contacts (e.g., consumer focus groups). Similarly, although R&D personnel in companies that sell to organizations visit customers more often, they frequently fail to exploit opportunities to get useful market information.

The usual excuse that R&D personnel (and R&D managers) use for not visiting customers often enough is that there is too much work in the laboratory. What this excuse does not take into account, however, is that the additional time that R&D personnel gain by not visiting customers is not worth much if the R&D personnel do not adequately understand the customers' needs and, as a consequence, work toward the wrong product objectives.

There are several approaches that companies have used to encourage R&D personnel to interact more often with customers:

- R&D personnel can accompany salespeople on many of their calls.
- R&D personnel can attend trade shows to meet customers.
- The company can arrange a conference concerning the quality of the company's products, inviting both company engineers and customers to attend.
- A company can hold exhibitions of its new products to help technical personnel meet customers.
- One or two technical personnel can be assigned to work in the field for a few years to learn customers' needs in depth and to pass on this information to the laboratory.
- The R&D department can arrange consumer panels to enable R&D personnel to learn what consumers want.

These approaches have all been tried in various combinations and have improved the commercial performance of new products.

Coupling R&D with Other Business Functions

Although the managers of R&D departments that direct their own efforts are not likely to consider their departments to be a problem, the managers of the other functions may see it as such. Many R&D departments are not linked to company operations, which frustrates non-R&D managers who have very little influence over the new product development process. If unchanged, this situation can result in high turnover of R&D managers and slashed R&D budgets.

Establishing greater discipline within R&D and making R&D personnel more commercially oriented forces the R&D department to become more accountable to business functions (e.g., marketing and finance). Better R&D project planning, control, and evaluation force R&D personnel to clarify objectives and priorities in terms of company goals. By becoming more commercially oriented, R&D personnel more readily see the need for the input of other company functions in directing R&D projects.

in becoming more commercially oriented, R&D personnel will more readily see the need for the input of other company functions to R&D projects

Strengthening the R&D Perspective

R&D departments are often not independent enough to function effectively; they tend to be dominated by other company functions, especially the marketing department. Once greater discipline is established in R&D and the department has become more commercially oriented, it will have the basic tools to assert itself and become more independent. By becoming both more independent and more commercially oriented, R&D personnel will also be in a better position to argue their case before marketing personnel. Regardless of how much technology contributes to the value of a product—that is, regardless of whether a company is in a technology-driven or a market-driven industry—what ultimately gives weight to opinions about the specifications of a product is what customers want.

Therefore, even if an R&D department has a great deal of influence in a company, it loses much of its influence if it cannot demonstrate that it has had direct contact with customers. On the other hand, if an R&D department has little influence within a company, it can lose whatever influence it does have if it has not had customer contact. By having its own information about customers' needs, an R&D department can participate effectively in company decisions about new product development.

STRENGTHENING MARKETING

Weaknesses in marketing usually stem from the absence of a definite marketing department in a company, as distinguished from a sales department. A marketing department should concentrate on analyzing customers' future needs, not on promoting a company's existing products. The R&D departments in companies without separate marketing functions typically select R&D projects without the benefit of any marketing information, which is simply unavailable. In companies that do have marketing departments, the department's analytical skills must be strengthened. In particular, the department's abilities to understand customer needs as well as the management of its marketing efforts must be improved.

The problem with marketing is that in most companies it barely exists. Before substantiating this assertion, however, it is necessary to define marketing, as opposed to sales. The distinction between sales and marketing was set forth by Lowell W. Steele:

> Sales represents the existing product line in meeting already demonstrated customer needs. The salesperson has clearly delineated objectives: to establish personal relationships that induce customers to prefer his company's products, to avoid anything that upsets or otherwise threatens the relationships he has so carefully fostered, to maximize the attractiveness of his products in meeting the customers' needs, and to achieve closure on sales.
>
> Marketing represents the customers in a more abstract sense: who they are, what they are like, what their needs are, and how they can be influenced. Marketing is focused on the future. Sales is focused on the present. It follows that sales is not a natural ally in innovation, but marketing is or can be.[3]

Before considering whether this distinction applies in most companies, I must make one more clarification. The question here is not whether a department called marketing exists. The needs of industries and companies differ, and as such, a department called marketing may be inappropriate—the important question is whether there is a group in the company that deals effectively with marketing issues. In a study of more than 200 R&D and marketing managers, it was found that not only did 60% of the R&D managers think that information from the marketing department lacked credibility, but 56% of the marketing managers themselves agreed with this assessment.

Two complicating factors make these weaknesses in marketing difficult to correct. One is that the methodologies of marketing and the body of knowledge in this area are far less impressive than the methodologies of science and engineering and the body of knowledge in those areas. In general, there is a low level of sophistication in marketing techniques.

In addition, the mission of a marketing department is always somewhat unclear. Part of the reason for this is that marketing skills can exist in another department (e.g., in engineering). Because people in other functions can have marketing skills and surely will have views about what the future needs of the company's customers are, a marketing department cannot have sole jurisdiction over marketing issues.

Another part of the reason is that marketing issues are too broad for a competent marketing department to deal with by itself. Because marketing issues are concerned with the very nature of a business and of customer needs, they involve other issues that senior management must address, including company growth and business diversification. Ultimately, marketing is of direct interest to senior management. Therefore, the mission of a marketing department is not easy for either the marketing department or the R&D department to define, and it is always subject to change according to changing business conditions.

Notwithstanding these obstacles, companies can take three steps to strengthen marketing: improve the marketing department's analytical skills; improve the marketing department's ability to understand customer needs; and improve the management of marketing.

because marketing issues concern the very nature of a business, they are of direct interest to senior management

Improving Analytical Skills

Although both marketing and sales departments focus on customers, marketing differs significantly from sales. As stated, salespeople work to establish personal relationships with customers in order to sell existing products; marketing people analyze customers' needs in order to guide the development of products. Because most marketing people start out as salespeople and spend much of their careers in sales, they usually have excellent skills in dealing personally with customers but have not developed the analytical skills necessary for marketing. In addition, because they have spent much of their careers selling existing products, they often have difficulty imagining products that have not yet been developed. Four tips to assist marketing people in improving their analytical skills are discussed in the following sections.

Focus on the Product. Marketing personnel, especially those who are relatively inexperienced in that function, should be encouraged to focus on the product itself rather than on the subsidiary activities involved in selling a product.

Guide Technology Development Effectively. This can be accomplished by considering the performance advantages of a technology, the possible product application of a technology, and the potential customers of a product in which a technology is embedded. The role of marketing is to translate the potential of a technology and of the applications of a technology into a product that customers want. To do this, marketing people must first understand the performance advantages of a technology. This does not mean that marketing people must become scientists and engineers and become capable of doing the technical work themselves; rather, they must learn what a technology can do. In addition, marketing people must identify and understand the possible product applications of a technology and the trade-offs among product features, prices, and schedules of development. In addition, the marketing staff must be capable of linking various product applications with various potential customers by understanding customer populations and their needs.

Improve the Rigor of Analyses of Market Potential. There are too many poorly written, unsubstantiated market analyses in companies. Regardless of how rudimentary marketing methodologies are, marketing people should be expected to analyze potential markets with as much rigor as possible.

Retain Some Skepticism. Although market research is necessary and often can be useful, marketing staff should be somewhat skeptical about what market research results signify. The basic assumption of market research is that customers can be asked straightforward questions and are able to give straightforward answers. Frequently, this is not the case; customers themselves may not recognize their own needs. The more novel a new product, the greater the limitations of market research will be.

Understanding Customer Needs

Marketing departments need to place a higher priority on requiring marketing people to understand the marketplace in depth. I have visited many companies in which the marketing people had only superficial knowledge of the market. In some cases, although marketing people did have adequate knowledge of the market at one time, conditions had changed. For example, in some industries, the users of a product have replaced the purchasing agents as the most influential party in buying decisions. Marketing people, especially those who had spent most of their careers in sales, often did not fully comprehend this shift in the marketplace.

Several approaches can be taken to improve the marketing staff's understanding of the marketplace. First, existing marketing people could be trained with regard to new marketplace conditions. Second, new marketing people could be recruited from customer organizations or from organizations that have close ties with customers (e.g., market research firms). Third, if the customers themselves are becoming more technically sophisticated, technical people from other parts of the company could be recruited for marketing positions.

Improving Management

Persistent criticisms of marketing management can be heard in many companies. One criticism is that no one holds marketing managers accountable for decisions that influence the use of R&D resources. Most of the resources spent in new product development are used in developing the product, not in analyzing and developing the market. Consequently, an R&D department spends much of the allotted money for a project on new product development, particularly during the early phases. If marketing managers do a poor job in planning products and determining product specifications, the R&D department will spend a large amount of money developing a new product that may never reach the marketplace.

Another criticism of marketing management is that marketing managers change positions and responsibilities so often that they do not have the incentive to support the develpment of new products. Marketing personnel and R&D personnel generally follow different career tracks. In most companies, R&D personnel spend their entire careers in R&D. R&D personnel usually remain with an R&D project or technical program for a few years, because their technical specialty tends to focus their efforts on certain product areas.

Marketing managers, on the other hand, often aim to establish a track record with a product line or business and then get promoted to a position managing a larger or more important product line or business. In many companies, marketing managers who are on this career track are moved to new positions every two years. Because marketing personnel may change once or twice or even more during the development cycle of a new product, many of these projects follow erratic courses and eventually fail.

The solution to both of these problems lies in the hands of senior marketing managers. What is required is to establish greater discipline in the way marketing operations and staff are managed. When marketing staff members make decisions on product specifications, they should meet certain standards with regard to the determination of customer needs and the technical experiments that are required.

no one holds marketing managers responsible for decisions that influence the use of R&D resources

In addition, senior marketing managers must recognize the trade-offs that exist between promoting effective marketing managers quickly and developing new products successfully. Although effective marketing managers should be promoted, they should be promoted in a way that does not hinder the development of new products.

STRENGTHENING INTERDEPARTMENTAL COMMUNICATION

Problems in the relationship between R&D and marketing are caused by several factors, including:

- Differences in the training and career experience of R&D and marketing personnel—R&D personnel usually have had technical training and have spent most of their careers in the laboratory. Marketing personnel, on the other hand, usually have not had technical training and have spent most of their careers in sales.
- Differences in their objectives and responsibilities—R&D personnel tend to focus on the technical performance of a product and on long-term objectives, and marketing personnel tend to focus on projected sales and market shares and on short-term objectives.
- Differences in their personalities—R&D personnel tend to be more comfortable working alone, whereas marketing personnel tend to prefer working in groups and generally are more socially oriented, gregarious, and assertive.

These differences between R&D and marketing personnel often result in serious communication problems. To improve communication between the R&D and marketing functions, five steps should be taken:

- The R&D and marketing departments should establish common goals for new product developments.
- The R&D and marketing departments should create a protocol for carrying out new product developments.
- R&D and marketing personnel should improve their communication with each other, paying particular attention to meshing a company's technical capabilities with market needs.
- The R&D and marketing managers should take organizational measures to improve coordination (e.g., by creating new product development teams).
- Senior management should help improve R&D and marketing coordination (e.g., by ensuring that R&D and marketing have equal power and prestige).

These steps are discussed in the following sections.

Establishing Common Ground Between R&D and Marketing

Departmental differences in objectives and priorities are frequently the cause of uncoordinated new product development efforts. A procedure that companies can use to resolve differences between R&D and marketing on a specific new product development was outlined by C. Merle Crawford.[4] Crawford recommended that both R&D and marketing sign a document at the time a new product is defined. In this document, the marketing department delineates what benefits or performance attributes it expects from the new product. In doing this, the marketing department commits itself to a new product that it will be able to market.

The R&D department agrees in this document that it will deliver the kind of new product that marketing wants within a certain time and budget. The R&D department, however, has the freedom to decide which approaches it will use to develop the product. In other words, by signing the document, the R&D department is not committing itself to respond to a set of product specifications that marketing may have defined too narrowly.

The value of this procedure is that it establishes a common ground in which the inputs of the R&D department are integrated with the expected outputs of the marketing department. The procedure by itself, however, cannot ensure that both sides will uphold their part of the agreement. As Crawford also mentions, this agreement must serve as the basis for later negotiations between R&D and marketing—for example, when the R&D department says that certain performance attributes cannot be delivered within the estimated schedule or costs.

Creating a Protocol for New Product Development

The R&D and marketing departments can also have different perspectives on what activities are required at what time during the development of a new product and on which department has what responsibilities. One way to resolve differences in perspective on how new product developments should be carried out is by creating a protocol—that is, an outline of what steps should be taken at what time. A protocol can serve as the guideline that the R&D department, the marketing department, and other company functions use to coordinate their activities. Protocols of new product development have been created at numerous companies. In almost every case, these protocols have been useful and have formalized the new product development process for those companies in terms that were meaningful.

The protocols that were related to new product development were not especially novel or insightful. They all dealt with the activities involved in new product development, including analyzing a new product concept, conducting a feasibility study, developing a working model, analyzing the market potential, developing a production prototype, and establishing a business plan. Some companies' protocols contained one or two activities that other companies' protocols did not, and each company had its own perspective on when activities began and ended. Nevertheless, all these protocols covered essentially the same issues.

What gave the protocols great value was that they served as a framework for discussion between the R&D department, the marketing department, and other company functions. R&D, marketing, and the other functions could all communicate more effectively with each other when they all were talking about the same things.

the aim of improved communication between R&D and marketing is not harmony but finding the right answers

Hurdling the Language Barrier

Even if R&D and marketing people meet regularly, communication problems may stem from the way in which they talk to each other or because each department has its own jargon. To improve communication, R&D and marketing managers can:

- Develop a common language for talking about the meshing of technical capabilities and market needs—For R&D and marketing people to communicate effectively, they need a common language; a set of common questions can usually serve as such a language.
- Require R&D and marketing people to communicate throughout the development process—R&D and marketing staff must communicate with each other at every step of the new product development process, and each must be a party to the other function's major decisions and analyses.
- Require both R&D personnel and marketing personnel to specialize—When R&D and marketing staff members know their respective specialties in depth, they find a great deal to talk about with regard to new products and new markets.
- Position R&D and marketing staff so they can work closely together—R&D and marketing managers should take advantage of any approaches that are available for allowing R&D and marketing personnel to see each other frequently.
- Promote the truth—The aim of communication betwen R&D and marketing is not a desire for harmony but a search for the right answer. The reason for having R&D, marketing people, and people from other company functions all involved in new product development is that each party contributes a necessary expertise.

The goal in bringing the two departments together is to use their skills to develop a new product that neither one of them could develop alone. The premise is that their discussions will function as a crucible and yield a new product that contains the best ideas of all parties. For this to happen, however, both R&D and marketing people must be prepared to challenge each other. To have constructive discussions, both sides must feel comfortable challenging the other side's arguments. Professional disagreements that focus on issues, not personalities, are what spawn many of the technical or commercial insights on which successful new product developments are based.

Organizational Techniques for Improving Coordination

R&D and marketing managers can take three organizational measures to improve coordination between R&D and marketing:

- Establish new product development teams.
- Create a separate marketing or new product group in the sales and marketing department.
- Assign R&D and marketing personnel to the other function as part of their career development.

These measures are discussed in the following sections.

New Product Development Teams. These teams can vastly improve coordination between R&D and marketing. By assigning specific R&D and marketing people and the representatives of other company functions to a team and giving them the responsibility for developing and commercializing a new product, R&D and marketing managers can create the conditions for effective new product development.

A Dedicated Marketing Group. In the sales and marketing department, a dedicated marketing staff will help improve communication between R&D and marketing. Such a group should be responsible for developing new products, rather than for promoting existing products.

Cross-Functional Training. When R&D and marketing personnel work in the other function, communication is improved in two ways. First, an R&D or marketing person carrying out an assignment in the other function can translate technical or marketing issues to that function. Second, when R&D or marketing staff members return to their original function, they bring with them understanding of the other function's concerns.

The Role of Senior Management

The senior management of a company greatly influences how much coordination there is between R&D and marketing. If a senior manager indicates that cooperation between R&D and marketing is valued, R&D and marketing staff will work more closely with each other. Unfortunately, in most companies, senior management does not play the constructive role that it could in improving interfunctional coordination.

There are two important things that senior management should do to improve coordination between R&D and marketing. First, senior management should establish the rules that dictate the way in which R&D and marketing will work together to develop new products—for example, ensuring that R&D and marketing establish a protocol that outlines what technical and marketing activities are to be carried out in new product development. Second, senior management should ensure that the power and prestige of R&D and marketing are equal so that both parties have incentives to cooperate. In other words, R&D should not ignore marketing, and marketing should not dominate R&D.

IT'S TIME TO ACT

in most companies, senior management does not play the constructive role that it could in improving interfunctional coordination

The aforementioned study by the Science Policy Research Unit of the University of Sussex, which pointed out the criticality of understanding customers' needs and paying sufficient attention to marketing, was carried out in 1972. Since then, dozens of other studies have confirmed that one of the keys to managing technology successfully is improving coordination between R&D and marketing.

Despite the fact that methods for improving coordination of these two functions have existed for 20 or more years, serious problems in coordinating them still exist in many companies. One conclusion that could be drawn is that the underlying problem that prevents improvements in coordination between R&D and marketing is not a lack of knowledge about what to do but an unwillingness to act. Many R&D and marketing managers feel that their senior management is not doing much, if anything, to improve coordination of the two.

The approaches described—for making R&D people more commercially oriented, for improving marketing people's analytical skills, and for improving communication with regard to meshing technical capabilities and market needs—are only tools. It is time for R&D managers, marketing managers, and senior managers in many companies to put these tools to use.

Notes

1. R. Rothwell, "The Characteristics of Successful Innovators and Technically Progressive Firms (with some comments on innovation research)," *R&D Management* 7, no 3 (1977); and A. Robertson, "The Marketing Factor in Successful Industrial Innovation," *Industrial Marketing Management* 2 (1973).
2. B. Little and R.G. Cooper, "The Role of Marketing Research in New Technology Ventures," *Journal of Product Innovation Management* 6 (1989), p 21.
3. L.W. Steele, *Managing Technology: The Strategic View* (New York: McGraw-Hill, 1989), p 294.
4. C.M. Crawford, "Protocol: New Tool for Product Innovation," *Journal of Product Innovation Management* 2 (1985), p 171.

SECTION

3

PROJECT MANAGEMENT AND TEAMWORK

Project management and teamwork constitute an important part of R&D management that, depending on the company, could also relate to R&D's relations with the rest of the company. Most R&D projects involve more than one person. Therefore, the need arises to set goals for two or more technical specialists and track the specialists' progress. Furthermore, in some cases, formal teams are created to carry out R&D projects. These instances involve the designation of a project manager who is then held responsible for the team's output. Finally, some companies create development project teams consisting not of technical personnel exclusively but including personnel from marketing, manufacturing, and other functions in the company. A project manager must face the challenges involved in coordinating people from several functions to develop a product.

This section presents four perspectives on this issue. Robert Szakonyi considers the creation and management of a new-product development team consisting of R&D and non-R&D personnel. Michael Grindel analyzes the particular challenge of managing interdisciplinary project teams within an R&D department. F.D. Galatola considers project management within an R&D department. Finally, William Guman looks at project management within an R&D department; in addition, he describes how this project management fits within a company's operations.

Szakonyi discusses the creation and management of a new-product development team that consists of members from R&D and non-R&D functions. He explains why such a team is useful and when it may or may not be needed. He delineates the qualifications, responsibilities, and authority of the team leader and the team members. He also highlights the issues that should concern the team manager.

Grindel depicts the efforts of the R.W. Johnson Pharmaceutical Institute (part of Johnson and Johnson) to establish effective multidisciplinary R&D project teams. The model of a multidisciplinary team that his company chose consisted of a project leader to coordinate R&D efforts and link R&D with the businesses, a project management professional skilled in project planning and negotiation, and other team members from various technical disciplines. In addition, Grindel analyzes the lessons his company has learned regarding multidisciplinary teams and methods that senior R&D managers can use to improve decision making.

Galatola describes an R&D department's first-time development of a system of project management. Explaining the rationale behind developing the project management system, Galatola then relates the handling of critical project management issues, such as statements of work, costing, and schedules. Throughout her chapter, Galatola explains how the staff group responsible for the establishment of project management cooperated with R&D personnel to accomplish the required results.

William Guman describes the use of project management at a defense contractor whose senior business managers controlled many details of R&D projects. Examining the strengths and weaknesses of this company's approach to project management, Guman concludes that company operations succeeded despite the senior business managers' considerable control. Technical innovation was allowed to progress, leading to evolutionary technical advances. However, Guman mentions, this approach to project management forced technical personnel to focus on short-term results neglecting major new technical advances that the company needed.

MISSION: NEW PRODUCT DEVELOPMENT

Creating An Effective Team

A primary concern of most companies, new product development can be a tricky process. Many companies repeatedly fail in their development efforts, despite plenty of good intentions and hard work. With effective exploitation of resources and strategic management of personnel, however, most companies can significantly improve their new product development efforts.

Robert Szakonyi

Most industrial companies place a high priority on developing and commercializing new products. Despite this emphasis, however, industrial companies fail significantly in their efforts to commercialize new products. According to a study by Merle Crawford, 30% to 40% of companies' new product marketing measures are ineffective.[1] Moreover, Crawford studied new products only after their introduction into the marketplace. Examining these companies' overall efforts to develop and commercialize new products would have yielded a much higher failure rate, as Crawford has acknowledged in his article.

It is a truism to say that a company needs great teamwork to develop and market new products successfully. R&D, engineering, marketing, manufacturing, finance, and others must all contribute to the development. The simple solution for a company, therefore, seems to be to create new product development teams composed of representatives from several functions.

However, most companies find it difficult to create effective new product development teams. In my work with more than 225 North American and European companies, I have found that only a small percentage were successful in this endeavor. Many of these companies have difficulty even setting up multifunctional new product development teams, regardless of whether or not they ultimately prove effective.

By assessing the various factors involved in creating such a team, most companies can improve their techniques of new product development. This chapter addresses the following specific questions:

- What is a new product development team and why is it needed?
- When should a company employ another method for new product development instead of a dedicated team?
- Who should be the leader and members of a new product development team?
- To whom should the team leader report?

ROBERT SZAKONYI, PhD, is the director of the Center on Technology Management at IIT Research Institute, Chicago. He has performed consulting work for many companies in a variety of industries and has written two books and more than 40 articles on technology management.

- What responsibilities should the team leader have?
- How much authority should the team leader have?
- Which responsibilities should fall to the other members of a new product development team?
- How should the team leader manage the team?

WHAT'S IN A TEAM?

A new product development team is a temporary group of specialists from several functions (e.g., R&D, marketing, manufacturing, and finance) and is led by someone typically called a project manager or a new product development manager. Typically, leaders spend all or most of their time developing the product, and the other team members spend one-third or less of their time with the same new product development.

The composition of a new product development team varies across industries. In a pharmaceutical company, for example, most members of a new product development team are technical specialists from the R&D organization, whereas an aerospace company's team usually consists mostly of engineers. Other industries, however, typically include more marketing people of various types (e.g., from market research, sales, and advertising) in a new product development team.

The mission of a new product development team is to develop a new product after most of the preliminary research has been completed (if any was needed) and to manage this new product until it becomes commercially viable.

Why a New Product Development Team?

Because industrial companies tend toward bureaucracy, with each function concentrating on its own mission, new product development—for which no function is totally responsible—often suffers. A few years ago, I visited a small specialty materials company, which then had annual sales of $10 million. This company exemplified the ideals of entrepreneurship more than any other company I have ever visited. The key functional managers were willing to consider the impact of their function's activities on the whole company and to manage their own operations accordingly. Moreover, these managers were willing to assist in other areas of the company's business. For example, the senior technical manager often acted as the company's primary salesperson with a major customer, because earlier he had been instrumental in obtaining that customer's business.

During the past few years, this company's annual sales grew to $25 million. I spoke again with the senior technical manager, who mentioned that he now concentrated on managing the engineering department and no longer became involved in other areas of the company. Even though his company was still fairly small, he felt that already it was becoming bureaucratic.

Leonard R. Sayles and Margaret K. Chandler, in *Managing Large Systems* (New York: Harper & Row, 1971), have highlighted this bureaucratic tendency of industrial companies and other organizations. In their discussion of project management, which significantly resembles new product development management, the authors note a common tendency to let lower-level, specialized routines and procedures govern larger program interests, often to the company's detriment.[2]

almost all companies, regardless of size, can suffer from a tendency toward overspecialization

All companies, regardless of size, can suffer from this tendency of functions to concentrate almost solely on their specialized tasks. A metal products company that I visited had one engineer, assisted by a technician and a part-time secretary. The engineer and his technician were the company's entire new products department. Although the company was quite small, there was a need for some sort of multifunctional team; the engineer and the technician found that the new products they developed constantly encountered obstacles when they were transferred to the production plant and the sales department.

In larger companies—which might have a laboratory with hundreds of scientists and engineers, nationwide or international manufacturing plants, and regional sales managers located all over the world—the problems of developing and marketing new products can be overwhelming.

Unifying the different company functions into new product efforts, the new product development team provides a counteraxis to the axis around which the various company functions usually revolve. This can be illustrated in three ways.

First, companies need multifunctional teams to address the nature of new product development. In the development of a new product, no company function can work effectively in isolation from the rest of the company. A new product cannot be designed without considering questions of its manufacture. A product cannot be priced effectively without assessing the product's performance characteristics or research results indicating potential customers' needs. Advertising for a product cannot ignore engineering analyses of quality, reliability, or stability. In new product developments, the discrete functions—regardless of their assigned mission and self-perception—are interdependent, relying on each other to do their own job well.

Second, I have observed that many companies without multifunctional teams often develop new products with great potential that travel with great difficulty to the marketplace or that die along the way, never even reaching the marketplace. One common cause of these failures was that, because the company had no champion committed to new product development, one function's blind spots went unchallenged. Perhaps a manufacturing organization was allowed to produce a new product by the least costly methods; perhaps a marketing organization was allowed to promote a new product through existing sales channels; or perhaps an R&D organization was allowed to design a new product without testing customers' reactions. Whatever the case, if a function's blind spots go unchallenged, as happens in many companies, failures in new product developments are likely.

Third, companies need multifunctional teams to compete more effectively. For instance, I visited a consumer electronics company that traditionally had concentrated on high-performance, high-price products. Because the company was sensitive to its niche, it recognized that competitors' products were taking away its business. Therefore, the company needed to find another way to compete. By developing systems products incorporating its earlier, individual products, the company provided much greater value to the customer. To meet the challenges of developing and marketing these new systems products, the company had to rely more on multifunctional teams to improve internal coordination.

In an article in the *Journal of Product Innovation Management*, Blair Little sums up the microcosmic nature of the new product development team. To improve new product development, Little writes, "there must be miniaturization of the corporate form, multiple innovation units within the overall unit."[3] In other words, functions represented in a company's management should be represented on a much smaller scale in new product development.

To create an effective multifunctional team, a company must strike a new balance between the individual functions' priorities and the priority of developing new products. In striking this balance, however, it is not necessary to eliminate or weaken individual functions. Functions are natural groupings within a company, providing centers of specialized skills and working together to fulfill the company's purpose.

Striking a new balance between these two priorities means exploiting the advantages gained by having a multifunctional new product team while capitalizing on the strengths of the various functional specializations. While providing focus for a new product development, a team can be held accountable for successfully developing and commercializing the product. Individual functions cannot effectively hold these responsibilities; they provide the complementary skills and resources needed to actually develop and commercialize a new product.

Team Benefits

An excellent example of an effective team's accomplishment was demonstrated by a food-processing company. Before a multifunctional team was set up, a particular new product development had been floundering for two years. The company was uncertain about how far to develop the product. The senior business managers in the company supported the product, but the next level of business managers had grave doubts about it; consequently, the new product development had been put on hold for many months.

Other factors also were stalling this new product development. For two years, the company's marketing organization could not decide among various courses of action related to the new product. The R&D organization, on the other hand, attempted to cover all possibilities in its technical activities, in preparation for whatever options the marketing organization chose. Thus, the R&D organization lacked a clear focus. Finally, the manufacturing plant managers and corporate managers were in conflict over repeated delays in this new product development and over the financing of the needed equipment.

The new multifunctional team that the company set up put this new product development on track within a few months and breathed new life into the entire new product development effort. Within nine months after the team was set up,

the advertising of a product cannot ignore engineering analyses of quality, reliability, or stability

all of the company's senior management supported the new product, and the company decided to invest heavily in capital equipment and expand the new business greatly from its earlier test-marketing phase.

To achieve these results, the new product development team had to assume many responsibilities that had formerly belonged to no individual person or function in the company. First, the team initiated action on issues concerning the new product development as a whole. For example, one of the things that the team soon noticed was that the new product had never been clearly defined. The team enlarged its charter and forced key functions, such as marketing and market research, to formulate a clearer product definition. When others were added to the personnel originally assigned to the team, the team also broadened its charter. In addition, the team introduced the idea of fast-track commercialization, which would have considerably shortened the new product development process. As a result, the company eventually generated a compromise approach to commercialization, significantly shortening the process.

Second, the team provided a focus for all company activities related to this new product development. As one company member, who was not on the team but worked closely with it, said:

> Before the team was formed, various functions within the company beat their own drums with regard to the way the new product development should proceed. Each function had its own agenda that it tried to push as the agenda for the new product development. With the team, all of this changed. Now, anyone working on the new product development could go to the team leader to find out what the highest priorities were.
>
> Having a focus has meant that everyone involved can work on something useful to the new product development—rather than work on what he thought would be useful, only to have to make major adjustments when the highest priorities eventually became known. Because of this, my own engineering group has been working twice as effectively as it normally does.

Third, the team helped clarify the priorities of the new product development among senior managers of various functions. The team had representatives from most of the key functions who were willing to disagree forthrightly about how the new product development should proceed. Thus, when the team agreed on a course of action, its recommendations carried great weight. Although each team member might have had difficulty challenging a senior manager alone, the team members gained a collective stature that they did not possess as individuals. Because company members recognized that the team's proposals had been tested through vigorous discussions with representatives of most functions, these recommendations gained legitimacy when presented to senior managers—who often disagreed among themselves. The team's clarity of purpose, consequently, helped to expedite these managers' agreement.

In summary, this new product development team accomplished what none of the individual functions with its own agenda could, breathing life into new product development with its commitment and sense of purpose. Many companies could gain a great deal by enabling their own new product development teams to follow this team's example.

When Should a New Product Development Team Be Used?

Although a new product development team can be quite effective, a team does not need to be set up for each new product development. Setting up and managing a team requires a great deal of work. In addition, the use of teams has disadvantages as well as advantages. As Sayles and Chandler pointed out, the longer a person from a function serves on a project team, the more his or her experience becomes defined by the limits of project requirements. In addition, requirements for skills and equipment fluctuate more within a project than within functional activities and thus often become costlier to manage.

Given the advantages and disadvantages of teams, a company must decide when using a team merits the additional cost. In a book on new product development, William Souder outlined many of the conditions that influence the decision to set up a new product development team. The common thread running through these conditions is the question of how much a new product development differs from a company's existing operations. If a company is aiming a new product toward a dynamic marketplace, is unfamiliar with the technology, or is making an innovative product, then it should use a team. However, if a company is aiming a new product toward a stable marketplace, understands the technology well, and is making only an incremental step, it should rely on its customary functions to develop the product.[4]

Even if it has a novel product development, a company may already have adequate organizational arrangements to handle the details of this task. For example, Souder mentioned three organizational arrangements that might already be adequate for the task. One such arrangement is a new product department staffed by technical people with extensive, open lines of communication throughout the company. Another arrangement, effective when used selectively, is a new product committee consisting of the managers who head the key functions. This arrangement usually is employed only for very costly or risky new product developments. Finally, if the key technical person and the key marketing person involved with a new product development have an exceptionally close personal relationship, this pair of committed specialists from R&D and marketing together may be equipped to guide a new product development to the marketplace.

Generally, in about 50% to 75% of a company's new product developments, the established functions can handle these more routine new product developments. A company may occasionally be fortunate enough to have an exceptionally qualified new product department, a new product development that can be handled by a senior management committee, or capable technical and marketing experts who are compatible with each other. Nevertheless, in most cases, the company will not be so fortunate. The remaining 25% to 50% of their new product developments will suffer unless they are backed by effective new product development teams.

the more time a person from a function serves on a project team, the more personal experience becomes defined by that project's parameters

In addition, these same 25% to 50% of new product developments (i.e., some of the most novel and often most promising new product developments) are frequently critical to a company's growth strategy. When these new product developments suffer, so does a company's growth. A company that wants to grow and hold its own in an intensely competitive marketplace should be able to create effective new product development teams.

WHO PLAYS ON THE NEW PRODUCT DEVELOPMENT TEAM?

The players on a new product team—its human resource—are vital to a company's achievement of its product development efforts. After making the decision to implement a team, the successful company will take great care in the selection of the people who constitute the team.

The Team Leader

The team's most visible and influential player, the team leader determines the amount of authority a team receives, the responsibilities it handles, and all the relations among the new product development team and various company functions. The leader also sets the direction the new product development takes and the degree of coordination among the competing viewpoints of R&D, engineering, marketing, manufacturing, and finance. The selection of a capable leader, in short, is essential to creating an effective new product development team.

Desirable characteristics of a new product team leader include:

- An understanding of the customers' needs.
- Knowledge of the technology involved.
- Knowledge of the operations of most of the key functions involved.
- Excellent management capabilities.
- Entrepreneurial skills.
- Stature with colleagues.

It is no coincidence that the desired characteristics for a new product team leader are similar to those desired in a company or business general manager. The leader of a new product development team is the manager of an embryonic business. As mentioned earlier, an effective new product team is a miniature company; its leader, therefore, is the general manager of a miniature company.

Only occasionally have I encountered new product team leaders with all of these desired characteristics. Of those who come to mind, one is in a defense electronics company; another, in a pharmaceutical company; and a third, in a communications systems company. All three are technical managers who have extensive experience with customers and with other functions in their company. All three are also effective managers and dedicated entrepreneurs, and they command great respect from their colleagues.

What really distinguishes these three exceptions from other team leaders, however, are two other characteristics usually not included in curricula vitae. First, all three leaders have a keen sense of the organizational politics in their company. Not only do they understand the organizational structure in which they operate; they also understand the people who make up the organizational structure—that is, the interests, prejudices, and objectives of key functional managers. Second, all three leaders rise above egocentricity in their dealings with the company's functional specialists. Uninterested in getting credit for making the right decision, these leaders ultimately are interested in getting the job done. Their colleagues, sensing this, trust their team leaders.

These two distinguishing characteristics are important because team leaders, though they may resemble general managers, do not have a general manager's authority. As mentioned earlier, a new product development team and its leader provide a counteraxis to the axis around which a company's functions typically revolve. However, a conflict can arise between a team leader and the heads of the various functions, should the heads of the functions have a higher rank than the team leader or should they control the resources and skills the team leader needs to develop a new product.

Given these conditions, a new product team leader cannot operate exactly like a general manager. For example, Sayles and Chandler explain that project managers or team leaders must operate according to a different theory of management and must have a different set of skills than general managers possess. Project managers or team leaders guide outsiders (i.e., the functional specialists who are not formal members of a new product development team or who are part-time members). Thus, team leaders accomplish their mission by monitoring others' work and by influencing decisions about a new product development; seldom do they give orders or make decisions like general managers.

Because a leader of a new product development team provides a counteraxis to the company's primary axis, a team leader must deal laterally, not hierarchically, with functional specialists. Furthermore, in these dealings, the team leader's lateral operations are very different from those familiar to most company members. Although many company members are accustomed to operating through an informal network of trusted colleagues to exchange information, a team leader operates laterally through functional specialists, many of whom the leader barely knows, to accomplish work and have decisions made. To accomplish anything under these conditions, the team leader must have a keen sense of organizational politics and be able to encourage functional specialists to do the right thing for the sake of the new product development, not just to fulfill the team leader's sense of authority.

I have observed how these distinguishing characteristics have made the difference. For example, a senior functional manager, who works with one of the three exemplary team leaders, mentioned that he would respond immediately to telephone calls from this exceptional team leader but not to those from other team leaders in his company. The functional manager knew that a call from this particular team leader evinced a critical problem that the leader had tried unsuccessfully to solve, and that now required the functional manager's help. In short, the manager trusted and respected this exceptional team leader more than he did the other team leaders.

companies should develop more effective procedures to nurture, train, and reward potential new product development leaders

Less effective team leaders sometimes attempt to manipulate the functional specialists contributing to their new product development. For example, one team leader admitted that he tries to manipulate functional specialists by giving them false or inadequate information, with the intent of spurring them to work harder on his own new product development. However, when the functional specialists sense his duplicitous approach, they take their time in doing the needed work.

Finding team leaders like the three mentioned is a constant problem in most companies. Companies can worsen their own situation, however, by neglecting potential candidates for new product development team leaders. Instead of initiating action, most companies prefer to react. Only when a new product becomes a possibility do most companies start looking for a leader, looking only to the currently available candidates. Because this attitude necessarily restricts the candidate pool to a small number, most companies have to settle for the best choice from among the available number.

What further complicates the problem of finding capable team leaders is that, in general, the state of knowledge about managing new product development is low. Industry publications, such as the *Journal of Product Innovation Management*, have only in recent years begun to pinpoint product development management as a long-neglected discipline worthy of study in its own right.

Finally, another obstacle in the way of finding and developing capable team leaders is the practice in many companies of transferring team leaders back to their functions after new product development projects have been completed, instead of assigning them to other new product developments. In addition, in some situations a person who serves as leader of a new product development team may have this position only as a part-time assignment.

In summary, most companies should develop more effective management procedures to nurture, train, and reward people who have the potential to be a new product development team leader. Because each company has different needs, there is no one set of management procedures that all companies can follow. Some companies may want to develop a cadre of full-time and permanent team leaders to serve as another company function whose responsibilities cut across those of all other functions. Other companies may choose to identify, more systematically, the pool of team leader candidates from various functions.

The Team

The members of a new product development team make up the remainder of this microcosmic company; their skills and abilities should reflect those elements necessary to run this miniature company effectively. Accordingly, functions other than

R&D, engineering, and marketing should be represented on the new product development team. For example, not only manufacturing engineering but manufacturing should be represented. As a manufacturing engineer who served on a new product development team once admitted to me, he could not adequately represent manufacturing operations on his team because he was unfamiliar with too many practical aspects of manufacturing operations.

A representative from the finance function should be included on a new product development team. An engineer for an automotive manufacturer once told me that a new product team in his company was much more effective when it included someone who could do cost analysis and return-on-investment calculations. Not only were the team's financial figures more accurate, but the financial officer of the company—and senior management in general—felt more comfortable with the team's efforts and therefore gave this team more support than they otherwise would have received.

Depending on the situation, other functions also may need to be represented, such as purchasing, quality assurance, sales, and legal services. A pitfall that a company must avoid in setting up a new product development team is enlisting almost exclusively people from the R&D organization. Among the numerous instances I have encountered of catastrophes largely caused by an R&D team's lone attempt to carry a new product to the marketplace, I am aware of four specific cases that vividly illustrate this type of failure: a consumer electronics company that introduced a new videotape recorder, a satellite communications equipment manufacturer that introduced civilian applications of space technology, a beer manufacturer that developed new products based on glass technologies, and a farm equipment manufacturer that introduced a process of coal gasification. In each case, an R&D group managed the entire new product development and commercialization—and failed. In some cases, the senior R&D manager responsible for the new product development was asked to leave.

A particular difficulty can arise when a company's R&D organization attempts to harness a technology with which they have little or no experience. In one illustration of this problem, a company that normally manufactured textile fabrics happened to develop a new membrane technology. The company discovered that it lacked people capable of the hands-on application work needed to develop new commercial products based on this technology. Another case involved a pharmaceutical company that developed new technologies, based on biological sciences, involving radioisotopes—significantly different from the chemistry-based technologies within the company's expertise. What the company lacked was the capability to translate these new technologies into manufactured products.

In both cases, critical skills needed to form a miniature company were missing. These companies—and others in similar situations—need to fill these gaps in skills as quickly as possible. If companies cannot do this, they should reexamine the risks and opportunities involved in developing a new product.

Companies also may have many functional specialists who do not understand the need for new product development teams. I have witnessed this problem at four kinds of companies: a dental products company, a computer company, a food-processing company, and a pharmaceutical company. In all of these, the functional specialists assigned to a new product development team misunderstood the team's purpose. When these specialists attended a team meeting, they thought of themselves as functional representatives who attended the meeting to report the meeting's proceedings back to their function; sometimes these functional specialists did not even want to be on the team. Failing to understand why they and their fellow team members were put together, they failed in their new product efforts.

when team members misunderstand the team's mission, the team will likely fail in its new product development efforts

Finally, when selecting members for a new product development team, the managers who set up the team, or the team leader, can encounter difficulty in getting ideal representation within a team. One such problem is related to the size of the team. Although it is desirable to represent all of the key functions, a new product development team of more than 10 members is unwieldy in most cases. To avoid this problem, someone from one function or discipline may have to represent two or more functions or disciplines. For example, someone from design engineering may have to represent both design engineering and packaging engineering. However, a problem can occur if, for example, packaging engineering believes that its needs are not fully taken into account by the team.

Another potential problem of team member selection concerns finding functional specialists with the appropriate combination of hands-on knowledge (e.g., about the technology or customers' requirements) and authority to manage effectively the other specialists in their function. Striking the right balance between the functional specialist's capabilities and the degree of cooperation he or she commands is often a difficult problem.

A third problem that can arise for the managers who set up a team concerns practical limitations with regard to the organizers' persuasive powers. New product development teams are not always popular with functional managers. Unless the operating division's general manager or the company president mandates that all key functions must participate, the team's sponsors or leader may run into difficulties in getting one or more key functions to participate. Like gaps in critical skills, gaps that appear because key functions do not want to participate must be filled as quickly as possible. If they cannot be filled, the risks and opportunities involved in developing a new product should be reexamined.

SETTING UP A NEW PRODUCT DEVELOPMENT TEAM

The organization's structure should include a system of checks and balances to ensure that the company's needs are met throughout the functions. The team leader's authority and responsibilities both within the new product team and throughout the rest of the company should be carefully monitored during the product development process.

Who Should Supervise the Team Leader?

Because a new product development team provides a counteraxis to the axis around which various functions revolve, the team's leader should report to that manager who supervises the function heads. This manager, usually the general manager of a business or company, must resolve conflicts between a function head and a new product team leader. If a team leader were to report to a head of one of the functions (e.g., the head of engineering), the team leader could be in frequent conflict with that boss.

In addition, having a team leader report to a function head creates problems for the function. For example, I know the engineering head of a medical instrument company who supervises all new product development team leaders. Because of this arrangement, he is caught repeatedly between representing the interests of his engineering department and those of a new product development. Furthermore, managers in other functions reproach him for inadequately representing their interests.

Therefore, the person to whom a team leader reports cannot hold direct responsibility for one of the functions represented on a team. Ideally, this supervisor ranks one level above the function heads and is a general manager. It is this general manager, in addition, who should appoint the leader of a new product development team. Specifically, this means that a general manager should not delegate responsibility for selecting a team leader; neither should the general manager first pick the members of a team, and then charge them with appointing their own leader.

First, if a general manager is going to depend on a team leader to provide an effective counteraxis, the general manager should take the initiative and find someone in whom he or she is confident. If the team leader is of the general manager's choosing, it is more likely that the leader will receive the private hearings with the general manager often needed during a new product development.

Second, a team has difficulty selecting its own leader, particularly if the team members do not know each other well. I have worked closely with a new product development team that had great difficulty calmly discussing which of the team members was best suited to be the leader. Because they had problems, they ended up choosing that member whose function would be making the most significant contribution to the new product development during its initial phases. In effect, the team members attempted to ignore personal issues in the voting process. What this process loses, however, are considerations of which team member has the best managerial abilities. Consequently, because team members often cannot evaluate themselves objectively, a general manager should select the team leader.

Responsibilities of a Leader of a New Product Development Team

Because new product development team leaders must focus the activities of various functions, they run the risk of being overwhelmed by the associated tangle of responsibilities. Therefore, the leader of a new product development team must have clear ideas about what is and what is not his or her responsibility. Melvin Silverman's book about project management succinctly lays out a general rule of responsibilities that could also apply to the leader of a new product development team:

> The general rule is that the functional manager is primarily concerned with the adequacy of the technical input (or the "how") and the project manager is primarily concerned with the task and when it will be done (the "what" and "when").[5]

to fulfill the role of an effective counteraxis, the team leader should report to that manager who supervises the function heads

Although there always is some overlap between the how and the what and when questions, Silverman's general rule applies to new product development.

I once formulated three questions for a particular team leader regarding responsibilities and concerns. Although these questions do not touch on the issue of when, they do point to three aspects of the what question:

- What technical results are needed to make the new product perform as well as desired?
- What will customers buy?
- What new product development efforts will be supported by the company functions and senior management?

According to Sayles and Chandler, a project manager or team leader should act as a metronome, keeping the various functions responsive to a central beat or common rhythm. In this regard, a project manager or team leader should not focus on the specialists' technical contributions. Because a project manager or team leader will have difficulty trying to second-guess a functional specialist on technical matters, the leader should concentrate on influencing how, how much, and when a function participates. Sayles and Chandler describe three specific responsibilities of a team leader:

- To make sure that problems related to a new product development are given their proper weight by the functions.
- To make sure that the functions tackle problems in the right sequence and at the right time.
- To shift the functions' decision criteria when necessary (e.g., to persuade a manufacturing organization to carry a costly inventory, to prevent possible distribution problems that another function in the company may later run across).

In other words, a team leader must look at a new product development as a whole, always challenging any function that carries out its activities with an overly narrow perspective.

To keep a new product development in perspective, a team leader must particularly beware of allowing the R&D organization's viewpoint to prevail. According to a study by Edward F. McDonough III and Richard P. Leifer, the function whose viewpoint is most likely to dominate a new product development is the R&D organization's. In the less effective new product development projects, the "project leaders allowed their engineers to pursue technological approaches and ideas that were interesting from a technical standpoint but were detrimental to the swift and successful completion of the project." On the other hand, in the effective new product development projects, the project leaders ensured that engineers and other team members always stayed aware of the needs of the new product development as a whole.[6]

Finally, another responsibility of the team leader is to force the team to make all of the decisions that it should, rather than refer controversial decisions to senior managers. There are decisions that a team should not or cannot make, such as those about buying capital equipment. Nevertheless, if a new product development team is to be a miniature company, the team should make all of the decisions that it can, within the bounds of its authority. The team leader holds primary responsibility for making this happen.

To elucidate to the company what the team leader's responsibilities are, the person who appoints team leaders (i.e., usually a general manager) should define them in writing. Otherwise, the position of team leader will seem vague to everyone. An unclear definition of responsibilities for team leaders can undermine the team leaders' authority because no one knows what the team leaders are supposed to do.

Another benefit of defining these responsibilities in writing is that the team leaders then know what is expected of them. Although individual team leaders each should have their own individual style within the defined guidelines of their responsibilities, they should not be allowed to delineate their own responsibilities. Not only might they then neglect certain key responsibilities, but the functional specialists who need to deal with these individual team leaders often will get confused about what to expect from them. A written definition of a team leader's responsibilities will help everyone involved.

The Team Leader's Authority

In view of the team leader's responsibility for focusing the activities of various functions, there arises the ongoing task of preventing the team leader from devolving into a mere clerk or paper-shuffler. After all, it is the various functions that actually carry out the product development work (e.g., conducting experiments, building prototypes, and organizing market promotions). In the midst of all of these activities, the team leader easily can end up operating simply as a conduit of information among the functions. For example, one team leader informed me that he had to make sure that he was informed about and attended function meetings concerning the new product development that presumably was his main responsibility. Because team leaders often must struggle constantly to make their presence felt, the extent of their authority is a critical factor in the success of the new product.

According to William Souder, this authority consists of a license to innovate. The team leader has the opportunity to create the means for developing and commercializing a new product; how to use this opportunity properly is up to the team leader.

A general rule is: The leader of a new product development team should have as much authority as needed to get the job done. Although this may sound like an exhortation to the team leader to build an empire, I have observed in my work with many companies that a team leader under this mandate seldom has too much authority. Given the various functions' control over resources and skills, the relevant concern is not whether the team leader will become a despot but whether the leader can become something more than a conduit of information among functions.

a team member must be more than a liaison between function and team: as a function representative, the member must participate wholeheartedly with the team

One way of helping a team leader progress beyond this conduit stage is to delineate initially, in concrete terms, those issues under the team leader's authority. For example, the team leader may be authorized:

- To have final say in the selection of team members.
- To evaluate the team performance of the functional specialists on the team.
- To control a budget that includes some activities related to a new product development.
- To direct outside subcontractors in their work for the team.

Even if formal authority is delineated in concrete terms, seldom will this formal measure guarantee the team leader's success. Far more important in this situation are informal factors, such as the team leader's stature, the extent of the leader's informal network of contacts throughout a company, and whether the leader has a mentor within the company who can tackle issues beyond the leader's influence.

Souder has identified a phenomenon connected with these informal factors. According to Souder, new product development team leaders who come from marketing have much greater success than those who come from R&D. A marketing person selected as team leader tends to elicit the desired reactions from the company—for example, senior management becomes more involved with the new product development, the team leader receives a broader license to innovate, and the marketing organization supports the new product development more than it otherwise would. In other words, when a marketing person is made team leader, many informal factors influenced by the marketing organization's position—at the core of a company's operations—suddenly come into play. An R&D organization, which in most cases holds a more peripheral position within a company, influences the team differently. Consequently, when an R&D person is made team leader, several informal factors in this situation tend to weaken, rather than strengthen, a team leader's authority.

This peculiar phenomenon aside, a company need not exclusively appoint marketing people to be team leaders. There are many other ways to ensure the leader's effective authority. Selecting the person with the desired characteristics, regardless of which function that person comes from, is the key to creating an effective new product development team.

The Team Members' Responsibilities

As mentioned earlier, one of the difficulties of selecting the other team members involves striking a balance between what specialists bring to the team and how much cooperation they can claim from others in their functions. These two aspects of a functional specialist's capabilities also are relevant to that specialist's responsibilities as a team member.

First, with regard to responsibilities, team members must be more than mere liaisons between their functions and the team. A function respresentative must be a real member of a team (i.e., must participate wholeheartedly). Team participation does not preclude the function representative from associating with the home function. Rather, the representative should maintain a double allegiance—one to the home function and one to the team.

Second, a team member must act as the team leader's surrogate in his or her home function. Although the team leader should take the lead in coordinating various functions' activities, he or she can become overextended. This can occur particularly when a function's responsibility for how and the team leader's responsibility for what and when overlap. Team members should get involved with these issues because they are likely to have greater expertise than the team leader with regard to how issues and because they are members of a team that must deal with the what and when issues.

MANAGING THE NEW PRODUCT DEVELOPMENT TEAM

The new product development team leader also has another set of responsibilities: to manage the team itself. The leader has four major concerns in this respect:

- Clarifying each member's role.
- Forging a team among functional specialists.
- Ensuring that the team is organized in accordance with the needs of the moment.
- Fostering effective communication within the team.

Each of these concerns is discussed in the following sections.

Clarifying Each Team Member's Role

When a new product development team is first set up, members will be unsure of what is expected of them. The team leader must take the initiative in clarifying each team member's individual role. For example, an R&D member might hold primary responsibility for designing the performance characteristics of a new product, a marketing member might be responsible for conducting market trials, a member from manufacturing engineering might have primary responsibility for developing pilot plant tests, and a manufacturing member might be responsible for conducting the initial manufacturing runs.

After the team leader has clarified the members' roles and all the members have become comfortable with their primary assignments, the leader should encourage team members to take more than one role in the team's activities. A marketing member should be able to contribute a great deal toward the design of a new product's performance characteristics, and a manufacturing member should be able to help a team member from manufacturing engineering develop the pilot plant tests. The more team members go beyond their specialty to address other issues within their knowledge, the more effective the whole team will be.

a project manager should send a meeting agenda to participants beforehand, inviting them to return suggestions for solving the problems to be discussed

Forging a Team Among Functional Specialists

The team leader must take a strong hand to forge a real team among the functional specialists who have been put together to develop a new product. In practice, the team leader should deal with a second tier of issues above that concerning the team members' individual roles. This second tier involves the whole team and should consist of such issues as:

- What is the team's specific mission?
- How should the team as a whole make decisions?
- How often should the team meet?
- How should the team communicate with company groups outside the team?
- How will the team hold itself accountable for its actions?
- How should strong differences of opinion among team members be resolved?

It is the team leader who must take the initiative in addressing these issues; otherwise, team members are likely to view themselves as a collection of functional specialists who happen occasionally to discuss a new product development—not necessarily as a dedicated team formed expressly for that purpose.

The potential strength of a new product development team stems from honest and open discussion among members from diverse functions. Effective communication results in better decisions regarding a new product development and a focus for coordinating the activities of various functions. This type of communication requires a real team. One sign of a real team is that its members refer to themselves as we in relation to them—that is, the rest of the company. It is this team spirit that makes it possible for the aggregated specialists to breathe life into a new product development. The team leader holds ultimate responsibility for this team spirit.

Ensuring That the Team Is Organized in Line with the Needs of the Moment

As a new product development proceeds from beginning to end, circumstances will change. The team leader is responsible for ensuring that the team is organized to meet the demands arising from the prevailing circumstances at any moment.

In an article on new product development, Frederick A. Johne points out the relationship between a team's effectiveness and the structure of a team's activities during the course of a new product development. With effective teams, activities are loosely structured during the early stages of innovation, then are tightly structured during the implementation stages. For example, the early stages involve a greater degree of brainstorming, whereas during the later stages, a development manual is used to monitor a new product development against a standard set of criteria. On the other hand, with less effective teams the opposite is true: activities are tightly structured during the early stages, when the team needs to innovate, and loosely structured during the later stages, when the team needs to implement.[7] Therefore, one of a team leader's responsibilities is to make sure that activities are structured according to the needs of each particular stage of development.

Another of a team leader's responsibilities in this area is to ensure that the various team specialists contribute the appropriate amount at the appropriate time. During the course of a new product development, team members from R&D, engineering, marketing, manufacturing, quality assurance, purchasing, and finance can be expected to play different roles at different times and to be more or less involved, depending on the needs at any stage. It is the team leader who must make sure that each team member plays the proper role at a given time and that each is as involved as he or she should be at each particular stage. For example, the team leader must make sure that an R&D member begins to play an increasingly advisory role as the new product development proceeds to market and that a purchasing member is involved sufficiently while design issues are still being considered.

Fostering Effective Communication Within the Team

A team leader also must foster effective communication among the various specialists on a team, which is not an easy task. Although many barriers can arise among various functional specialists, I have noted that one of the most difficult barriers to overcome is that between marketing people and technical people, when *technical* is defined in the broadest sense (i.e., R&D, design engineering, manufacturing engineering, quality assurance, manufacturing, and purchasing). Marketing people often see themselves as piloting a new product development, whereas technical people see themselves as implementing it. From marketing's perspective, technical people address only issues of secondary importance and therefore should follow marketing's guidance. Technical people generally deem marketing people incapable of handling the nuts-and-bolts issues involved in developing and manufacturing a new product. The backgrounds of the two groups are usually so different that they frequently find it difficult to talk meaningfully with each other. This barrier between marketing and technical personnel is one that a team leader must overcome in promoting effective communication.

A team leader also must be able to conduct team meetings well. In this regard, Melvin Silverman offers useful advice in preparing for a meeting:

> A properly designed format for a problem-solving meeting requires that most of the individual problem-solving work be done prior to the meeting itself. Meetings are very expensive. They use people's time, and the only reason to have them should be to provide an interaction among people that doesn't

happen in a one-on-one meeting. Therefore, the contributions of individuals invited to the meeting should be coordinated beforehand. The meeting is then necessary only to achieve group interaction.

to be effective, a team leader must concentrate on the what and when questions involved in the development

Silverman also recommends that a project manager send a meeting agenda to participants beforehand, inviting them to return written comments outlining possible solutions to the problems that will be discussed.

CONCLUSION

A multifunctional team provides most companies with a focus for a new product development. Such a team is, in effect, a miniature company that coordinates the various functions' activities related to the new product.

The leader of the new product development team is essential to achieving such a focus. To be effective, a team leader must concentrate on the what and when questions involved in the development. The functions, which provide the skills and resources, must then concentrate on the how question.

The selection of a team leader and team members is critical to the success of the new product. When the team is set up, care should be taken in designating the person to supervise the team leader, in clarifying the responsibilities of a team leader and team members, and in delineating a team leader's authority. Finally, the team leader must fulfill his or her own responsibilities in managing the team.

Multifunctional teams can improve a company's new product development. Before companies can take advantage of such teams, however, they first must improve the methods for creating their new product development teams.

Notes

1. C.M. Crawford, "New Product Failure Rates: A Reprise," *Research-Technology Management* 30 (July–August 1987), pp 20–24; and C.M. Crawford, "New Product Failure Rates—Facts and Fallacies," *Research Management* 42 (September 1979), pp 9–13.
2. L.R. Sayles and M.K. Chandler, *Managing Large Systems* (New York: Harper & Row, 1971).
3. B. Little, "Significant Issues for the Future," *Journal of Product Innovation Management* 1, no 1 (1984), p 59.
4. W.E. Souder, *Managing New Product Innovations* (Lexington MA: Lexington Books, 1987).
5. M. Silverman, *The Art of Managing Technical Projects* (Englewood Cliffs NJ: Prentice-Hall, 1987), p 147.
6. E.F. McDonough III and R.P. Leifer, "Effective Control of New Product Projects: The Interaction of Organization Culture and Project Leadership," *Journal of Product Innovation Management* 3, no 3 (1986), pp 152–154.
7. F.A. Johne, "How Experienced Product Innovators Organize," *Journal of Product Innovation Management* 1, no 4 (1984), pp 216–217.

GOING GLOBAL
Establishing R&D Project Management Teams

When Johnson & Johnson's pharmaceutical research and development efforts were consolidated recently, its project management approaches were reassessed. In the process of implementing a standard project team approach across previously diverse and distributed R&D activities, the author, who was instrumental in the change, encountered complexity and resistance in many quarters but ultimately enjoyed success.

J. Michael Grindel

A highly decentralized health-care corporation, Johnson & Johnson is organized into three business lines: professional, consumer, and pharmaceutical sectors. The pharmaceutical sector, in turn, has three major components: the Janssen Group based in Beerse, Belgium; the ICOM Group comprising McNeil Pharmaceutical, Ortho Pharmaceutical, Cilag International, and the R.W. Johnson Pharmaceutical Research Institute; and the Diagnostics Group. This structure is detailed in Exhibit 1.

To maximize the leverage of the new pharmaceutical company's research capabilities, the R.W. Johnson Pharmaceutical Research Institute (PRI) was formed in 1988 by consolidating the R&D divisions of McNeil Pharmaceutical, Ortho Pharmaceutical, Ortho Pharmaceutical (Canada) Ltd, Ortho Biotechnology, Cilag AG, and the J&J Biotechnology Center. These operations were then located in Pennsylvania, New Jersey, California, Canada, and Switzerland.

To facilitate the merger of these R&D units, a project team was assembled to assess the status of the new company and to make recommendations to its incoming president on issues related to the management of information technology, worldwide clinical trials, and project portfolios. As a member of the implementation team, I was responsible for assessing the project management practices in the merging divisions and recommending an appropriate future course for this management approach.

THE NEED FOR PROJECT MANAGEMENT

The process of new drug development has become increasingly longer (the average cycle is 12 years) and more expensive (the average cost is $231 million). Consequently, pharmaceutical companies have looked increasingly to project management as a means to control their investments in new product development

J. MICHAEL GRINDEL, PhD, is the executive director of Project Planning & Management Worldwide at the R.W. Johnson Pharmaceutical Research Institute in Raritan NJ.

Exhibit 1
The Location of the R.W. Johnson Pharmaceutical Research Institute in the Organizational Structure

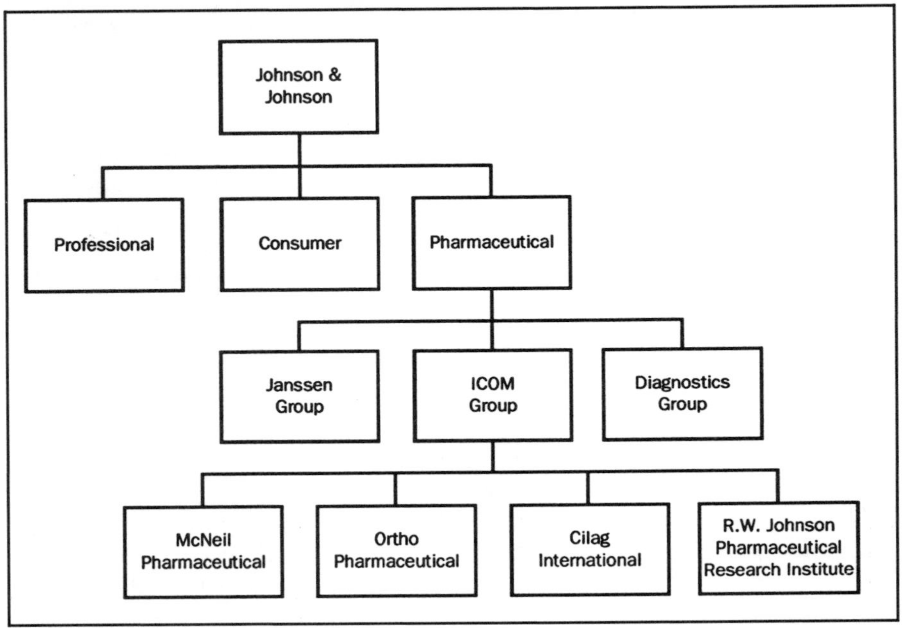

more efficiently and effectively. Proper management of risk, cost, and development time can determine the outcome of the race to the marketplace by decreasing the time it takes to bring a new drug to regulatory filing and by speeding the regulatory approval process. Successful project management should minimize the resources needed for product development, thus allowing other resources to be allocated to discovery research and product support opportunities. Such optimal use of available resources can be accomplished by minimizing the duplication of clinical trials and by focusing R&D efforts on new products with the greatest potential for success.

The complexity and length of the product development cycle combine to present a unique and formidable management challenge. Integration is required across multiple phases and functions (e.g., preclinical drug safety tests, pharmaceutical formulations and analysis, medical analysis, regulatory filing and review, marketing, and manufacturing) as well as across geographic locations (e.g., the US, Europe, Latin America, and Japan). In addition, the drug development process is subject to strict regulation in each of the phases through which a new product must pass, including preclinical trials, clinical trials, regulatory filing, regulatory review, and product launch.

Other factors contribute to the complexity of new drug development. High attrition rates are standard in drug development: only 20% of the new chemical entities that undergo preclinical safety evaluation ultimately reach the marketplace, and most (i.e., 70%) fall by the wayside before or during the clinical trial period. Therefore, it is essential to make the right decisions about new products at the earliest possible point in their development cycle.

successful project management should minimize the resources needed for product development, thus allowing other resources to be allocated to discovery research and product support opportunities

Globalization of the pharmaceutical industry requires an R&D organization to operate in a multiregulated environment. In the US, the Food and Drug Administration (FDA) requires access to clinical data developed worldwide, the European Community harmonizes its registration requirements through the CPMP, and adverse drug experience must be reported worldwide. In addition, the industry has seen a shift in needs for new drugs. The demand for drugs that treat acute diseases (e.g., bacterial infections) or that relieve pain has given way to the need to develop pharmaceutical treatments for chronic diseases (e.g., arthritis, cancer, and schizophrenia). This trend has contributed to increased development time and costs.

Companies that have successfully met these challenges have focused their efforts on three areas:

- Organizing and managing for functional and global integration.
- Making trade-offs among timing, cost, and market position.
- Achieving the flexibility to respond rapidly to changes in the market, regulatory, and development environments.

PROJECT MANAGEMENT PRACTICES AT PRI

Project management and planning were perceived to be important to the success of PRI. At the time PRI was formed, project management practices varied throughout the divisions of the organization. Each site considered itself effective at drug development, and each believed that its approach should be adopted for the new organization.

One site had a well-established project management department with multidisciplinary project teams that were linked to the business unit. Together with management, these teams set directions for their projects and actively recommended ways to advance them most effectively. By comparison, other sites had project teams for some projects but no infrastructure (e.g., no project planning or management department). In these instances, the project teams simply coordinated activities and communicated problems. They were not encouraged to recommend improvements to the project plan, nor were they supported.

In yet another case, the international division had used project teams and project management methods to coordinate its work with the activities of the medical departments of each affiliate company in Europe. These project teams were effective until changes in the research center's senior management upset the delicate balance. Decreased management support for and understanding of the project management process coupled with the lack of a formal infrastructure to support the teams created gaps in the process.

It was evident that formal project management procedures had to be introduced carefully in the new organization. Much resistance to their implementation was encountered, and many questioned the need for them as well as their potential effectiveness. At those sites where project management systems and structure existed, the environment was more open and line managers were accustomed to grappling with cross-functional issues. At the sites where project management practices were to be introduced for the first time, the line management organizations were very strong and there was little cross-functional communication and problem resolution except at the very highest management levels. The challenge was to determine the direction PRI's project management should take to establish PRI as a worldwide R&D organization.

Choosing a Model for Project Management

The models for project management that have been applied successfully to the pharmaceutical R&D environment include:

- Project teams led by part-time project leaders who also perform line-management functions.
- Full-time project managers with multifunctional project teams or task forces.
- Project leaders with no project teams but with access to all of the pertinent functional managers.
- Therapeutic area teams and therapeutic area divisions managed by a full-time divisional director or vice-president.

Any of these approaches can succeed provided it meets an organization's needs for project integration and implementation.

PRI adopted a multifunctional project team headed by a part-time project leader and assisted by a project management professional. The project team approach was judged most suitable for PRI's need to meet the challenge of pharmaceutical product development. Project teams allow the R&D manager to maintain the skills and specialization required to deal with rapidly evolving technologies. They also provide the mechanism for integrating a large number of independent but related projects that are in various stages of development. A project team's success depends on its ability not only to bring the product to the market but, more important, to identify critical junctures in the process and shepherd the project through the many decisions that must be made during its development life.

THE PROJECT TEAM

The project team consists of representatives of specific organizational units in the R&D department and of each business unit—marketing, manufacturing, and sales. The exact composition of the team changes as the project progresses through various stages of development. At a minimum, the typical PRI project team includes a project leader (usually a senior scientist or physician), a project management professional, and representatives from the participating domestic and international business units. The early involvement of business unit representatives in the development cycle has been critical to the teams' success. Additional team members represent such other functions as manufacturing, preclinical trials, medical analysis, statistical analysis, and regulatory filing and review, as required.

a project team's success often lies in its ability to make a well-informed decision to terminate a project at the earliest possible stage

Forming the Team

The first challenge in forming a project development team is to designate an effective and knowledgeable person as the project leader. The leader and the project management professional must form a partnership to run the team effectively. The team members should be well-informed, senior technical personnel who understand their respective roles, and they must be actively supported in these roles by their managers. The project team's effectiveness suffers if management does not allow time for team meetings or if communicating team issues before and after the meetings is not important to the manager. At PRI, it has been an ongoing challenge to orient team members to their roles and to drum up senior and middle management support for the team leaders and members. Another significant issue in developing effective project teams at PRI has been the scarcity of personnel with experience in global drug development. Although many PRI staff members were expert in dealing with the US FDA, few were conversant with European or Japanese regulatory issues.

Ensuring Team Success

At PRI, project team success is measured by a team's ability to see a drug successfully through all phases of development. Progress must be routinely observable in the form of tasks completed as planned and interim objectives accomplished. The team must identify key issues confronting the project and discuss them with the appropriate PRI and business unit managers. At major milestones in the development cycle, the team must critically assess the project's potential and issue unequivocal recommendations to PRI senior management regarding the investment of future resources in the project. A project team's success often lies in its ability to make a well-informed decision to terminate a project at the earliest possible stage.

Success is also judged by a team's ability to get the product approved and on the market around the world. The need to develop drugs for a global marketplace was a primary impetus for consolidating research capabilities in PRI. Project teams are necessary to ensure that drugs are developed rapidly and efficiently on a global basis. The teams' goals are not just to file registration documents on schedule but to create documents of such quality that the drugs will meet with approval in several countries in a short period of time. To do so, the requirements of health registration authorities of major countries around the world must be well understood.

A core program of clinical studies needed to fulfill the drug approval process in all of these markets must be undertaken. To obtain universal filing, those studies and their supporting preclinical safety evaluations must meet the highest standards.

Other prerequisites for a project team's success include:

- Consistent hard work and a high level of motivation and dedication from all team members.
- Well-defined roles, responsibilities, lines of authority, and information flow to minimize conflict with respect to project procedures and administration.
- Clearly defined project objectives, plans, schedules, budgets, and reporting and monitoring systems.

It was generally believed that project management at PRI would be effective only if a consensus on the roles and responsibilities of the project teams, their leaders, and their members was reached. To this end, discussions were held between the project team leaders and PRI senior managers in 1988 to clarify these responsibility issues, the outcome of which is summarized in Exhibit 2. The product of months of discussion, this list compares the responsibilities of the project team to those of functional management. Although recent changes in senior management necessitate revising this list to make it consistent with the goals and understanding of current PRI management, the ongoing dialogue between senior managers and the project team leaders is critical to the continued effectiveness of project management practices.

Exhibit 2
Project Team Versus Functional Management Responsibilities

Project Team	**Functional Management**
Integrate the overall project strategy and direction for senior management approval	Develop project strategy and direction for the functional area
Focus on the needs of a single project	Balance the needs of multiple projects
Develop the project phase objectives and plans	Develop functional area plans
Integrate and reconcile functional plans and resource budgets for the project	Assign resources to the projects as negotiated in the approved project plan
Coordinate and integrate activities among functions, sites, or licensors	Conduct project activities and studies using functional area expertise
Evaluate progress according to the approved plan	Ensure quality, timeliness, and effectiveness of project work
Anticipate, identify, and resolve project problems	Anticipate, identify, and resolve problems
Communicate accomplishments, status, and plans	Communicate study data and information
Coordinate technology transfer	Transfer technology in the functional area
Perform as the project expert and advocate	Perform as the functional expert

Selecting Project Leaders

To be successful, project leaders must accept the pivotal role they play in the management process. To be effective, a project leader should possess as many of the following characteristics as possible:

- Leadership abilities.
- Interpersonal skills.
- Communication skills.
- Negotiation skills.
- Problem-identification and problem-solving skills.
- Credibility with senior management.
- A scientifically informed understanding of the project.
- Sound judgment.
- Organization and planning skills.
- Confidence and a positive attitude.
- The willingness to accept responsibility.
- The ability to cope with an uncertain environment.

because open communication is critical in project environments, electronic mail, telephone conferencing, videoconferencing, and face-to-face meetings have been vital to effective work teams

At PRI, project leaders are selected from all divisions and all locations. They are selected typically from the ranks of senior scientists and physicians and occasionally from among experienced managers in other areas (e.g., regulatory requirements and statistics). As a rule, they have high levels of technical competence and excellent communication skills. Although many have needed help to improve their team management skills and their ability to influence others, most have grown into the role over several months with some coaching from an experienced project manager or senior corporate manager. Although some individuals were initially reluctant to invest the time needed to become project leaders, management's visible support for the project leaders and the organizationwide acceptance of project teams helped them to overcome their reluctance.

The project leader should be a major force for integration within the R&D department and between R&D and the involved business units. An effective project leader exerts personal influence to induce the performance expected from project team members and functional managers. The leader must foster an environment in which the collective talents of all team members contribute to the success of the project. Therefore, in addition to strong problem-solving skills, the project leader must also be able to resolve conflicts and conduct efficient, effective meetings.

A project leader's responsibilities also include acting as the spokesperson for the project and ensuring that the project has clear objectives as well as an acceptable plan and schedule and that it progresses according to plan. The leader must get the team members to identify and resolve issues before they become crises. Though lacking the formal authority to make decisions, project leaders exert a great deal of influence in the recommendations they and their teams make to senior management about developing drugs for the company's global markets. Initially, support for the teams' active role in drug development was lacking because line management perceived such involvement to be a challenge to its authority. As these managers grew accustomed to the global scope of the project managers' responsibilities and as the number of projects increased, however, many of their fears were put to rest. Indeed, project leaders are now welcomed for the support they lend to management in developing drugs for global markets.

Selecting the Project Management Professional

The project management professional plays a key role on the project team. Administratively, this person ensures that meeting agendas are established, issues are raised and discussed, action items are assigned to team members for follow-up, and the necessary follow-up occurs. After PRI's team meetings, the project management professional prepares the minutes of the meeting and sends a report of critical issues to the management board and to other senior managers. The project management professional also sees to it that all presentations and reports are prepared in a professional, timely, and accurate manner.

Most project management professionals have a background in R&D, clinical research, or regulatory affairs and later join the project planning and management organization, a branch of PRI. As members of that organization, they develop the knowledge and skills requisite for project planning, scheduling, and developing drugs for worldwide markets as well as for influencing, communicating, and negotiating with management. With this background and training, they are well equipped to support project leaders in achieving project goals and objectives. The following sections discuss the project management professional's planning and scheduling responsibilities in greater detail.

Developing the Strategic Background Document. The project management professional's most important function is to critically evaluate drug development plans and to establish with the teams efficient, effective, and realistic plans of action. For each project they support (most are involved in four to five simultaneously), the project management professionals write a strategic background document. This document contains critical information about issues that are fundamental to the project plan. Exhibit 3 outlines the generic topics covered in the strategic background document that apply to all projects.

The first documents took months to write. The services of outside consultants were needed to impress upon PRI staff and managers the significance of the documents in establishing project plans. Each section of the document had to be negotiated with the appropriate line managers. The sections on goals and risks were especially enlightening because they pointed up inconsistencies in those risks and goals identified by middle management and those acceptable to senior management in the same function. Such discrepancies underscored the need for these documents as a means to clarify such issues, and they dramatically demonstrated the communication gap that professional project managers needed to bridge.

Exhibit 3
Outline of Generic Strategic Background Document

A. The project's background
   1. Project objectives
   2. Scientific rationale
   3. Importance of the drug to the business franchise
   4. Chemical name, structure, and patent status

B. Anticipated claims structure
   1. Indications, dosage forms, and routes of administration
   2. Potential comparisons with existing and future therapies
   3. Contraindications and warnings as they become known

C. Worldwide development strategy
   1. Goals
   2. Risks
   3. Pivotal studies and rate-limiting activities
   4. Project milestones
   5. Registration filing targets

D. Operations and manufacturing evaluation

E. Business evaluation by indication or dosage form

F. Appendixes
   1. Chronology of significant regulatory agency actions
   2. Chronology of major project decisions
   3. Patent expiration dates by country
   4. List of registration filings and approvals
   5. Development schedule summary

Since their first use, strategic background documents have been simplified to eliminate redundancy and to improve readability. Updated semiannually—unless a more frequent revision is required—they now serve as excellent information sources and as a means for new PRI managers to get up to speed quickly with the project portfolio and the major project issues in their areas of responsibility.

Preparing Project Schedules. On the basis of the strategic background document, the project management professional prepares a detailed schedule of project activities for each phase of development. The schedules clearly detail the interrelated project activities across departments, functions, and sites. Dates are derived for all performance milestones from the schedule and then negotiated with and approved by management. If necessary, schedules are modified to agree with the established milestone.

Project schedules are maintained in a consolidated data base known as the project scheduling system. The detailed project schedule, the clinical operating plan, and the strategic background document form the basic project plan. Information from the strategic background document is also entered into a data base called the project reporting system to aid in the preparation and maintenance of the appendixes that are a part of that document. Although only project management professionals had ready access to these systems previously, PRI has now made the project reporting system and the project scheduling system available to all active managers.

Selecting Team Members

Successful project team members at PRI have the following key characteristics:

- The ability to learn quickly and to adapt to changing conditions.
- Endurance, drive, and high energy.
- An even temperament and consistent performance.
- Technical proficiency in their area of responsibility.
- Satisfactory skills in organization and communication.
- Optimism about the project.
- An interest in their own personal development and project's success.

Like project leaders, most team members are selected on the basis of technical competence, familiarity with the project, a demonstrated interest in furthering their careers as well as the project, and management's concurrence that involvement on a project team would be a growth experience for all participants. Most scientists have expertise in their specific work area but usually need to broaden their knowledge of ancillary areas in their own functional areas. They need to be allowed to develop loyalty to the project and the project team and therefore must be encouraged early and often to establish ongoing lines of communication with middle and senior managers within their functions.

To a large degree, the project team's success depends on the team members' ability to represent their functional areas to the team and, conversely, to represent the team to their respective functions. Team members must communicate departmental plans, budgets, activities, and issues so that decisions made by the team have the benefit of complete functional area input. In addition, they must be able to identify problems in their functions and must help to achieve their resolution. Team members are expected to act as project advocates within their respective areas.

The team members must participate actively in meetings and willingly provide information and support to the project leader and the project management professional in the preparation of project plans and management presentations. The project leader and the project management professional should consel and coach team members to improve their effectiveness. In addition, human resource consultants may work with the teams on team-building and group-interaction skills.

To demonstrate the value management attaches to project team members and to reward them for their contributions, PRI has used group lunches, small gifts, personal notes from the president, and similar gestures to express appreciation for jobs well done. Team managers are continuously on the lookout for other appropriate rewards for team members.

Clarifying Decision-Making Responsibility

despite its accomplishments in opening lines of communication, PRI has found a major stumbling block to be middle managers' reluctance to share information

In addition to defining the roles and responsibilities of the project team and members, PRI recognized early in its conversion to a project-based management approach that other issues had to be clarified before changes in the way projects were coordinated could be put into effect. Unlike a functionally oriented operation in which individuals have responsibility to one manager and one functional area only, a project-based organization means that individuals may have more than one supervisor depending on the needs of the project. Therefore, it was crucial that decision-making and problem-solving responsibilities be clarified for the project teams. The best decisions in terms of project success are made when the bounds of authority are clearly delineated between senior management and project management and when the project team is responsible for making decisions within its realm of expertise. To that end, senior management and project teams must agree on three basic issues:

- The need to provide the project team with the information it needs to manage the project successfully.
- The need to acknowledge the project leader's and team members' technical expertise and to apply it in ways and in places that will yield the greatest benefit.
- The need to grant decision-making authority to project team members representing their departments.

To be effective, project teams need to operate in an open environment in which they are free to discuss issues and problems with whomever they need to. If such an atmosphere could be created, it would allow for informal discussions of critical issues at team meetings and elsewhere without causing alarm among senior managers. Major decisions and actions would be taken only with the agreement of line management, and no issues would become critical before senior management heard about them.

Establishing the Management Board and Project Reviews

To ensure that there would be no surprises in store for senior managers, PRI's management board initiated monthly reviews—a process that allowed senior management to take stock of each project once or twice a year, depending on its importance and the need for high-level review. The board actively encouraged open discussions of issues and was careful not to shoot the messenger when bad news surfaced during the course of these meetings. Despite the board's well-meaning intentions, however, the reviews were ultimately unsuccessful because they did not involve a large enough portion of the management group to clearly establish the open lines of communication that were sought. In addition, many interpreted the review process as an attempt by the board to manage the projects more actively and in greater detail.

An alternative to management board reviews was recently established to allow for discussion of projects in an open forum—namely, project reviews. The purpose of these reviews is to focus attention on important project issues and to present data relevant to these issues. The project teams are responsible for coordinating preparations for these meetings. Attendance at these reviews includes the management board, all directors or vice-presidents with worldwide responsibility, patent attorneys, and business development representatives from the business units.

The forum has been notably successful so far; the presentations have been professional, ample time has been allowed for open discussion, and attendees have participated actively in constructive discussions. The forums have ensured that actions taken by the various line functions are appropriate and consistent with each other. Because each project requires a lengthy review, major projects are reviewed only annually and smaller projects are reviewed only as needed.

In addition, PRI's executive committee (PRICOM) meets quarterly with senior managers of the business units to set the direction and strategy for PRI and the units it supports. PRICOM consists of the company group chairperson; the presidents of PRI, McNeil Pharmaceutical, Ortho Pharmaceutical, Ortho Biotechnology, and Cilag International; and the management board of PRI. The committee approves new projects, reviews the overall project portfolio, and establishes priorities among projects. It also makes go/no-go decisions at major project milestones and resolves business issues between domestic and international business units.

PRICOM reviews individual projects at the completion of each development phase, or roughly every one to two years. Critical decision points occur after clinical trials and the preregistration filing. During periodic reviews at each stage of development, the project team provides a comprehensive overview of accomplishments and presents plans for the next development phase. Also discussed at these reviews are key issues confronting the project and the project's economic outlook, with particular emphasis on market competition. Finally, the team makes specific recommendations for executive management action.

Despite its accomplishments in opening the critical lines of communication, PRI has found a major stumbling block to be middle managers' reluctance to share information with project teams. Habit forces these managers to seek clearance for releasing information from everyone further up the chain of command. In addition, many senior managers have had difficulty adapting their management styles to the new culture of openness, and their words and actions are often inconsistent. Their message of openness is often met with skepticism by subordinates who remember these managers' past behavior. There are no quick solutions to these problems—only consistency of deed and word can provide proof of better intentions.

TOOLS FOR SUPPORTING PROJECT MANAGEMENT

With the project team approach being used for 10 of its more than 50 development projects, PRI has acquired new tools for managing large multifunctional teams at multiple sites. Because open communication is critical in project environments, electronic mail, telephone conferencing, videoconferencing, and face-to-face meetings have been vital to effective work teams.

the project team approach has provided a critical link between the business units and the R&D function

Managers are learning to accept (but not always like) the easy access that project leaders have to senior management and vice versa through electronic mail and voice mail communication. The organization has learned ways to rapidly inform large numbers of people at multiple locations about critical issues. In the process, people have had to overcome the ingrained habit of withholding bad news until the last minute to protect their departments from any negative consequences.

Other ways in which PRI supports project management include training and development programs that meet the needs of project leaders and project management professionals. In addition, the PRI management board established a formal mentorship program to support project leaders in developing their skills. Board members are responsible for providing coaching on strategy and direction and for focusing attention on critical issues and decisionmaking. These mentors contribute their substantial expertise in drug development for global markets to the project management process by meeting regularly with project leaders. Board members who have been involved in this program have responded positively to the responsibility and feel limited in the impact they have only because of the multitude of other demands on their time.

A CRITICAL LINK

The rapid adoption of project management practices at PRI in the recent past has led to the acceptance of this management approach as an integral part of the R&D process. The project team approach has provided a critical link between the business units and the R&D function. It has brought to senior management's attention many issues that affect the line functions' ability to perform effectively. Most important, the project teams have proved to be an integrative force not only for project activities but within the entire research organization.

RUNNING INTERFERENCE

Experimenting with Program Management

At a defense-contracting corporation, a new director of the corporate research center decided to bring R&D into line with the corporation's goals by instituting a program office. The office oversaw R&D progress, trained the technical managers in program management, and guided researchers in their preparations of contract bids, presentations, and business management.

F.D. Galatola

Several years ago, a major computer manufacturer hired an individual from outside the company to direct its East Coast research center. After undertaking this job, the new director found that the center comprised several groups of highly talented individuals. Each group was pursuing its own agenda, with allocated company resources, but in some cases with little hope of any long-term payoff for the corporate lines of business.

A visionary with the ability to turn his dreams into reality, the new director quickly established two relatively straightforward goals: obtaining additional outside contracts that would allow for growth and test the validity of the center's ongoing research, and making the center a critically valuable resource for the corporation.

To accomplish these objectives, the director decided to introduce program management to the organization, starting with the idea that a research program should have well-defined goals and a well-defined life cycle, characteristics that might have appeared to be antithetical to traditional approaches to research.

Program management also would have required the appointment of a program manager to be responsible for cost, schedules, and team performance. Although discipline and accountability existed in varying degrees in each R&D subgroup and were adequate for many of the internally funded programs, the existing research managers had the priorities and prejudices of a traditional technical manager: they believed that the technical aspects of the research program came first. Although it would take some effort to get the researchers to accept the concept of total program management, the director believed that the long-term success of the organization depended on gaining acceptance for this management approach.

The new director could have introduced program management techniques to his staff in several ways: by working directly with the staff on a daily basis, through extensive training for senior staff, or by creating a separate program office. Be-

---

F.D. GALATOLA is president of MG Management Group, a computer consulting firm based in Valley Forge PA. She spent more than 16 years in the defense industry, most of that time at General Electric Co, leading large-scale software development projects. She subsequently worked at UNISYS Corp, where she successfully transferred aspects of management techiques used in production to the R&D environment.

cause the director's agenda for the East Coast research center included growth and technology transfer, he opted not to work directly with the staff on a daily basis. He felt that his time would be better spent marketing the center's new role rather than supervising the transition to program management. In addition, he felt that if he could delegate the supervisory responsibility, he would be able to act as a sounding board and intermediary for the research staff as the new methodology was imposed, thereby guiding its progress but not playing an instrumental role.

Extensive training did not appear to be the answer either. Although several senior staff members had already been through corporate program management seminars, most of them had never worked in a contract environment and therefore did not have the ability to translate the information obtained from these courses into skills that they could use in their own daily management experiences.

The director decided to adopt the third approach—setting up a separate program management office. He hired an individual from outside the research department who had extensive technical and program management experience to set up a program office in the center.

THE PROGRAM MANAGEMENT ENVIRONMENT

At the time that its new charter was introduced, the East Coast research center had many of the characteristics common to an academic or small start-up research organization trying to expand its base. The research center's history can be traced to the 1950s, the era of the ENIAC (electronic numerical integrator and calculator), the first all-purpose, programmable, electronic computer. By the 1980s, however, its research focused on developing software, particularly artificial intelligence (AI). The center was part of the defense-contracting division of the corporation, and most projects were defense- or government-related contracts. Because of its contract base and because the Ada AI programming language must be used in all Department of Defense software programs, the center had become nationally recognized by 1985 in Ada environments and software reusability and by 1987 in natural-language processing. The center had one of the largest concentrations of Prolog programmers in the US, which it put to use in knowledge-based maintenance systems.

The East Coast research center grew from approximately 55 persons to 85 in the three years under the new directorship. More than 20% of the staff held doctorates; 70% held master's degrees or higher. All researchers had their own workstations or had terminals connected to a mainframe and were able to access a network through the mainframe. Communication by way of electronic mail was standard. Only a few research staff members had ever worked in a nonresearch environment; most came from academic positions or had joined the research center immediately after receiving their degrees.

The Program Office Structure

In many ways, the East Coast research center functioned as an independent entity in the corporate organization (see Exhibit 1). The center reported at the same

Exhibit 1
The East Coast Research Center Within the Corporate Structure

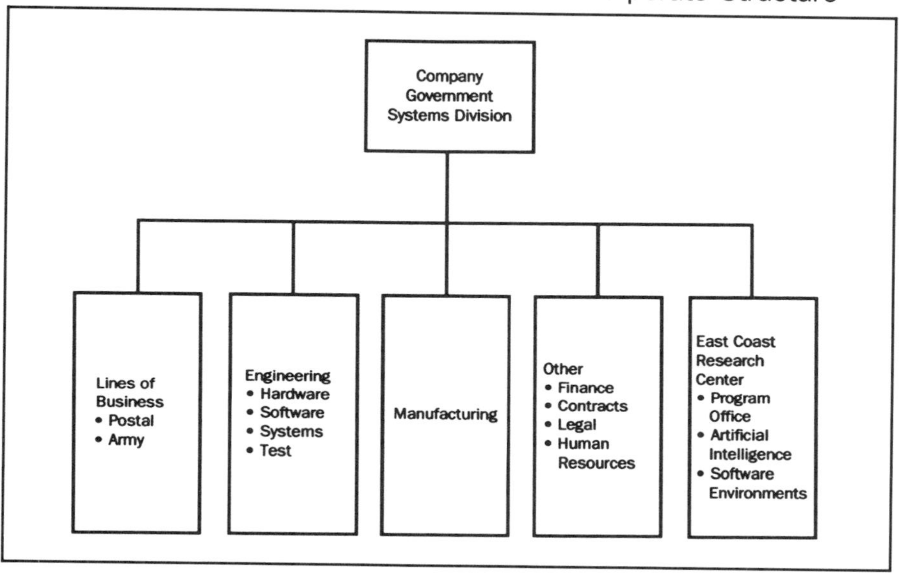

level as did the lines of business, the engineering group, and the manufacturing group. It had minimal need for support services until the arrival of the new director, who placed the responsibility for obtaining these services with the program office. The program office handled this responsibility in two ways: it hired personnel into its own organization in a few critical areas, including project engineering, accounting, and program control; and it obtained support from the division for such areas of need as legal concerns, contract issues, subcontracts, finance, and technical publications.

In addition, the program office manager was given full responsibility for all contracts and proposals, including cost, schedule, performance, customer contacts, and marketing. Technical management, additional customer contact, and some marketing remained with the research staff.

THE OBJECTIVES OF THE PROGRAM OFFICE

A primary goal of the program office was to train the research staff in program management to the degree that the office itself could cease to exist as structured. It was believed that most of the East Coast research center's programs would be small enough to allow the technical managers to handle both the program management and the technical aspects of the contracts. The program office would ultimately become a support group.

Two other major goals of the office were to win contracts and to perform successfully on these contracts. All three goals were met, with varying degrees of success. In addition, the director had one more hidden goal: to succeed at imposing structure on the program without losing any researchers. That, too, was achieved.

The Preparation of Proposals

One activity that would appear, on the surface, to be routine but is absolutely critical in successful contract procurement is the preparation of proposals. The following examples are intended to illustrate some of the typical problems arising in the proposal-writing process and to offer some possible solutions.

Three months after establishing the basics for the program office, the corporate postal line of business informed the East Coast research center that it had received a request for proposal (RFP) on which it wanted the center to bid. The postal line of business had been marketing the research center's capabilities in AI to its customers and had requested to be placed on the bidders' list for research programs that might benefit from the center's expertise. Unfortunately, when the RFP arrived, it sat on someone's desk for weeks; by the time the research center staff was finally aware of the bid, it had only two weeks in which to respond. The possibility of writing a winning proposal within the limited time and using inexperienced personnel appeared to be remote. The director of the center, however, thought that it was worth the gamble, so work was begun.

A Statement of Work. Experience had shown that the first draft of a proposal seldom made sense. During this particular effort, however, there was a refreshing surprise: the technical people not only could write but could write well. In fact, over the next several years, top-quality writing was the norm on all proposals.

Unfortunately, the problem in this case was of a more fundamental nature: the researchers did not know what to write about. The RFP dealt with a study focusing on the use of AI technology to help maintain postal equipment; it would subsequently be used to develop a prototype. The entire first draft of the proposal was a tutorial on knowledge-based maintenance assistance written in Prolog. The proposal team had to be convinced to change its approach to the write-up from one that was tutorial to one that addressed the requirements of the RFP by detailing the cost, schedule, and deliverables.

The second draft proposal was considerably improved; it fully addressed the needs of the study, but addressed only the internal configuration of the prototype. The proposal authors needed to appreciate the importance of the user interface in any prototype as well as understand how the customer might use it. Because researchers had the mistaken belief that the prototype would be seen only by the technical staff of the organization issuing the RFP, they also needed to be made aware that they would have to convince less technical readers of the long-term benefits (particularly the cost savings) of the proposed system. In addition, the prototype had to be of such quality that the customer and its management would instantly recognize that they had not been given just a high-tech toy.

Because competing proposals would be submitted for this project, the researchers had to be persuaded of the importance of their prototype's appearance when compared with those of the competition. (There would be three contracts for further study awarded initially, with the potential for a follow-up contract at the end of the first one). Although the original prototype concept called for a simplistic user interface with text only, this was subsequently amended to position the proposal more competitively among its competitors.

most of the center's programs were small enough to allow the technical managers to handle both the program management and the technical aspects of the contracts

By the third draft of the proposal, a complete package was developed that presented a minimum of tutorial material, definite technology solutions, a graphics interface using the latest hardware, and a strong case for the proposed approach. The East Coast research center was selected as one of the three winners and ultimately won the follow-up contract.

After this first experience, the program office worked with all of the senior staff on several proposals. In most cases, oral explanations and previous training did not appear to be enough to guarantee that the writer would produce a responsive proposal. Most researchers' first attempts at proposals resulted in first drafts similar to the one that emerged in the preceding case. In subsequent proposals, however, much less input was required from the program office. What became evident from this experience was that it was crucial to the ultimate success of any proposal that the first draft be reviewed critically and in detail by an expert.

Costing New Projects. A significant component of any proposal—estimating cost—is no more or less difficult in contract research than it is in an engineering environment. The only noticeable difference in this case was the continued reluctance of some researchers to bring in outside help from areas in which they lacked expertise. Costs in these areas would generally be determined from the bottom up, with few allowances made for the fact that the solution might already be at hand in the form of an existing technology or an experienced engineer available to perform the task.

There were apparently two reasons for the staff's reluctance to seek advice: the belief that there was no one on the outside qualified to do the job, and the belief that the East Coast research center had to know every detail of the job in order to guarantee success. These beliefs may have resulted from the center's general lack of experience with large-scale projects and from the staff's lack of specific management experience, especially on projects that required more than five or six people.

The program office was responsible for helping the center's researchers understand that they could use their general technical and management expertise to determine whether individuals assigned to specific jobs were performing adequately. The East Coast research center staff was encouraged to monitor the progress of contracts, answering such questions as:

- Were there constant delays?
- When problems were reported, were they clearly defined?
- Were goals and milestones clear, realistic, and measurable?
- Were there problems with technical exchange?

The program office also worked to help the staff understand the importance of using contracts as a means of technology transfer. Not only could the engineering department provide expertise to the center (e.g., in the form of standard data base techniques or graphics capabilities), but the researchers could have given the engineers an opportunity to gain hands-on experience in current research done at the center (e.g., knowledge-based techniques for maintenance problems).

Theoretically, an engineer who had worked directly on a research project would return to the applications environment with a basic knowledge of the East Coast research center's activities and would be sufficiently sold on the merits of such research to advocate it to engineers in general. This argument was helpful but did not totally convince the research staff to allow engineers to participate in its proposals. After several engineers actually worked on several contracts, however, the staff saw the benefits of such technology transfer and became more willing to embrace the concept. Experience was again the best teacher, and the program office was responsible for ensuring that the researchers were afforded the opportunity for such experience.

The argument that proved to be decisive in persuading researchers to include their engineering counterparts in bids was the increased potential for losing the contract if they did not include them; an otherwise excellent proposal could lose out because of excess cost or because it did not convey an understanding of an underlying technology. For example, an entire technical proposal could be called into question if one part proposes to do research (when it is not needed) on an existing, appropriate technology. This line of reasoning was usually most effective in helping researchers put costs in perspective.

THE PREPARATION OF CHARTS AND PRESENTATIONS

Formalizing a procedure for preparing charts and presentations proved to be an enormous challenge to the program office, which had no direct authority over the research center's technical managers. For years, these managers had been preparing monthly, preformatted management review charts; they also made presentations (usually no more frequently than semiannually) to their customers on the few outside contracts they had. To the technical managers, it was clear that everything was under control; it was equally clear to the program office that these managers needed help.

The Management Review Chart

The basic content of the management review chart remained the same over the years, detailing the program description, accomplishments, management concerns, cost, and schedule. Minor format changes that had been introduced tended to emphasize program objectives and risks. The research center's management review charts were usually reviewed by the technology branch of the defense-contracting group and not by the actual lines of business, though the lines of business had stakes in the money invested in the research. Because the reviewers were involved in related research, the strong technical emphasis on the research center's charts went unchallenged.

it was crucial to the ultimate success of any proposal that the first draft be reviewed critically and in detail by an expert

Also uncontested were the inconsistencies in the material presented. Program descriptions usually did not clarify the relevance of the specific program to the general corporate research program but instead delved in great detail into the specific tasks currently under study. The key accomplishments cited were almost always of a technical nature (e.g., finding a correct algorithm) and seldom reflected the overall results of the program (e.g., positive customer feedback on the latest delivery). Management concerns also focused on technical issues (e.g., the algorithm does not lend itself to an easy solution) rather than program-level problems (e.g., not enough funding, not enough staff, or not enough computer resources). Because the program office's objectives were to increase the center's visibility, foster technology transfer, and secure additional funding, it had to work with the research staff to make the charts more useful and more presentable to the line-of-business managers.

Guidelines for Program Descriptions. The program office issued guidelines for updating the program description, directing researchers to use the summary of a program as it was described in the contract or the appropriate internal research publication. The description was to be brief, could not change fundamentally over the life of the program, and could not contain jargon. The definition of jargon, however, remained open to debate.

For example, one such debate involved the description of a program on application-specific languages (ASLs) and whether to clarify what ASLs were and to emphasize that the program was concerned with the development of specialized software languages. This additional explanation was important because most managers outside of the research group had a background in hardware, not software. The program office felt that it would be preferable to be explicit rather than to place senior management in the position of needing to ask for basic definitions. The research staff was confident, however, that there would be no question on this point; the acronym was viewed by researchers as standard terminology. When the charts were presented, the vice-president of the division posed the first question: What are ASLs?

Although written and oral instructions were provided for the preparation of charts, the most effective method of implementing the required changes was the East Coast research center management review process. Charts were presented by the staff to the director of the center every month. At that time, either the director or the program office would ask questions about the programs strictly on the basis of the information presented on the charts. The staff would then be able to spot potential problems with the charts and update them accordingly.

After several review cycles, the program descriptions were acceptable. The review process also made the staff more sensitive to the need to report regularly on program-level issues (e.g., customer feedback, cost, and schedule). Program objectives were also defined after the researchers answered three questions: What problem were they solving for their customer? What was the benefit of the program to the corporation? and How did the program benefit the East Coast research center?

Customer Presentations

Four areas were addressed in customer presentations, two involved content and two related to the format of the charts. These are discussed in the following sections.

Problems with Content. The issues relating to chart content in customer presentations were similar to those encountered with proposals and in-house management reviews. The first problem was that the charts lacked overall perspective and program definition. The presentations often had no introduction and began with current technical issues. This approach was based on the assumption that, because the East Coast research center manager had been in close and frequent contact with the customer, the customer would therefore be knowledgeable about the general background and status of the program.

There were several problems with this approach. The main problem was that the omission of a brief statement of program definition and requirements left the objectives and directions of the program open to misunderstanding. The inclusion of an overview for customers would eliminate potential misunderstandings as well as provide all necessary information to customer representatives, who were not as directly involved in the program as the primary participants. After definition charts were included, minimal time was spent on them generally; however, they were available as part of a complete package for the customers to use as they wanted.

A more serious problem with presentation content became evident on at least one program in which the research center faced significant technical problems, schedule delays, underspending, and the real danger of losing funding for the program. The technical manager in charge of the program prepared a series of charts that detailed the program's technical and staffing problems, tracing the source of schedule and budget difficulties back directly to his inability to hire the right personnel. The presentation material essentially made customers responsible for finding a solution to the technical problems and implicitly asked for their understanding and forbearance.

After 24 hours of work with the program office, the content of the charts was turned around completely; they now presented possible solutions to the technical difficulties and made a strong case for the center's preferred solution. The charts were also modified to reflect a recovery plan for the staffing and schedule problems. The results made a positive impression on customers, who continued funding for the program. In addition, the center's researchers arrived at a better understanding of the program, and the technical managers were able to gain more control over it.

when the researchers were exposed to the broader objectives of their programs, they soon realized the importance of moving the programs along on schedule

Problems with Format. Although the two format-related issues in customer presentations were minor compared to the content-related issues, they bear mentioning. The first concern was the amount of information the charts should contain. One presenter would invariably turn the charts into full-text descriptions, and another would use them only as written reminders of the topics to be covered.

Neither format was appropriate for technical presentations. The first had to be edited down to bulleted entries that contained sufficient information to communicate but not so much as to distract from the presentation. The second had to be augmented to provide sufficient information so that the customer could use the charts for later reference. In this case, the presenter preferred to continue using charts only as cues but supplied the customer with backup charts containing the pertinent details of the presentation.

The second format problem was one of style: the corporation required the use of a company template and general format style for outside presentations. This requirement either had not been introduced to the research department or had been ignored. It was a firmly held belief among the researchers that informality was preferable. Although all charts were typed and some contained excellent graphics, there was no consistent style among presenters at any one meeting. In addition, corporate affiliations were seldom identified.

The significance of the presentation format is aptly demonstrated in a case in which a researcher who had been asked to give a paper at an international meeting failed to investigate the audience and arrived with a set of unassuming charts to accompany a highly technical presentation. After listening to several papers, she realized that she had made two mistakes: she was too technical for the audience, and her charts lacked quality and all the hallmarks of showmanship. Fortunately, she was able to change the focus of her presentation, but she did not have the facilities to change the format of her charts. The lesson was painful but worthwhile.

Maintaining Schedules in Research Programs

Many researchers showed considerable reluctance to generate schedules to present to customers or to management, let alone schedules that were well-planned and could be used to evaluate progress. Each set of charts discussed in the section on management review contained milestones, and slip dates accompanied each milestone. When the program office began to review charts on a monthly basis, it requested copies of the schedules that corresponded to the milestones; none existed. When the technical managers were asked to generate these schedules, their response was simply, We know the schedules will change, so why produce them? The program office's rejoinder was equally simple: How will you know that they've changed if you have nothing to measure them against?

It is debatable whether it is more difficult to get schedules from researchers than from engineers; the major difference appears to be that engineers tend to acknowledge the importance of schedules. Early reviews of research projects by the program office found that milestones were mentioned only peripherally, if at all. Although the researchers were asked pointedly whether these had been met, their answers were never direct; instead, they usually offered long explanations of program status and discourses on general research activities in that area.

Their focus on pure research and their limited or nonexistent relevance to applications had allowed these researchers the freedom to work on problems no matter how long the research took and to change the direction of that research whenever they thought it appropriate. Another earmark of the pure-research mind-set was that, given the choice between meeting a contract schedule with reasonable results that satisfied performance requirements and missing the schedule to find the right technical answer, these individuals usually opted to miss the schedule. When the researchers were exposed to the broader objectives of their programs and, in one significant case, after working closely with a competitor, they soon realized the importance of moving the programs along on schedule.

During the three years in which program management techniques were used at the East Coast research center, the practice of scheduling and schedule reporting was gradually accepted. The primary inducements to these changes were the increased visibility of the programs and the potential for recognition of research efforts by other parts of the corporation.

With prodding from the program office, schedules and clearly defined milestones began to appear and technical managers began to initiate discussions about milestones. The inevitable slips in schedule and changes in program direction still occurred but far less frequently than in the past, especially on contract deliverables. The center began to establish a reputation with several customers for prompt, professional deliveries.

Support Services

In many ways, support services were easier to deal with than any of the other issues discussed here. Perhaps this is because most of these services (e.g., financial reporting, legal counsel, and marketing) were peripheral to the researchers' primary responsibilities and, to a certain degree, remained so even after the introduction of a program office.

With the increased number of outside contracts, however, certain changes had to be effected. The program office attempted to establish firm links with the marketing, contracts, and legal departments in the reasonable expectation that the technical managers themselves would gradually take over these liaisons. After two years, the program office served less as a liaison than as a directory that pointed individuals from either side to their appropriate counterparts.

More significant changes occurred in the areas of resource planning and program budgeting. Resource planning was required on a monthly basis, and program budgeting on an ad hoc basis. Before the advent of the program office, resource planning and program budgeting reports were created manually and bore limited resemblance to reality. The program office automated both reports and linked them; consequently, the turnaround time to replan resources and costs for an entire department went from more than a week to approximately a day. The software packages used for such replanning ran on microcomputers and were not commercially available to run on the workstations used by the research staff. The researchers found the automated system helpful but inconvenient to use because it was not available at their desks; therefore, they wrote programs that duplicated some aspects of the system to use on their own computers. Although this duplication effort obviously resulted in additional development cost, the subsequent involvement of technical managers in the planning process provided some justification for that cost.

resource planning was required on a monthly basis, and program budgeting on an ad hoc basis

BECOMING A CONTENDER

The program office was a catalyst for changing the East Coast research center from a pure research organization to a competitive contender in the open research market. Was the experiment successful? Certainly some aspects might have been handled differently. For example, more authority should have been granted to the program office to institute required changes. A different set of problems would surely have arisen as a result. In the long run, however, the increased autonomy would have allowed for quicker feedback on the program management methods introduced, which in turn might have resulted in quicker acceptance of the changes.

Budgeting for the program office could also have been handled more effectively. When the center's director created the program office, he believed that most of its work would be tied to specific programs and that the costs could be charged accordingly. Contrary to expectation, the program office spent much of its time on problems that affected the center across all programs (e.g., management reviews, marketing, and staffing). As a result, most of the office's time was charged to overhead. This caused an unanticipated increase in overhead rates, even though the higher rates did not appear to significantly affect the center's ability to bid competitively on new programs. The center might have been in a better (i.e., less defensive) position in the corporation, however, if it had projected its overhead requirements more accurately. These are implementation issues, however, and are only indirectly related to the objectives of the program.

The true measures of the program's success would have to be its contribution to the center's long-term competitiveness and its efficacy in instituting program management concepts and practices. Unfortunately, the results cannot be known because the corporation that the East Coast research center is a part of underwent a major restructuring and downsizing after experiencing serious financial problems across the board.

As might be expected, this major change had a significant impact on the research center; the director was replaced (actually promoted), and internal corporate funds were cut significantly. The funding cuts resulted in staff reductions and a lack of resources to bid on new programs. The new director moved the center back in the general direction of pure research, downplayed the importance of program management, and eliminated the program office.

Although the long-term success of the experiment cannot be measured, it was an unqualified success during its brief existence; competitive contract research programs were won, follow-up contracts were awarded, and in the process, the researchers incorporated program management issues and techniques into their daily work. Program management was shown to be an effective means of ensuring the competitive success of a strong research organization.

KEEPING A TIGHT REIN

Aligning R&D Efforts with Business Objectives

Managing in an R&D-driven industry is often a balancing act in which trade-offs must be made between creativity and control. Similarly, managers are often confronted with situations in which their short-term objectives seem to be in conflict with their long-term goals.

William J. Guman

Senior managers of large, multidivision corporations use various methods, structures, and controls to ensure that the efforts of all divisions are directed toward achieving the strategic goals of the corporation. In some cases, executives have adopted a philosophy of empowerment, seeking to tap the creativity and entrepreneurial spirit of employees and achieve levels of innovation typically associated with small, start-up companies. In other cases, senior management has placed extremely tight controls on the activities of all divisions in their company—that is, they have created a microcontrolled environment—to ensure that all business units are working toward a common set of objectives. This chapter details the experiences of an organization in which senior corporate management created a microcontrolled R&D environment to ensure that the R&D activities of each division supported the corporation's overall business objectives.

SETTING THE STAGE

Understanding how the company described in this case study microcontrolled R&D to achieve its business objectives entails a review of some company background. The following sections of the case study describe the company's primary business, its R&D program and investment strategy, its organizational structure, and the parent company's strategic plan for R&D, control and administration of R&D resources, and documentation of R&D plans.

The Company's Business

The Alpha Co was a major profitable division within Beta Corp. (Both *Alpha* and *Beta* are fictitious names.) Beta's corporatewide business was diversified and extended into the following major market segments: government (defense and nondefense) aerospace; communications and electronics; space; commercial aviation; and commercial industrial products.

Approximately 85% of Alpha's business was related to defense aerospace. This portion of the business provided between \$100 million per year and \$570 million per year of Alpha's total annual sales from 1972 to 1986. The peak occurred

---

WILLIAM J. GUMAN, PhD, is involved with corporate strategic planning at Grumman Corp. He was director of R&D Contracts in Grumman's Corporate Technology organization and held several R&D management positions with a major aerospace company, including director of R&D, director of Advanced Product Development, and director of Technology & Operations. He has published more than 60 technical papers in various professional journals and is the author and coauthor of numerous company and in-house R&D contract reports.

during 1981. These sales came from the defense-related product lines within each of its four major strategic business units (SBUs). Beta dissolved Alpha as a division in late 1987, and Alpha's military aircraft program and associated assets were acquired by another major aerospace corporation.

The Company's R&D Program

In the defense industry, R&D is called independent research & development (IR&D), which refers to contractor-initiated efforts relating to basic and applied research, product development, or systems and concept formulation studies. IR&D is distinguished from R&D in that it is independent of any contractual or grant commitments to the government. In essence, a company uses IR&D to technically and competitively position itself in the defense industry. The major characteristic of the bid-and-proposal (B&P) effort in defense work is the intent to use the results of IR&D directly for preparing a specific bid or proposal to the government.

Both IR&D and B&P are indirect costs carried in the overhead account of a contractor's accounting system. IR&D and B&P are the only overhead costs, however, that are required by law to have a separately negotiated advance agreement. This places a ceiling on the annual amount of such costs that may be included in the overhead allocated to all work performed for the DoD. To determine these IR&D/B&P cost ceilings, the government uses the following evaluation factors:

- The contractor's commercial and government business activity.
- The technical quality and the potential military relationship of the planned IR&D projects.
- Past contractor performance.
- Business base projection.
- Other forecast business and financial data.

The IR&D/B&P Investment Strategy. As a major defense aerospace contractor, Alpha negotiated annual IR&D/B&P advance agreements with the DoD's Tri-Service Negotiating Group. Alpha's annually planned IR&D/B&P budget and program were heavily influenced by this annual cost recovery process. For annual planning purposes, the company's IR&D/B&P work focused on sustaining and generating new aerospace defense business. The IR&D/B&P effort was to be performed within the limits of the annually negotiated cost ceilings. These annual ceilings were approximately 2% to 3% of Alpha's annual sales to the DoD. Supplementary R&D funding was provided by Beta management and was generally allocated to Alpha for new non–defense-related business.

A primary aim of Alpha's senior management and Beta's management was to maintain the IR&D/B&P program within the negotiated ceiling. In effect, this ceiling produced an implicit management objective and also defined Alpha's operational environment for the IR&D/B&P program. The management strategy that was used to achieve this implicit objective was to microcontrol the IR&D/B&P program and its expenditures.

deviations from the business plan were permitted only after Alpha received approval from Beta's corporate vice-president for aerospace development

Under the terms of the agreement with Tri-Services, Alpha could fully interchange the individual IR&D and B&P ceilings that had been negotiated. In other words, the company was legally permitted to increase recovery of costs for either IR&D or B&P above the individually negotiated ceilings, provided that recovery of costs for the other would be decreased below its ceiling by a similar amount and provided that the total IR&D and B&P amount would not exceed the amount negotiated. During the 16 years that Alpha negotiated IR&D/B&P ceilings, the company overspent the annual IR&D/B&P ceiling approximately 50% of the time. Such overrun expenditures came out of profit. Over the same period, the company underspent the IR&D portion of the total ceiling dollars 75% of the time. Alpha senior management elected to invoke this interchangeability option and to use the larger amount of B&P funding thus made available to enhance the company's position to win major, new short-term defense business. In short, Alpha gave B&P higher priority than it did IR&D.

The Company's Flexible Matrix Organization

For the specific purpose of managing the company's IR&D/B&P program, the company was viewed as a flexible matrix organization in pursuit of its key business (i.e., product) areas, supported by its functional directorates. The company set up key functional directorates for: business development; engineering; manufacturing; materials; finance, contracts, and legal; and integrated logistics support. These directorates supported the IR&D/B&P program directors and managers in the matrix.

Corporate Control over the IR&D/B&P Program

Beta had a corporate strategic plan that defined the direction the corporation was taking to seek new business, retain and increase current market shares, and increase technical skills. The plan was coordinated annually with the senior management of all of Beta's divisions and subsidiaries, after Beta received suggested programs from them.

The business opportunities assigned to Alpha were an outgrowth of agreements between Alpha and Beta senior management. The programs to be pursued were designed to permit Alpha to meet its goals, which supported Beta's goals for profit and growth. Deviations from the plan by Alpha were permitted only after Alpha received approval from Beta's corporate vice-president for aerospace development.

Beta had a documented corporate policy entitled "Bid & Proposal and Independent Research & Development Programs" that strongly affected how Alpha's IR&D/B&P program was managed. As part of Beta's strategic planning effort, each division and subsidiary within Beta formulated its own long-range plan for IR&D/B&P, consistent with its own strategic plan. The corporate vice-president of product development approved all such IR&D/B&P plans and changes. Furthermore, if planned annual expenditure for any IR&D/B&P project at Alpha would exceed $25,000, Beta's approval of detailed project documentation was required before the project could be initiated. It was also Beta's policy that the following written reports be submitted to Beta:

- Monthly reports, presenting the status of resources expended on IR&D/B&P.
- Quarterly reports, with a brief narrative description and a milestone chart detailing the progress made on each IR&D/B&P project.
- A detailed year-end IR&D/B&P progress report. Furthermore, Beta conducted quarterly oral reviews of the IR&D/B&P program.

Control and Administration of IR&D/B&P Resources

Within Alpha, the engineering directorate administered and monitored the efforts, expenditures, and progress of the company's entire IR&D/B&P program. However, the business development directorate controlled how the B&P budget was used and applied for all proposal activities that focused on winning major new defense aerospace business. Specific IR&D projects were dedicated to developing and providing the requisite focused technology for the product lines within each SBU. The engineering directorate, however, controlled the use and application of a smaller percentage of the total IR&D budget, to develop the technology base in the technical disciplines that were applicable to all SBUs in the immediate future (i.e., in less than two years). Such technologies included manufacturing technology, advanced concepts, and operations analysis and simulation.

Furthermore, the engineering directorate was also allocated a B&P budget of approximately 2% to 5% of the company's total IR&D/B&P budget. This budgeted B&P activity was dedicated to winning government-funded R&D contracts in newly emerging technologies or new system-enabling technologies.

Company Leadership Styles

Alpha had five company presidents during a nine-year period, and each president's management approach and leadership style varied somewhat with respect to the IR&D/B&P program. However, Alpha's IR&D/B&P program was always microcontrolled and consistent with Beta's published policy on IR&D and B&P programs.

Planning Documents for the Annual IR&D Program

During preparation of the company's annual strategic plan, each division within Beta Corp followed a generic outline provided by Beta (see Exhibit 1). Specific details of the company's plan were based on its business assignments. The IR&D/B&P program was one of the company's implementation plans in its overall strategic plan. The IR&D/B&P effort addressed the strategic issue of how Alpha would attain its objectives while staying below the IR&D/B&P cost ceiling.

Alpha's requirements for short-term and intermediate-term development of technology under IR&D were linked directly to its specific defense aerospace product

Exhibit 1
Generic Outline of Alpha's Annual Strategic Plan

1. Where are we today?
   - A. Mission statement
   - B. Strategic issues
     - Impact
     - Actions required
   - C. Situation analysis
     - External environment and competition:
       - — Products
       - — Companies
     - Internal environment and capabilities:
       - — Strengths
       - — Weaknesses
       - — Threats
2. Where do we want to be . . . when?
   - A. Objectives (five-year)
     - Industry position
     - New business development
   - B. Assumptions
     - External environment
     - Internal environment
3. How are we going to get there?
   - A. Analysis of market opportunities
   - B. Strategies, programs, and goals
   - C. Financial measurement
   - D. Resource deployment
     - Facility requirements
     - IR&D/B&P
     - Labor resources
   - E. Implementation (i.e., action) plans
     - Strategy, interim goal, and action:
       - — Individual, due date, and expected result

lines. Much of the associated IR&D planning documentation pertaining to the technical aspects and resource requirements was generated annually by Alpha's IR&D project managers and middle management. Various versions of this documentation were required by Alpha management, Beta management, and the DoD Tri-Services. The rationale underlying this documentation is depicted in Exhibit 2.

Implementation Plan for Technology R&D Contracts. In addition to the IR&D plans that were generated annually, planning and implementation documentation was generated for technology-developing R&D contract orders. These contracts were to be won using the small budget (approximately 2% to 5% of Alpha's total annual IR&D/B&P budget) that had been allocated to Alpha's engineering

Exhibit 2
Linkage Between Strategic Plan and IR&D Projects

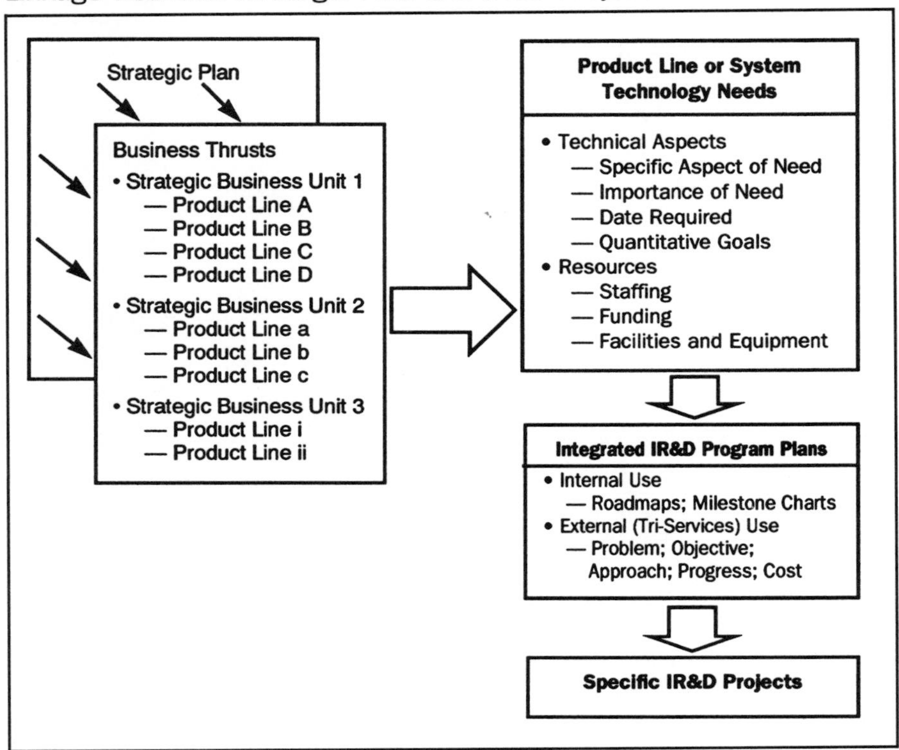

directorate. In general, these government-funded R&D contracts entailed conducting R&D in newly emerging or new system-enabling technologies. Technology-developing contracts in manufacturing technology, advanced composites, and operations analysis and simulation were also pursued under this contract program. The annual goals and five-year goals for this program were always expressed in terms of dollars. Senior management closely tracked the annual goal of contract R&D dollars won.

A formal status review of each contract R&D project was conducted once a month at a meeting with the principal investigator of that project. During this meeting, a small review group—with representatives from the engineering, contracts, and finance departments—completed a one-page status report of each project. The report included assessments of technical performance, financial status, scheduling, staffing, and any other relevant factors. Contract R&D projects that had a definite problem were brought to the attention of the company's president by the finance department during monthly management review meetings.

ANALYSIS OF THE STRATEGY OF MICROCONTROLLING R&D

The experiences of Beta and Alpha can help technology managers who are considering the use of a microcontrolled R&D environment in their own organizations. The following sections review the factors that contributed to the success of Alpha's R&D environment and the lessons learned by Alpha's management.

a microcontrolled environment merely defines a scenario in which a program manager must produce results despite the environmentally imposed constraints

Success Factors

The microcontrolled R&D environment at Alpha was merely part of Beta's corporate culture. In this environment, several factors were important to achieving the company's short-term objectives. The presence of one or a few of these factors within the company did not necessarily guarantee success; however, when most of the factors were present in a given situation, their cumulative effect significantly enhanced the probability of success.

Strong Program Leadership. A microcontrolled R&D environment merely defines a scenario in which a program manager must perform and produce expected results despite the environmentally imposed constraints. R&D program managers who performed well in this environment were able to advocate and successfully defend their programs to senior management. Their charisma and ability to be perceived as winners usually enhanced their teams' morale and motivation to succeed. Certainly, a strong ego drive for achieving success contributed significantly to the performance of program leaders in the microcontrolled R&D environment.

In addition, such strong leaders sought, hired, and retained the best-qualified people to work on their programs. The presence of several equally capable program managers led to strong internal competition for priority in the ongoing use of the limited number of key technical specialists in Alpha's matrix organization. In conjunction with this personnel competition, these program leaders also emphatically sought to obtain and control the amount of the IR&D/B&P budget they perceived necessary to do their job, regardless of the effect this action would have on other ongoing or planned R&D projects.

Despite the microcontrolled environment, the successful leaders usually also ensured that their program area could draw readily available resources for use during unforeseen contingencies. This resource reserve was usually either a management reserve budget they established or a funded special project they initiated and controlled.

Support from Senior Management. R&D projects that had either the CEO's or company president's personal attention and endorsement were always allocated an operating budget. However, some endorsements caused discomfort for R&D middle managers when they discovered that a visibly endorsed program was technically tenuous but knew that it had been sold to the highest levels of management. The microcontrolled R&D environment could not prevent the occurrence of such an anomalous project; fortunately, in the few isolated instances encountered, such projects eventually tended to fade away.

Senior management's attention to contract R&D projects also contributed significantly to the progress of these projects. When technology-developing R&D contracts were pursued, won, and performed specifically to implement some aspect of Alpha's short-term strategic plan to enter a new business, senior management followed the progress being made by the supporting IR&D/B&P contract R&D effort.

Adequate Resources. In general, the R&D projects that achieved their objectives received the budget, technical personnel, and capital assets consistent with the results expected of the program by senior management. At Alpha, these projects always directly supported short-term business plans. It was not uncommon for IR&D/B&P funds to be reallocated from several lower-priority projects to a high-priority project within a matter of days. The microcontrolled R&D environment at Alpha provided management with the timely information needed to assess the likely impact of such program interruptions and resource reallocations.

Contract-Funded Technology Development. By law, IR&D/B&P funds cannot be used to support contractually required work on R&D contracts awarded by the government. Because they provided Alpha with R&D funding above the level of the negotiated IR&D/B&P ceiling, management usually viewed these R&D contracts as a way to augment its R&D budget for developing technology. Seeking to obtain this larger R&D budget, senior management established annual goals to win a specified amount of contract dollars provided by these R&D contracts. To meet this financial goal, technical personnel could pursue, win, and perform R&D contracts if they met the following three conditions:

- R&D management assessed the probability of winning the contract to be relatively high.
- The B&P funding required to win the R&D contract was not excessive.
- The B&P funding could be provided by the relatively small B&P budget for contract R&D projects (i.e., 2% to 5% of Alpha's total IR&D/B&P budget).

Contract R&D activity provided an outlet for technical specialists to conduct the type of research they preferred. The program also provided the opportunity to establish formal relationships with customers in anticipation of future R&D business. Alpha's contract R&D program successfully established several technology-developing programs in addition to those funded by the IR&D program. Alpha did not develop a formal strategy, however, regarding the potential of the contract R&D program to enhance the process of developing future new major business, nor did it fully exploit its potential. New enabling technology that was developed for the customer by these contracts was rarely used to strengthen a particular long-range business plan. In addition, the monthly review process of all contract R&D programs caused such discomfort to some talented technologists that they preferred not to pursue and win R&D contracts in the microcontrolled R&D environment that prevailed at Alpha.

contract R&D programs provided the opportunity to establish formal relationships with customers in anticipation of future R&D business

Assignment and Tracking of Employees' Annual Performance Goals. Alpha conducted formal annual performance appraisals for exempt personnel. (*Exempt* legally signifies a job function that is excluded from the application of the Fair Labor Standards Act and the Walsh Heally Public Contracts Act, primarily as they relate to payment for overtime.)

Individuals being appraised would meet with their immediate manager. During the review, the individual's performance, potential, career interests, and development needs would be discussed. The performance of the individual's major goals and special assignments would be rated numerically. Furthermore, the employee and manager would jointly agree on a set of major goals and tasks with measurable criteria for the next appraisal period. These were usually stretch goals, instead of goals that could be easily reached. The scored assessment of how well the goals were met became a factor during the annual salary review process.

R&D personnel usually had goals related to an identifiable improvement of some specific operational or technical aspect of Alpha's technical capability and competitiveness. Another goal commonly set by an R&D program manager was to win a specified dollar amount of R&D contracts with a specific customer in a particular area of technology. Sometimes Alpha would set a goal to formally establish a team arrangement with another company for a future joint effort. Frequently, an individual's goals would be a subset of a major goal of the company. By setting and tracking goals at various levels of the R&D organization, management had a means to anticipate the possible results of R&D projects if all individual goals were met. Furthermore, tying the goals to the annual salary review process motivated R&D personnel to achieve their goals.

Dedication of IR&D Projects to the Short-Term Technical Needs of Specific SBUs. At Alpha, IR&D projects were dedicated to help the business development directorate meet its short-term business objectives. The focus of this R&D was systems oriented, and it had several enduring objectives.

One objective was to provide the technical foundation for the task of verifying that significant modification of one of the company's major products (e.g., military aircraft) would result in a product variant that would either perform a new military mission or significantly expand its present capabilities. Although the R&D organization successfully accomplished this objective, the business development directorate failed to obtain government funding for producing any of these derivative military aircraft, despite intensive marketing efforts.

A second objective of the systems-oriented R&D endeavor was to provide the technical foundation for winning major new system business in the SBUs. The R&D program was performed to provide the technical foundation to win defense business that was planned for and budgeted by the government. In general, this

R&D activity successfully supported the business development directorate in winning such new short-term business.

A third objective of Alpha's R&D program was to establish the feasibility of some special technical enhancement of an existing product. In this case, the research activities were primarily part of Alpha's engineering directorate's IR&D program. The engineering directorate also succeessfully developed new analysis tools and simulation capabilities to support the feasibility-verifying efforts of the systems-oriented IR&D mentioned in the discussion of the first objective. The engineering directorate was also responsible for developing new manufacturing technology along with assessing the applicability of new structural material to existing and future company products. Senior management's interest contributed to the success of this latter R&D effort to enhance Alpha's competitive manufacturing capability.

Results of the Microcontrolled R&D Environment

The overwhelming drive to win major new short-term business caused a greatly unbalanced distribution of the type of technology that was developed under IR&D. As can be expected, people worked within the company rules and were productive. However, they also learned how to play the games deemed necessary for them to function in the microcontrolled environment. The following sections examine the results from Alpha's strategy of using a microcontrolled R&D environment to achieve business objectives.

Technical Innovation Was Not Smothered. Senior management required that the main thrust of Alpha's IR&D be linked to its short-term business objectives. The net effect of this focus, along with the planning and tracking of results, was that the IR&D program became primarily development oriented; however, technical innovation was not smothered. Instead, Alpha's R&D personnel used their creative talent to seek evolutionary modifications, improvements, and derivatives to existing Alpha products or processes.

In addition, the microcontrolled environment did not prevent R&D personnel from seeking out entirely new applications of Alpha's technical expertise. In one case, this particular type of lateral technology transfer allowed Alpha to rapidly establish itself at the forefront of an emerging area of business within the defense industry. This lateral transfer of technology was facilitated by the attitudes of technical personnel and their interaction with potential customers.

Two additional outlets that fostered technical innovation were acceptable to Beta and Alpha's senior management. One was the technology-developing R&D contracts awarded by the government. The other was the highly focused corporate-funded and directed R&D program (i.e., new aerospace defense business).

Management Awareness Was Enhanced. Frequent reporting provided management with timely project status information for both the IR&D/B&P contract R&D projects and the Beta-funded R&D projects. Two types of reports were generated and issued: numerical data on the financial and human resources used during the reporting period and progress reports based on the input received from each project manager. In addition to those reports that were used within Alpha, Beta management received monthly, quarterly, and year-end written reports for all Alpha's IR&D/B&P projects with annual expenditures that would exceed $25,000.

Alpha's R&D personnel used their creative talent to seek evolutionary modifications and improvements to existing products

The usefulness of the weekly financial expenditure and monthly open financial commitments data was significantly enhanced after it was graphically displayed against time. These graphs were generated for each IR&D/B&P project as well as for the entire IR&D/B&P program. Extrapolating and analyzing the trend of expenditures at a central point within the company usually caused potential problems to surface sufficiently early to initiate inquiries and, if necessary, corrective action.

In the microcontrolled environment, project management was characterized by minimal communication. The written progress reports and the associated milestone charts did not necessarily reveal what was really happening in a project. Therefore, some Alpha presidents conducted monthly oral progress reviews of major IR&D/B&P projects. Dialogues initiated between the president and the project managers sometimes uncovered issues not discernible from the printed status reports.

Beta managers also conducted quarterly oral reviews of Alpha's major IR&D/B&P programs and issued the results of these reviews in a memo to the CEO of Beta Corp. In addition, permanent consultants to Alpha occasionally worked with IR&D/B&P program managers and provided assessments and recommendations.

The microcontrolled environment did reduce passive or static performance of R&D personnel. However, surprises occurred. For example, a vendor persistently delayed delivery of an essential item and the delivery milestone was not being tracked by the formal system. This situation eventually produced the surprise that prevented the project from being completed on schedule—despite the project manager's diligent efforts.

In addition, management became aware that inappropriate goals were often set and tracked. New technology-developing R&D contracts were planned for and tracked by Alpha primarily for the dollars they added to the IR&D/B&P budget. Thus, contracts that cost little to win and that were relatively easy to win were pursued, instead of contracts that would enhance the company's intermediate- to long-term strategic technical position.

Finally, management realized that the true intent of a goal can be misunderstood. Project managers and R&D personnel must have an unambiguous understanding with their supervisors of the results that are expected of them. The dialogues that took place during employees' annual goal-setting sessions and performance reviews could occasionally fall short of communicating what was really expected from each employee.

Two cases can be cited to illustrate this point. In one case, the goal was to develop a specific new capability for the company in the short term. However, the R&D program manager who was responsible for the project did not know that this new capability also had to be transferred out of R&D and put into practical use within the company. In another case, the individual's goal was to win an R&D contract from a particular customer and for that contract to have a targeted contract dollar value. When that planned procurement was canceled by the government, the R&D program manager was initially unaware that the targeted dollar value of the contract being pursued would remain a goal. The R&D program manager subsequently learned that receiving a contract from the originally targeted customer was not considered the real pertinent issue of the goal.

The Microcontrolled Environment Linked the Potential of New Managers. In addition to being exposed to the different leadership styles of various presidents, Alpha's engineering and R&D personnel were also exposed to the different leadership styles of various vice-presidents and directors who were hired and brought into the company from outside Beta Corp. The professional experience of these individuals augmented the in-house experience, and this new source of experience introduced different insights and approaches to R&D planning and problem solving.

However, their talent and this additional outside experience focused primarily on meeting short-term business objectives and achieving victories soon after they joined the company. Microcontrolling R&D primarily for this purpose essentially prevented the full potential of this new technical management—and of the company's inherent R&D talent—from being used to prepare Alpha to win new business that would use emerging technologies.

CONCLUSION

Microcontrolling R&D was an effective implicit strategy for realizing the particular business objectives to which this strategy was applied. At Alpha, the primary thrust of the IR&D/B&P contract R&D efforts to win short-term profitable major new business was too sharply focused and unbalanced in scope. Despite excellent R&D management, high morale, and select islands of technical expertise, the secondary R&D activities did not position the company to win contracts for the technically more-sophisticated, next-generation products that eventually replaced the company's existing products. Indeed, of the last three major prime military aircraft programs that Alpha won over a period of approximately 20 years, each new program was for a technically less-sophisticated Alpha product line than the one it replaced.

What cannot be determined is the percentage of the total annual IR&D/B&P budget a company must dedicate and manage strategically to prevent its product line from becoming technologically obsolete. This problem is further complicated by the need to manage technology-developing contract R&D projects and to sustain interest and motivation from senior management. The threats and opportunities of emerging technologies must be integrated carefully into the overall long-range business development strategy and must be an integral part of the scope of a microcontrolled R&D environment.

SECTION

4

NEW PRODUCT DEVELOPMENT

New product development concerns almost all companies. To ensure growth, a company generally needs to develop new products to sell; to simply maintain existing sales and profits, a company often must develop new products to compensate for the shrinking sales and profits of mature products.

Various perspectives are presented on new product development. Hector Cochrane and Colin MacPhee both analyze their companies' strategy in new product developments. Thomas Fidelle examines new product development from the viewpoint of an applied researcher who must coordinate the many company functions involved in the process. James Doherty then examines his company's efforts to revive an existing product line—that is, to think anew about customer needs and product attributes.

Cochrane compares his company's present method of new product development with the method it used 10 years ago. Although the earlier new-product development approach was successful, the company learned lessons in the process. Cochrane points out that today four issues would receive more emphasis: defining a new product concept in terms of market needs, creating a product team, maintaining close ties with senior management, and recognizing the importance of time.

MacPhee examines his company's execution of a new product development from another point of view. In its development efforts, MacPhee's company emphasized risk minimization. MacPhee outlines the six phases of his company's plan. At the end of each of the first four phases, the company conducted design audits that enabled it to spot trouble early; a design audit in the sixth phase involved a pilot run. MacPhee shows how his company proceeded methodically through these phases to transfer technology successfully to manufacturing.

Fidelle looks at new product development from the applied researcher's viewpoint, explaining how applied research bridges the gap between R&D and customer requirements. He also points out how the applied researcher plays an important coordinating role among R&D, sales and marketing, manufacturing, and the customers. In addition, Fidelle describes the requirements for an applied researcher and relates the success stories of applied researchers who helped develop new products.

Doherty recounts how his company revived an existing product line through a variety of approaches, many of which could be applied to developing the requirements for a new product. Recognizing that much of its existing product line was unprofitable, the company took steps to boost productivity. These steps included establishing a customer focus panel, analyzing the value of various product attributes, and restructuring the company to focus product engineering, purchasing, inventory control, and manufacturing on a common product.

IN SEARCH OF A MARKET

Improving the New Product Development Process

During the course of the development, introduction, and commercialization of an actual new product in the specialty chemical market during the period from 1977 to 1987, several changes were made to improve the product development approach. These improvements and the reasons behind them are discussed in detail.

Hector Cochrane

The key to developing a commercially successful new product is to first get market acceptance for it and then rapidly penetrate the market before competitors succeed in matching the product and regaining or establishing a significant market share. In this case study, the development of a new product in the specialty chemical market is traced, and the product development program is evaluated at every stage to assess the effectiveness of the strategies applied.

In the course of developing this product, a change was made in the basic approach to product development. The initial product development process was divided into four stages:

1. Birth.
2. Exploratory research and development.
3. Commercial development.
4. Full commercialization.

Throughout the project, however, several difficulties were experienced in managing the process; consequently, a new product development protocol has been instituted to overcome these problems. The new process is divided into five stages:

1. Product concept.
2. Concept checks.
3. Product concept testing.
4. Product scale-up and optimization.
5. Commercialization.

The discussion in the following sections focuses on the critical management differences between these two approaches that can affect the success of the product development process. This particular case involves a small, relatively independent division of a medium-sized chemical company. It is anticipated that larger companies would have a more formalized approach to new product development.

HECTOR COCHRANE, PhD, is the technical director of the CAB-O-SIL Division of Cabot Corp, Tuscola Ill.

THE BASIC APPROACH TO NEW PRODUCT DEVELOPMENT

The Birth of the Product

In 1977, Cabot Corp was approached by a small firm (referred to in this chapter as Company A) with the proposition that Cabot market an experimental hydrophobic fumed silica that Company A had developed. Company A was involved in the toll production, manufacture, and sale of treated fillers and felt that its expertise lay in the manufacture of these products rather than in applications development and marketing. The company's management believed that Cabot Corp could perform these functions profitably.

Although no special market niche for the new silica product had been identified, the product concept intrigued us at Cabot. At that time, the US market for hydrophobic fumed silicas was small but was predicted to grow—a forecast based on market developments for similar fumed silicas in Europe. In addition, although Cabot's competitors could offer hydrophobic grade materials to their customers, Cabot had not yet entered this market.

At that time, Cabot did not wish to invest the capital to build a plant to manufacture either a copycat product with limited sales in the US (its major marketplace) or a new product without proven market potential. The treatment agent Company A used to make the silica surface hydrophobic differed from those used on existing commercial products being sold by Cabot's competitors. Therefore, there was a potential opportunity to develop a unique fumed silica product with new applications in an established market. These factors all added up to a promising new product opportunity, so a secrecy agreement was signed between the two companies to evaluate the commercial potential of the experimental product.

Exploratory Research and Development

Preliminary laboratory screening studies showed that the new hydrophobic fumed silica was an excellent thixotrope (i.e., it became liquid when shaken or stirred but reverted to a gel when left standing). It was, therefore, a potentially useful ingredient in epoxy resin adhesives, sealants, and coatings. This market was already dominated by hydrophilic, or wettable, grades of fumed silica. The new silica solved an existing industrywide problem with the use of hydrophilic silicas in epoxy resin systems—namely, the loss of sag control after long-term aging before use and during high-temperature cure conditions.

Performance studies were carried out, and data showing the utility and cost-effectiveness of the new silica as compared to other existing thixotropes was developed. The results showed that the superior properties of the new silica were related to the specific treating agent used and the level of treatment. After discussions with Cabot Corp patent attorneys and representatives of Company A, it was decided not to patent the process used to make the new treated silica or its use as a thixotrope for epoxy resin systems. Instead, it was decided to retain the treatment process as a trade secret in order to secure a competitive edge.

Commercial Development

it was decided not to patent the process; instead, the companies opted to retain the treatment process as a trade secret

Identifying the first market niche for the new silica was the next major step in the development of the new product. Because Cabot supplied products (i.e., epoxy resin products) to this division of the specialty chemical industry, its marketing personnel knew the size of the market and the potential customers for the new silica. Although they recognized that, initially, the new silica would replace some of Cabot's existing hydrophilic grade materials, the firm felt that the superior performance of the new value-added product would support a significantly higher price. Cabot would then gain market share from its competitors, initially in the US and later in Europe and Asia. After the new product had been established in the epoxy product market niche, further effort could be spent in developing additional new applications for the new silica.

The commercial development program began in May 1979 with the appointment of a product development manager. The individual chosen for this position had been intimately involved in the exploratory research and development program and was convinced of the technical utility of the new silica in the epoxy product market niche. Before commercial development began, he had already emerged as the product champion for the new material.

A commercial development plan was made that was reviewed and approved by Cabot's senior management. The areas covered by the plan were:

- Market analysis.
- Financial analysis.
- Price list.
- Customer list with targeted accounts.
- Toll production agreement with Company A and cost.
- Product name.
- Packaging.
- Preliminary product data.
- Preliminary applications data.
- Material safety data.
- Product specifications.

A low-profile, low-cost strategy was chosen to introduce the product to the market for such reasons as:

- The product properties might need to be modified slightly on the basis of customer feedback.
- New products typically require five to seven years in the market before they start to generate a positive cash flow; hence, Cabot wanted to minimize costs as much as possible during the initial negative cash-flow period.
- The new silica would sell well without expensive packaging, brochures, and advertising if it solved existing customer problems cost-effectively.

- There was a need to avoid attracting Cabot's competitors' attention until Cabot was ready to expand sales rapidly during the full commercialization stage.
- Company A's capability to produce the silica was limited, which meant that full commercialization would have to be staggered so that Cabot could match sales and production volumes.
- Limited human resources were available for development and sale of the new product because of other commitments by Cabot.

After the commercial development plan was approved, the first small commercial run of new silica was produced. A few epoxy sealant manufacturers who were using hydrophilic fumed silicas were selected as targeted customers on the basis of their ability to provide Cabot with laboratory and production-scale feedback on the performance of the new silica.

Visits were scheduled at each of Cabot's targeted accounts so that the product development manager and the regional sales manager could introduce the product and explain its potential utility to the customers. Close contact was maintained with the customers during the several months that it took them to evaluate the product in their laboratories and then carry out plant trials. In many cases, these companies then had to wait for several months while their customers (the end users) evaluated their products containing the new silica.

In many cases, a total of two years elapsed between the initial visit to a customer and the time that customer would start to purchase the product regularly for commercial use. It was gratifying that nearly all the companies that evaluated the new silica in liquid epoxy systems reported beneficial results similar to those that were stated in Cabot's literature. These findings often encouraged these companies to carry out further evaluations of the product in their formulations.

During the commercial development phase, the product development manager controlled all aspects of the commercialization program, from order packaging materials and controlling the issuance of samples to customers to introducing the product to customers, coordinating production to match sales, and monitoring product orders.

The key to successful market entry was to respond quickly to all customers' needs, whether large or small. One such large problem involved modifying the treatment process so that the new silica was easier to disperse. In another case, Cabot was asked to maintain a truckload of the new silica in inventory for a manufacturer that was developing a new sealant to be used by an automotive company in a new car model. This case proved to be quite a problem because the truckload represented several months of Cabot's typical sales volume at the time. Fortunately, the sealant was approved for the new car, and very large sales have resulted since then.

Another interesting development during this time was that Company A relocated its plant, resulting in its inability to produce the new silica for several months. Fortunately, the inventory of product that Cabot had built in anticipation of the move lasted until Company A could resume production. Cabot did, however, quickly develop the capability to produce limited quantities of the new silica in its own plant.

The commercial development phase of the program lasted approximately four years. By the end of the first year, customers were buying very small quantities of the silica. At the end of the fourth year, approximately 40 customers were established and sales had grown by a factor of 400.

the key to successful market entry was to respond quickly to all customers' needs, whether large or small

During this time, additional preliminary laboratory studies were continuing the search for other potential applications for the product. These included the product's use in corrosion-resistant coatings and water-resistant greases and as a free-flow agent. Additional extensive studies were also conducted to generate toxicity data for the new product. Prior information used in the material safety data sheet was based on the knowledge that both the starting silica and the treating agent, a silicone fluid, were nontoxic and that other commercially available hydrophobic silicas, which would be expected to behave similarly to the new product, were also nontoxic. Additional studies confirmed the nontoxic nature of the new treated silica.

Full Commercialization

On the basis of the success of the low-profile commercial development program and the promise of the new silica's wider utility in markets other than epoxy products, it was decided in 1983 to commercialize the product fully. The major strategy for full commercialization was to maximize the penetration of the US markets for epoxy adhesives, sealants, and coatings as quickly as possible—before Cabot's major competitor could commercialize a product to match the new silica—and to widen the customer and applications base for new silica sales through a commercial development program for new applications that had been discovered for the product.

An expanded marketing program aimed at increasing market acceptance for the new product included these activities:

- Building product inventory at warehouses around the US.
- Training regional sales managers on the properties and applications of the product.
- Presenting technical seminars at national technical meetings (e.g., the Adhesives and Sealants Council conferences and the National Lubricating Grease Institute's conferences) and publishing papers demonstrating the utility of the product.
- Advertising in national technical and trade journals.
- Conducting laboratory studies to support customers' applications development projects.
- Developing a new, more formal product brochure that would include extensive data comparing the new product with existing products in such applications areas as liquid thickening, corrosion-resistant coatings, epoxies, greases, free-flow agents, and silicone rubber—All these investigations emphasized the unique properties and cost-effectiveness of the new silica relative to other products.
- Improving packaging to service customer needs more effectively.

In 1985, the product was turned over to Cabot's distributors for sale to a wider range of customers and exported for sale to companies in Europe, Japan, and elsewhere. Care was taken to continue to expand the production of new silica to match the increase in marketplace demand for this versatile product. The rate of growth in the sales of new silica in the US from the birth of the product until 1987 is shown in Exhibit 1. By 1987, the worldwide demand for the product had reached a volume at which it was decided by mutual consent to terminate the manufacturing agreement with Company A and build a much-expanded facility to manufacture the product in Cabot's US plant.

THE REVAMPED APPROACH TO NEW PRODUCT DEVELOPMENT

The new product development process described in the previous sections does not represent the optimum product development path based on Cabot Corp's most

Exhibit 1
The Relative Growth of New Silica Sales in Three Stages

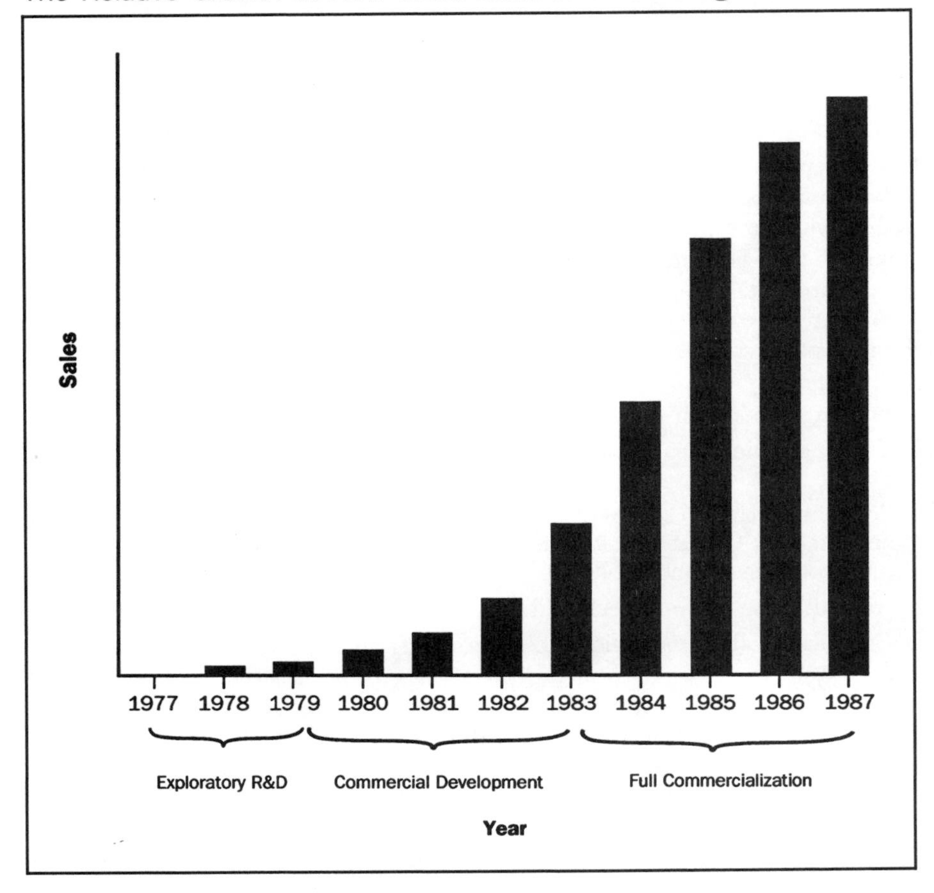

recent experience. An outline of Cabot's current new product development process is shown in Exhibit 2. It may be useful to compare the basic areas in which the former and current approaches differ. The product described was developed successfully in a relatively unstructured process described here in detail; recommendations are then made on how this process could have been improved.

Product Concept and Concept Checks

In the case of the new treated silica, the new product was technology driven. Company A was a manufacturer of a specialty treated silica, and the company had

Exhibit 2
Cabot Corp's Revamped New Product Development Process

1. Product Concept
   a. Market need or potential size
   b. Ideas to satisfy need
   c. Product profile
   d. Competitive activity review
   e. New product development team formed
2. Concept Checks
   a. Existing technology modified or radical new technology developed
   b. Regulatory considerations
   c. Technical, production, and sales inputs
   d. Time estimate
   e. Economic evaluation
3. Product Concept Testing
   a. Microplant manufacture
   b. Product characterization
   c. In-house application evaluation
   d. Patent application
   e. Economic feasibility rechecked
4. Product Scale-up and Optimization
   a. Pilot plant scale-up
   b. In-house trials
   c. Preliminary data and material safety data sheet
   d. Customer trials selected
   e. Further product modification conducted
   f. Economic estimates refined
   g. Acceptable market price confirmed
5. Commercialization
   a. Commercial development plan approved by management
   b. Customer base increased
   c. Marketing literature developed
   d. Training of sales personnel
   e. Distribution
   f. Full commercialization

experience converting inorganic pigments with silanes and silicones to meet its customers' needs. The new hydrophobic silica was the result of an experiment that involved treating a fumed silica with a silicone fluid. Apart from producing a new, unique hydrophobic silica, no specific commercial utility for the product was envisioned when the product was conceived at Company A.

The later, very successful commercialization of the product was attributable to Cabot's application studies, which discovered that the product solved important deficiencies exhibited by hydrophilic fumed silicas in a large, well-established market. Although other markets were subsequently found for the product, the decision to commercialize the product was predicated on its wide potential use as a thixotrope in epoxy resin coating, adhesive, and sealant systems. In addition, Company A had already experimented with commercial-scale production of the product in its small batch reactors. This eliminated the need for microplant and pilot-plant scaleup during Cabot's initial trials.

In the author's experience, it is frequently very difficult to commercialize new products successfully when they are developed as technical curiosities rather than as solutions to known market problems (i.e., market-driven products). The most effective approach to successful new product development is to identify technical problems that exist with the use of existing products within a given market. Although this can be achieved through reading technical articles, attending seminars, and reading patents, it is ideally accomplished through direct technical discussions with industry peers. Having such discussions reduces the incidence of misinterpretations of the problem. It also allows for the collection of information on the size of the market and, in some cases, helps to develop an appreciation of the value of a new product to the customer in solving a problem.

After the problem has been properly defined, ideas can be developed through individual and group brainstorming sessions. As indicated in Exhibit 2, the revised new product development process involves extensive data collection and examination of the possible solutions to the problem in light of the technical feasibility, competitive activity, regulatory considerations, projected cost, and the time to bring the new product to the marketplace.

In contrast, the case described here was technology-driven; the new product probably would never have been commercialized without the discovery of its utility in epoxy systems. Because no market need had been identified initially, no thought was given to alternative potential products.

In essence, the product concept and concept check stages are the critical planning phases of the whole new product development process. Insufficient consideration of the available information may result in the process moving in the wrong direction. It should be noted that during the initial phases there may be more than one possible solution to a problem. If this is the case, all options should be examined until all but one becomes eliminated to lower the technical development costs and other overhead.

Product Concept Testing

The earlier approach to new product development focused on selecting a highly motivated individual—often someone who had been involved with the project from the beginning—to lead the project as the product champion. The timely success of the new product development process would depend primarily on this individual, who would be required to coordinate all aspects of the process, including seeking senior management's continued approval, obtaining the daily support from peers and superiors in other departments for research equipment and human resources, setting up customer visits and sampling programs, and negotiating for development time on production equipment.

because initially no market need had been identified, no thought was given to alternative potential products

Designating one person to led the development process is an undesirable approach to project management for several reasons:

- Most organizations have a very limited number of personnel who have the personality, the competence, and the stamina to take on this responsibility.
- The project will be severely disrupted if the product champion becomes incapacitated for any reason.
- It may be difficult for the product champion to obtain adequate human resources and equipment to support a project's needs—Negotiating for resources from those with a peripheral interest in a given project can be time-consuming and can detract from the product champion's efficiency.
- To be successful, the product champion and the project must have continual exposure to senior management and the active and continued support of at least one member of the management team.

The Team-Based Approach. A more effective management approach is to appoint a new product development team with a team leader to coordinate the team and new product development activities. The selection of the team members and leader is critical to the success of the project. The team members should be chosen on the basis of their knowledge, the contributions they can make to the project, and their ability to work effectively in a team situation. The team leaders should be chosen on the basis of their knowledge, leadership abilities, and leadership styles (i.e., persuasive rather than dictatorial). The team leader must be fully committed to the success of the project.

The advantages of the team approach are that it:

- Provides a wide diversity of knowledge to tackle problems.
- Provides much easier access to support across departmental lines—This should expedite the collection of data and performance of studies.
- Allows a new product development program to be set up without a champion.
- Allows the program to continue smoothly even if the team leader leaves the project.

The Interface with Senior Management. Interaction between senior management and the new product development team is critical. In the case described here, the product champion was directly involved with solving the initial product variability problem and collecting the applications evaluation data. The product champion's only prior experience in new product development was as part of a team that attempted to find new uses for a product initially made to US government specifications. The marketing approach was to sell the potential utility of the product, on the basis of its hydrophobic properties, to selected markets with the expectation that potential customers would generate the actual data demonstrating the product's use in their specific applications. This approach was not highly successful, so the product champion was careful to generate extensive applications data demonstrating the utility of the new product in specific target markets.

To ensure an effective and timely new product development process, it is essential that the new product development team have regular discussions with at least one member of senior management of the company or division. Ideally, these meetings should be with the entire senior management team, the purpose of which is to keep management apprised of the progress of the project and to seek its help if the need arises. These high-level management contacts can be especially useful in providing business advice and financial and resource support to speed the new product development process. Astute senior managers will be careful not to let their own ideas and interests overshadow the development path taken by the team.

In the case cited here, the product development manager had infrequent meetings with senior management, especially during the early stages of product development, before senior management realized the utility of the product. This shortcoming was later rectified in new product development processes using the team approach.

Product Scale-Up and Optimization and Commercialization

The commercial success of a new product development program depends not only on having the right product but on being able to bring it to market ahead of potential competitors. Currently, major efforts are being made by US companies—especially the automotive and electronics industries—to reduce lead times to market for new products.

In the development of the new treated silica discussed in earlier sections, significant time could have been saved during the exploratory R&D and initial commercial development phases of the project by focusing more human resources and effort on the project after the utility of the product in epoxy resin systems had been identified. Fortunately, this lapse was not a problem because of the low-profile market development process chosen and the failure of our competitors to appreciate the potential utility of the product and quickly develop competitive products.

One factor that was not anticipated by the product champion was the significant time (up to two years) required by potential customers to evaluate the product before purchasing commercial quantities. The time lag was attributable to several factors, including the customers' need to:

- Evaluate the utility of the new product in their specific formulation.
- Evaluate the shelf-aging characteristics of systems containing the new product.
- Obtain their customers' acceptance, in some cases, of modified products containing the new treated silica.

Once an opportunity for a new product has been identified and the new product development team has been formed, it is important that the team determine a time line for developing and bringing the new product to the market. The time frame specified for the process as a whole should be broken down into shorter segments for each phase of the program. Once the development timetable has been developed and approved by senior management, great effort must be made to adhere to it as closely as possible. To maintain a time advantage over industry competitors during the crucial exploratory and commercial development stages, great care must be taken not to alert competitors to the firm's activities. For this purpose, customer trials should be carried out under joint secrecy agreements.

high-level management contacts can be especially useful in providing business advice and financial and resource support

BRINGING THE RIGHT PRODUCT TO MARKET ON TIME

The original new product development process used to develop the new, commercially successful silica had a low profile and was not very efficient—but it worked despite its flaws. The initial phase was technology driven, not market driven. The product champion had stamina, and the new product was commercialized. The path taken depended very much on mutual goodwill between Company A and Cabot. The project took longer to commercialize than was desirable because of limited management support until significant customer support was obtained. Fortunately, there was no strong competitive threat until most of the initial epoxy market for the new product had been captured.

In contrast, the revised new product development process is much better planned. It is market driven, the product needs are defined, and much more time and effort are spent considering alternative new product options to solve the problem. The competitive situation is continually monitored, and the projected costs and estimates of necessary development time are considered from the concept check to the full commercialization stages.

Widespread support for a project is obtained through the formation of the new product development team and through the frequent meetings with and support from senior management. Although the team leader is an important individual in the revamped process, product development can continue successfully in the leader's absence. In contrast, if the product champion is removed from the original development process, the new product activity will quickly die. The revised process is designed to bring the right product to the marketplace in the shortest time possible. Most important, the revised process has proved to be very successful in practice.

Acknowledgments

The author wishes to thank Richard Geiger, PhD, for useful discussions, Roberta Porter for typing the manuscript, and Cabot Corp for permission to publish this case study.

COVERING ALL THE BASES

Reducing New Product Development Risks

Managing technology to meet the conflicting needs of the marketing, sales, manufacturing, quality control, and engineering functions and reconciling those needs with the essential demands of senior management is an art, not a science.

Colin MacPhee

The term *new product* has different meanings for different functions throughout a company. To the sales department, a new product is one that overcomes customers' problems with existing products. To the marketing department, a new product is one that differs demonstrably from preceding ones and therefore is easy to promote, though it need not be radically new in concept.

In contrast, a new product provides the typical engineer with an opportunity to explore a new design concept or to redesign an existing product to incorporate new technology and improve performance. The manufacturing department usually views a new product as a redesigned product with reduced component costs and assembly labor. To senior corporate management, a new product is an opportunity to improve the profitability of the company; if this profitability is not forthcoming, the new product will probably not be around for long.

Managing technology to satisfy all these conflicting views is an art, not a science. This case study details one company's attempt to formulate a more disciplined approach to new product development than it had previously used.

The need for the particular new product had been established on the basis of several factors, one of which was that the company knew that its competitors had introduced or were planning to introduce similar products. Consequently, to improve its competitive position, the company felt the need to be perceived as state of the art. In addition, some problems were associated with the existing product—specifically, high production costs and difficulties in field service—that could be overcome by a new product.

RECOGNIZING THE NEED FOR CHANGE: THE EVOLUTION OF A PRODUCT

The product discussed in this case study is the overcurrent relay component of a molded-case circuit breaker typically used in industrial and commercial electric power distribution systems. The power system is designed to meet users' needs. Still, because it is connected to a public utility that is capable of supplying excess power, the system must be protected from currents that are greater than the capacity for which it is designed (i.e., it must be protected from overcurrent). Damage can be caused by persistent small overloads or very high instantaneous

---

COLIN MACPHEE is the principal of the technical consulting firm Albion Enterprises. Previously, he spent 22 years in R&D and engineering management for Federal Pioneer Ltd, Toronto, Canada.

overcurrents; in such cases, the results can be catastrophic, leading to system breakdowns (e.g., insulation failure).

The relay component monitors the current flow in the circuit and opens the circuit breaker if a fault occurs, thus isolating the section experiencing a fault. Because power distribution systems break down power into progressively smaller blocks as it flows from the utility to the user, there are often several overcurrent relay components in the system. Therefore, the more precise the relay components are and the more accurately they can be set, the better the system can be controlled.

Correcting the Problems of the Existing Product

Molded-case circuit breakers were invented during the 1920s to provide a low-cost alternative to the air-break circuit breaker. The overcurrent relay component used exclusively since that time makes use of thermal effects for overcurrent control and electromagnetic effects for catastrophe control. Because the action is mechanical, the relay component combines thermal, mechanical, and electromechanical elements. All of the components in the relay component are subject to various types of tolerance errors, so both the precision and the setting accuracy of the relay component became problems as electric power use escalated rapidly during the past 20 years. No satisfactory, low-cost solution had been found to compensate for the errors, and the problem was compounded by the high currents that had to be used during product testing and adjustment.

As is the case with many product innovations, the company in question anticipated that there might be strong resistance to new products or components because the industrial users had already developed methods for coping with the deficiencies of existing products and might be reluctant to exchange the problems they understood for a new set of potential problems. To overcome such resistance, sales personnel had to assure customers that their complaints concerning the existing product had been considered in the design of the new product, and the marketing department dealt realistically with those complaints in its product introduction program.

Introducing the New, Improved Product

Improvements that the company sought for the new product included ease of manufacture, ease of installation, and field testability as well as use of a minimum number of close-tolerance components. It was also vital to maintain the product's performance requirements. In addition, it was believed that the new relay component should be calibrated before installation in the breaker, should be easily adjustable for different breaker ratings, and should not require high-power testing during production. Finally, it was critical to produce a new product that was cost-competitive with existing relay components.

This last point is often a subject of acrimonious debate; some parties assume that a factory-cost comparison is sufficient. In fact, the lifetime product cost is more meaningful because it considers such critical factors as:

- The cost of overdesigning the breaker and the power system to compensate for the inaccuracies of the relay component.
- Warranty costs caused by lack of precision in the relay component and the difficulty of performing overcurrent tests.
- The need to remove the old products from service for routine testing.

to overcome market resistance, sales personnel assured customers that their grievances with the existing product were considered in the design of the new product

SPECIFYING THE NEW PRODUCT

Preparing a commercial specification that accurately defines the desired characteristics of a new product is the single most important step in the development process. As such, it is central to technology management, and it is therefore essential to take whatever time is necessary to prepare a fully detailed specification before the development program is started. This conviction is fully supported by studies on the causes of failure in new product introductions.

Writing a new product specification is too important a task to be assigned to any single function (e.g., marketing or engineering). It must be a truly cooperative effort involving all interested parties—from the customer (represented by the sales department) to the senior managers of the company. Unfortunately, this is precisely the area in which technology management is weakest in most companies.

The commercial specification for a new product should consider only what is required, never how it is to be achieved. If consideration is paid to how to achieve the specification, there is a significant risk that the design will be constrained by existing technology and that the designers will be unable to give their imaginations free rein.

The specification process begins by laying down general guidelines based on the desired characteristics of the new product. The details are developed through dialogues between the product developer and the potential buyers until the document is fully detailed. Only then should the development process be allowed to start. The specification process is time-consuming and often regarded by the participants as a tedious and unnecessary impediment to getting on with the job. Without it, however, the risk of failure is extremely high. If the product is allowed to evolve through continuous fine-tuning in response to the perceived needs of the customer after the specification stage, development costs will usually be higher than necessary.

Incorporating New Technology

The commercial specification for the new relay component dictated that it should use solid-state devices. Therefore, it was determined that the relay component should be composed of:

- A current transformer to allow each power conductor to monitor current and provide power to operate the relay component.

- A solid-state circuit to carry out the logic and control functions.
- A tripping device to unlatch the breaker when needed.
- A frame to hold these parts in proper relationship for easy assembly, calibration, and insertion into the circuit breaker.

In addition, the low power required by the relay component suggested that a battery-powered, truly portable test set could be designed to permit in-circuit field testing of the relay component and its associated circuit breaker. The testing equipment was added to the list of requirements.

CONTROLLING THE DEVELOPMENT PROCESS

It is characteristic of a properly executed new product development process that the level of uncertainty over the result is at its greatest when the project commences and that this level diminishes as the project progresses. It is also true that the costs increase rapidly as the project nears completion; accordingly, the risk increases at the same rate.

The development process must be designed to control risk at all stages. Therefore, it is vital to begin with a fully detailed commercial specification and to manage the development process so that the specification requirements are met at each stage of the process before authorization is given for the next stage.

The Phases of Product Development

To ensure that the commercial specification requirements were met in the case at hand, the development process was broken down into discrete phases, as described in the following sections.

Phase 1: Reducing Uncertainty. During this phase, the first steps were taken to reduce uncertainty. The development team studied the commercial specification, decided on a preliminary approach to solving the technical problems, and scoured technical journals and other sources for information about competitive products and patents. After the team conducted a feasibility study, the project was ready for the first design audit.

In general, the purpose of the design audit is to review the work of the development team, examine its conclusions, and resolve any conflicts, always keeping in mind the requirements of the commercial specification. Only when each member of the audit team is satisfied is the next phase authorized.

The design audit team includes the engineering area managers (usually designated as chairpersons) and representatives from field service, manufacturing, manufacturing engineering, marketing, purchasing, quality control, and test equipment engineering. Each departmental representative is responsible for ensuring that its concerns are addressed and resolved at each design audit.

At the first design audit of this product, the project engineer presented the development team's report on feasibility, drawing attention to possible conflicts with patents or competitive products and stating the team's opinion on the probability of meeting all the commercial specification requirements. Problems that arose were discussed by the audit team and resolved. In some cases, resolving a problem required further work during phase 1, so authorization to begin phase 2 was withheld and the design audit was rescheduled accordingly. Thus, the audits were designated as first-, second-, third-, and fourth-phase audits, because each phase included more than one.

the prototype specifications were reviewed with particular attention to test requirements, which significantly affect manufacturing and capital costs

Phase 2: Constructing and Testing the Working Model. Phase 2 began with the construction of a working model of the product, followed by detailed testing to ensure that the performance requirements were met by the model. The second-phase design audit considered the results of these tests and paid particular attention to the proposed method of construction to determine the product's ease of assembly and ease of service. With the consent of the appropriate members of the audit team, it was agreed that some minor changes would be made in the construction to simplify assembly but that these changes would be incorporated in the prototype design scheduled for phase 3 in order not to delay the project by repeating part of phase 2.

Phase 3: Specifying the Prototype and Choosing Components. Phase 3 opened with the specification of the prototype and the task of choosing components for reliability. Preliminary bills of materials and engineering drawings were prepared and detailed cost estimates were made. Test requirements were established, test points were incorporated into the design, and test specifications were prepared. During the entire process, the various departments represented on the audit team were kept fully informed of developments affecting them, and their input was sought to minimize conflicts and delays during the audits.

During the third-phase design audit, the prototype specifications were reviewed with particular attention to test requirements because they significantly affect manufacturing cost and capital costs for test equipment. Before authorizing phase 4, the audit team scrutinized the development team's ability to meet commercial specification requirements up to this point—an indication of the probability of success in future phases. It was crucial to proceed only if this indication was favorable because the development costs—and hence the risk—would escalate very rapidly in the following phase with the purchase of tooling and sizable quantities of components for prototype sample manufacture, along with the development of production test equipment.

Phase 4: Releasing the Preliminary Design. During phase 4, the preliminary design was released to the manufacturing, purchasing, test equipment engineering, and quality control departments to permit those departments to gear up for production. The tasks these departments had to perform at this stage included:

- Designing jigs and fixtures.
- Modifying the circuit-breaker tooling.
- Developing new labeling for the circuit breakers.
- Determining labor and space requirements.
- Sourcing components.
- Discussing quality control requirements with subcontractors.
- Designing in-house test equipment.

Several prototype samples were made and tested exhaustively for compliance with the commercial specification as well as additional manufacturing and quality control specifications. Subsequently, certification testing was carried out as the samples were distributed to other departments represented on the audit team to determine how well their needs were met. The results of certification testing were reviewed during the fourth-phase design audit, with each department reporting on any difficulties it encountered in preparation for production that might affect the design. Because these departments were extensively involved in the development process, no problems requiring design modifications were reported, and phase 5 was authorized. The release date for the product was set for February 1988 to coincide with the Electrex 1988 conference, the biennial electrical engineering showcase held in England.

Although the risk to the company was very high at this point, it was offset substantially by the certainty that the product would perform as required, given satisfactory manufacturing procedures. The possibility remained that some unforeseen factor would emerge to affect the performance of the product in the field, but this risk was adjudged minimal because of the stringent procedures followed in the development process.

Phase 5: Completing the Engineering Drawings and Bills of Materials. In phase 5, the engineering drawings were completed and the final bills of materials were drawn up. This information package was released to the marketing, manufacturing, purchasing, quality control, and test equipment engineering departments to allow them to complete their preparations for the pilot run and for full production. The stakes increased rapidly with the purchase of tools, equipment, and subcomponents necessary for the pilot run.

The development team had not yet completed its job at this point, because it was charged with overseeing the pilot run and evaluating the relay components produced. It was also responsible for determining the failure rate of the subcomponents used and the efficiency of the production testing method both in detecting failed subcomponents and in minimizing testing labor. After these tasks had been completed, the team documented the technology used in developing the product and prepared generalized procedures for troubleshooting failed products and for transferring technology to production.

Phase 6: Performing the Pilot Run. The purpose of the pilot run was twofold: to evaluate the product as made in a manufcturing environment (which differs markedly from that of the development laboratory because it produces prototype samples) and to evaluate the production process to permit fine-tuning before actually commencing production. Phase 6 was designed to control the pilot run and to ensure that these evaluations were carried out.

improving technology management became more important as the solid-state products became more complex, causing more difficulties in transferring technology from engineering to manufacturing

The manufacturing process proved satisfactory, and the product met the full requirements of the commercial specification. However, problems arose with the supplier of the current transformers, whose interpretation of the component specification differed from what was intended. It was therefore necessary to design and manufacture a simple test unit to inspect incoming components until more satisfactory supply arrangements could be made.

Systematic analysis of failed subcomponents in products fabricated during the pilot run suggested that these subcomponents should be purchased only from suppliers prepared to guarantee their quality. The engineering department was directed to modify the bills of materials to list only approved suppliers in order to minimize subcomponent failure. After the pilot run evaluations had been completed and the modifications described had been made, a final design audit was held and the product was released to the manufacturing department.

ANALYZING THE RESULTS

The new product development case described here represents a concerted effort to improve technology management, with the objectives of reducing product development failures to a minimum, controlling risk at all times, and facilitating the transfer of technology from the development group to manufacturing. The need to improve technology management became more important as the solid-state products developed by the company became more complex. Greater product complexity initially caused considerable difficulties in transferring technology from the engineering function to the manufacturing function, which in turn caused significant delays in scheduled product release dates. These delays led to a loss of confidence on the part of the sales force and disappointing results from new product introductions.

The company's senior management faced the challenge of trying to obtain the active participation of all the major departments in its efforts to improve the management of technology. In the past, other departments had regarded new product development as the engineering department's job; however, the engineering department had engineering jobs to do that kept them fully occupied. As a result of these ingrained attitudes, not all members of the audit team attended meetings or carried out the necessary detailed reviews at every stage of the process. Consequently, there was some overlap between the phases. Nevertheless, the process was successful in controlling risks by not permitting the product to proceed to the next phase without the appropriate audit.

The technology management improvement effort was successful in most respects:

a successful product was introduced, the risks were controlled during the development process, and the product developers were able to transfer to the manufacturing department detailed information on the product technology as well as the results and the pitfalls to be expected during the production process. Success could be attributed to the unremitting efforts of the engineering department to make the process work and the recognition by other departments represented on the audit team that this process (or a similar one) was the way of the future. All participants came to realize that the traditional method of developing products was unworkable in the current environment of short product lifetimes and unceasing pressures to reduce manufacturing costs. Eventually, all the appointed departmental representatives participated actively in the team.

As anticipated, the most difficult part of the entire process was to establish the detailed product specification quickly enough to meet the intended launch date. At the Electrex 1988 conference, the centerpiece of the company's display was to be a new line of molded-case circuit breakers incorporating the solid-state overcurrent relay component. Despite the pressure to succeed that this expectation exerted, difficulties were experienced in completing the specification before the project began. Immediate everyday operating problems sidetracked some project team members. As a result, several minor details of the specification were not made available until they had become absolutely essential. No solution was found other than persistent efforts by the engineering department to pressure the delinquent departmental representatives to provide the necessary input.

The product was first shown to the general electrical engineering community at its planned release at Electrex 1988, where it was judged the best new product at the show. And as planned, the relay component was subsequently incorporated into higher-rated circuit breakers; although an unforeseen technical problem was detected in this application, it was subsequently eliminated with a minor change in the product.

IMPLEMENTING THE GAME PLAN

During the new product development program described, the senior management of the company was changed and a major effort was begun to reduce overhead costs by increasing the efficiency of all departments, with particular reference to those associated directly with product development. This effort has taken precedence over all others; as a result, the benefits arising from the attempt to improve technology management described in this chapter have been deferred.

This case history has bearing on any product development effort. Any small or medium-sized company (i.e., up to $250 million in annual sales) with an extensive product line in an established industry can use the process outlined as a formal mechanism for developing and screening new ideas. Without such formalized procedures, any new idea—regardless of its merit—will probably be developed only if it has some appeal to management. In such precarious circumstances, the only way to improve the chances of launching a successful new product is to manage the risks actively, using the approach described here or a comparable methodology. Otherwise, the risk of failure is great indeed.

FORGING A CRITICAL LINK

The Role of Applied Research in Corporate Development

Applied research is governed by the same general principles that govern less practice-oriented forms of research. However, the methods that work best in applied research, the way goals are set and accomplishments are measured, the skills that applied scientists need, and the resources that applied scientists must cultivate and use effectively are different.

Thomas P. Fidelle

Applied research is usually associated with industry and basic research with academia; however, this is an oversimplification. Much excellent applied research is conducted in universities as well as in industry. Conversely, relatively little basic research is conducted in industry, and that's probably how it should be. Industry can draw on academia for the scientific foundation necessary to support its applied research programs, and academic institutions can draw on industry for financial backing and guidelines to ensure the relevance of their work.

The case studies discussed here are based on industrial applied research simply because my experience is centered almost exclusively in the industrial arena. The business issue as well as the technical ones are covered because, quite bluntly, the ultimate goal of industrial research should be to make money for the company.

It comes as a surprise to some observers that most companies make products or offer services that the consumer never sees as finished products. These products or services are called intermediates and will pass through one or more stages of development before reaching the consumer. This is especially true among chemical and petroleum companies, which make up a substantial portion of the capitalization in the US. For example, Dow Chemical realizes close to $15 billion in annual revenue, yet less than $2 billion of sales are directly related to consumer products. It behooves a company that supplies mostly intermediates to understand as much as possible about its customers' technologies, given the limitations that may be imposed by secrecy. That is the principal responsibility of applied research in industry. As a supplier of intermediates, the company cannot value its products, especially new products, unless it can establish the value of those products as the customer sees them.

Sometimes this evaluation can be achieved through rigorous specifications. More often than not, however, specifications alone cannot predict downstream performance, especially when unexpected and unusual developments come about (as is

THOMAS P. FIDELLE, PhD, is the manager of applications research at Great Lakes Chemical Corp, West Lafayette IN. Previously, he worked for Celenese Corp, where, among other responsibilities, he served as the applications engineering project manager for two of the PBT-based product development projects: carpet fiber and stretch denim.

frequently the case with substantially different new products). It may be necessary for the applied research department to be familiar with a broad variety of complex technologies if the products they sell are to enter several markets.

HOW APPLIED RESEARCH FITS

Because of the uniqueness of its role, the applied research group may be an integral part of the R&D organization or a separate group outside R&D or, in some companies, part of the marketing department. For example, DuPont coined the term *end-use research* for its applied science department and had several specialized groups occupying several buildings at the Wilmington Experimental Station. Dow Chemical still uses the term *Technical Service & Development* for its applied technology area. Before its acquisition by Hoechst AG (Frankfurt, Germany), Celanese Fibers Group had a marketing and technical department that reported directly to the president of the Fibers Marketing Co. As the name suggests, the group performed marketing as well as technical services.

Exhibit 1 depicts the spectrum of research and development in a specialty chemical company. This medium-sized company annually spends between 5% and 6% of gross revenue on R&D. Some synthesis among the various research areas is accomplished, which is aimed primarily at new products. Because of limited resources, the synthesis effort must be directed at technologies at least somewhat familiar to the company. A major share of R&D resources is dedicated to process development that is aimed at cost reduction, efficiencies, environmental requirements, quality, or product development.

There is room for knowledge building when the need exists, both from a fundamental and an applied perspective. When new products or processes emerge, they must first be demonstrated at the pilot level, then on the production-scale level. The R&D department is likely to be involved in all phases of scaleup.

The freedom to innovate must be present throughout the entire R&D structure; the organization is feckless without it. Frequent interaction among R&D staff members and other functions outside R&D is also essential for success.

The R&D applications group commands a unique position in the organization. It is the key link between R&D and sales and marketing and—directly or indirectly—with the customer base. Ideas may originate from within the R&D organization or from an outside source. Several potential sources are shown at both sides of the R&D spectrum in Exhibit 1.

Sources of Invention

It is important for R&D managers to note that R&D departments do not have a monopoly on ideas. Many bright ideas emanate from other sources, and it is the responsibility of R&D, led by the applications group, to tap those outside resources effectively. Nothing can isolate and ultimately destroy an R&D organization as readily as a narrow development philosophy that fails to recognize the rich outside resources available for obtaining useful ideas.

In responding to either opportunities or problems, the applications group has

Exhibit 1
The R&D Spectrum

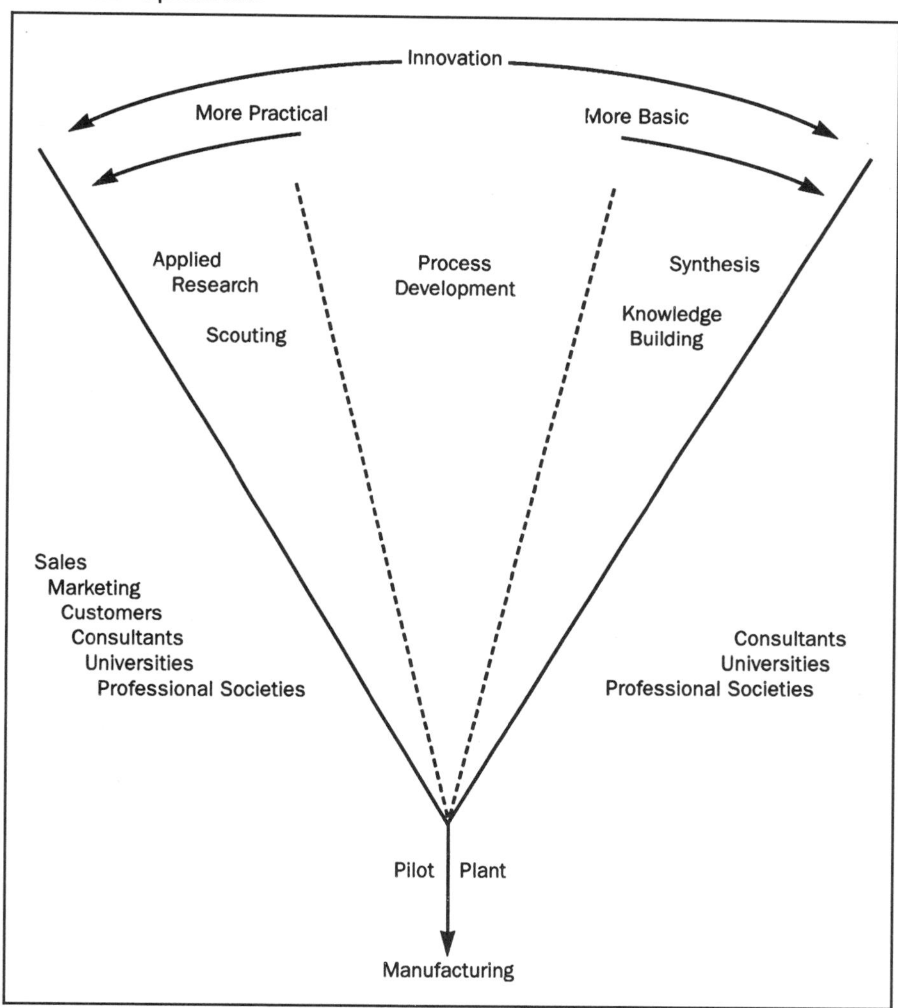

a relatively short-term focus—that is, between two years and five years for developing significant new products. If there is no separate technical service department, the applications group is likely to be the first line of technical support for the sales force. In such cases, however, they must constantly strive to avoid getting bogged down in daily battles at the expense of longer-term objectives.

The Tasks of Applied R&D

A typical distribution of effort for the applications group is:

- Routine technical service (30% of effort).
- New product development and introduction (40% of effort).
- Finding new applications for existing products and exploring new technologies (30% of effort).

Routine technical service consists of troubleshooting, particularly regarding quality problems, and specifying the use of products to customers. Troubleshooting typically may require some detective work or laboratory evaluation. Advising customers on product use may be accomplished with existing information or it may be necessary to go to the laboratory or pilot plant to derive the data.

In new product development, the applications role is to take the new product offering at the early stages of development, to determine the value of the product in the customers' eyes, and to participate in all the subsequent stages of development, from introduction through commercialization. This includes preliminary evaluation, defining performance criteria and specifications in the interest of the customer, reviewing cost and selling-price scenarios, making product presentations to prospective customers, and sometimes conducting field trials on location with the customer.

The tasks of the applications function are the most difficult to define and perform. Finding new applications for existing products sounds easier than it is, especially for mature product lines. However, it is the responsibility of applications research to constantly seek new markets for existing products as well as new technologies for the company to grow into. If the applied research department is large enough, it makes sense to organize the department around functional product lines.

An Organizational Example. One applied research department was divided into three groups, each consisting of six to eight scientists and a group leader. Two of the groups were aligned with the core businesses of the corporation. These were well-defined markets that had not only some mature products but also some opportunities and the need for new products. The objectives for these two groups were reasonably well defined, and the projects generally had development time frames of between one year and three years. There was a respectable amount of routine technical service provided by the two groups.

The third group in the organization had lofty goals, much more poorly defined objectives than the other groups, and considerably more freedom to diversify outside the company's core business area. Little technical service was provided, and only a limited contact with the traditional customer base was required.

This group's resources for new product ideas included available literature, technical conferences, university contacts, start-up companies, consultants, entrepreneurs, and others involved in state-of-the-art technologies. On average, projects took from 5 years to 10 years to complete, from idea to commercialization. It is often difficult to justify such expansive and risky undertakings in R&D, given the bottom-line mentality that currently besets many US corporations; however, technology-based companies that fail to make sacrifices today for long-term growth are doomed to atrophy as their product lines mature and are replaced by new products with new technologies.

THE ROLE OF THE APPLIED SCIENTIST

The applications researcher, or applied scientist, functions as the focal point for new developments and must possess certain skills in order to be successful. Over the years, I have interviewed hundreds of people for applied research jobs and hired many of them. In describing the position to the applicant, I often use the diagram in Exhibit 2, which depicts the four key elements of a business and the way they must interact to be effective. Most large companies are organized along the lines of Exhibit 2. The larger the company, the greater the premium placed on communication and cooperation.

For the business to be successful, communication among the business functions must be excellent, particularly if new product development is meant to play a significant role in that success. A key link in this chain of communication is that between R&D and manufacturing. R&D must be able to scale up new products in the plant, and plant personnel frequently must rely on the R&D department to solve problems.

In my experience, the most difficult communication bridge to build and maintain is that between the R&D and the sales and marketing functions (as indicated

Exhibit 2
Functional Interaction

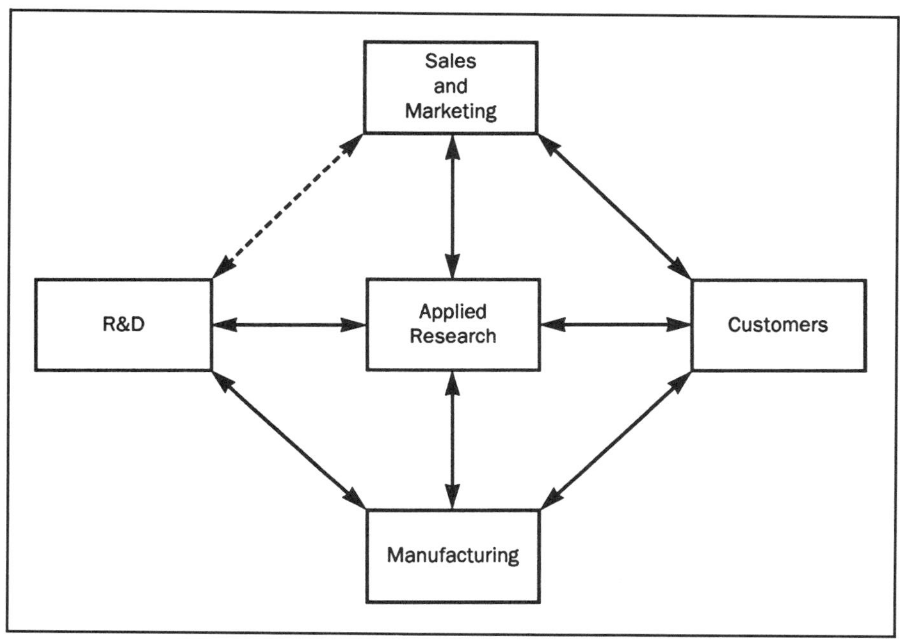

by the dashed line in Exhibit 2). Weak links in the communication chain tend to be caused by mutual misunderstanding of the other's professional goals and point of view. A central figure is needed who understands the diverse perspectives of each party and can speak their languages.

The applications researcher is equipped to play such a role, coordinating the overall effort in introducing a new product. As mentioned, the idea for the new product may originate in R&D or come from any number of outside sources. The first decision on whether or not to proceed with the idea is likely to be made by the applications researcher. If the idea originates in R&D, the decision will be based on the idea's commercial feasibility; if the idea comes from an outside source, the decision is likely to be based on its technical feasibility and estimated cost.

Likewise, the research applications department is usually the first gate for a new product to pass through when it is synthesized or formulated on a laboratory (i.e., prototype) scale. At this stage, the product is evaluated in terms of its ability to perform its ultimate end use. As the project unfolds, the applications group evaluates several versions of the product as it is scaled up and begins to formulate specifications on the basis of its process capability and market needs. At this stage, the sales and marketing function identifies potential customers, projects volumes and prices, and plans initial field trials.

If significant start-up capital is needed, engineering and manufacturing are brought in at this stage to aid in planning and specifications. Patents need to be filed before any public disclosure is made. Environmental and toxicological issues may need to be addressed early on. An applications researcher is likely to go along on the initial calls to prospective customers to promote the merits of the product and discuss any special handling requirements. If the customer allows, the applications researcher may be heavily involved during initial stages of field trials.

During the scale up to manufacturing, an applications researcher is responsible for setting specifications according to customer requirements. If any major changes become necessary, some or all of the foregoing steps may need to be retraced to obtain continual feedback as the results of such changes. In short, the applications department serves as the focal point for new product introductions, and the applications researcher is well suited to act as the project manager.

A PROFILE OF THE IDEAL APPLICATIONS RESEARCHER

When interviewing candidates for applied research, look for the following characteristics:

- Social skills—Does the applicant speak clearly and concisely and express thoughts logically? Does the candidate maintain eye contact? I expect eye-to-eye contact for at least 80% of an interview, which reassures both parties that they are having a meaningful, honest exchange. Does the candidate make a favorable first impression, dress acceptably, and exhibit a commanding presence? Is this person likely to be a leader?
- Technical qualifications—At a minimum, a bachelor of science degree is required, in a subject ideally but not necessarily related to the company's products.

it is the continuous responsibility of applied research to look for new markets for existing products and for technology that the company can grow into

Advanced degrees are welcome but not required; in my view, an advanced degree can be a handicap if the applicant is too narrowly focused. I'd rather have a person with average formal technical training who is innovative and has initiative rather than a technical genius who can't carry ideas to fruition and is inflexible in the subject area. Experience and a successful track record in applied research are highly desirable, especially in related fields.

- Innovative ability and hands-on style—People who can generate their own ideas and build on the ideas of others are well suited to applied research. In my experience, fewer than 1 out of 20 technical people are naturally innovative; the rest of us have to work at it constantly. In addition, the job requires a hands-on person—that is, someone who enjoys actual lab work and interacting with the physical product and with colleagues.
- Winners and achievers—The candidate should exhibit a few significant accomplishments as a professional and as a student and should also exhibit a positive attitude when faced with challenges.
- Multifaceted—Can the applicant effectively handle two to three projects simultaneously? In general, candidates who would rather work on several short-term projects that last from 1 year to 3 years than on one long-term project lasting from 5 years to 10 years make better applied researchers.
- Goal orientation—The candidate's career goals may reveal a general attitude. For example, does the candidate intend to remain in research or to move into management, sales and marketing, or another function? Because of the variety of their experience, people who have worked in applied R&D for a few years often make excellent candidates for sales and marketing or general management.

HOW APPLIED R&D PERFORMS ITS JOB

The following two case studies illustrate different aspects of the applied R&D department's function.

Grappling with the Difficulties of Corporate Growth

During the early 1970s, Calloway Chemicals was a small specialty chemical company. A public company traded on the American Stock Exchange, in 1973 it reported sales of $30 million and an 11% after-tax profit. Calloway started as a regional supplier of dyestuffs and finishing chemicals for the textile industry in the southeastern US. As an alternative to the large multinational companies, Calloway offered a broad range of products, attractive prices, and excellent service. None of its products was proprietary or unique; the company simply followed its larger competitors with me-too products. The R&D department's spending was less than $1 million (less than 3% of gross sales).

Throughout the 1970s, the company grew at a compound rate of 22%, and by 1980 it was generating $120 million in sales annually with an R&D budget of $2 million. Most of the R&D budget was being spent on process development aimed at cost reduction, because at this point there was increased competition in the industry. In addition, the company began to experience slower growth and loss of market share to smaller, more flexible companies that were doing exactly what Calloway had done to become successful: copy its large competitors.

A new, three-pronged corporate strategy for the 1980s was designed to solve this problem:

- In R&D—Increasing the R&D budget to 5% of sales, implementing more new product development (synthesis), and finding new opportunities outside the core business (i.e., textiles).
- In corporate management—Aggressively pursuing acquisitions and joint ventures to facilitate growth.
- In sales—Establishing a global sales posture by branching out to other US regions and to Europe and the Far East with novel and proprietary products.

The first leg of the strategy was to beef up the R&D function. Previously, the company had never had an applied research group. Dominated by textile chemists, the R&D department had a strong internal focus, with most of the outside influence coming from direct sales and customer service. The company hired an R&D director with a strong background in applied research and made him accountable to the corporate president. (The R&D department had previously reported through operations and did not have strong support or contact at the senior management level.)

The new director formulated a five-year plan with the following goals:

- Doubling R&D spending in the first full year (a budget of $4 million was still less than 4% of sales).
- Committing $250,000 to a general-purpose organic specialties plant for the small-scale manufacturing of high added-value products.
- Creating an applied research group and a small organic synthesis effort, and recruiting aggressively to staff these positions.
- Bringing technical service into the R&D group.
- Establishing a steering committee for new products that is chaired by the corporate president and consists of high-level representatives from R&D, sales and marketing, and manufacturing.

Creation of the steering committee was a key element, as the new director could not make sweeping changes without broad-based support from senior management. First, the president had to be sold on the plan; the new director then had to actively solicit support at key levels throughout the organization. As might be expected, the most resistance came from the sales and marketing staff members, who expected their power and influence to erode after the changes in the R&D department went into effect. Eventually, the new director was able to convince them otherwise by painting a vista of new products and worldwide business opportunities that would result in resumed growth for the company and enable it to recapture a significant market share.

between R&D and sales and marketing, a central figure is needed who understands the diverse perspectives of each party and can speak their languages

The turnaround in organizational positions was time-consuming. It took a year to reorganize R&D and to staff the new positions with high-caliber people. It took another year before the group began to produce workable ideas and new product road maps began to emerge. Three years from that stage, new products were finally established in the marketplace and began to earn respectable revenues. Thus, the constant investment of time and resources paid off.

By 1988, the corporation had grown to just under $500 million in annual sales, putting it back on the 20% annual growth curve. More than $100 million of the growth resulted from products that were developed in the 1980s. By 1987, 10 new product patents were being issued to the company every year, and new technology had resulted in two joint ventures totaling $50 million annually. Nearly 25% of R&D costs were offset by annual licensing fees paid for by technology developed after the reorganization.

By 1987, the R&D directorship had been elevated to the corporate vice-president level and the R&D department had close to 100 scientists. Twenty-five percent of R&D personnel were involved in applied research. Much of the success of this corporation was attributed to the emphasis on applied research and the leadership exercised by the applications researchers who acted as project leaders in developing new products.

A Technology Looking for a Use

During the 1950s, Celanese Corp, in a joint venture with UK-based Imperial Chemicals, launched a company called Fiber Industries Inc. Imperial Chemicals brought polyester and nylon fiber technology to the relationship, and Celanese brought expertise on marketing fibers in the US. Eventually, the joint venture became one of the largest polyester fiber producers worldwide, with sales in excess of £1.5 billion annually. In chemical terms, the polyester of choice for synthetic fibers was (and still is) polyethylene terephthalate, or PET, the simple condensation polymer of terephthalic acid and ethylene glycol. This fiber offers the right combination of chemical and physical properties for many apparel, home furnishing, and industrial end uses, ranging from T-shirts to tires.

Another simple condensation polymer, based on terephthalic acid and butanediol and called polybutylene terephthalate, or PBT, could be made easily with the equipment used to make PET. Although PBT has unique physical properties compared to PET and is used extensively in injection-molded plastics, the material never seemed to have any value as a fiber during the early years of applied development of polyester.

Celanese prided itself on fiber technology and employed some of the best applications researchers in the industry, who doggedly tried to find a use for this new fiber. They first looked at hosiery yarn. Nylon was then the miracle fiber for hosiery. DuPont had chosen hosiery to introduce nylon because it was cost-effective; the new polyester polymers were quite expensive in small volumes, however, so this use seemed impractical though possible. In retrospect, fabricating a knockoff of nylon for this end use would have been a Herculean task.

PBT had a few minor advantages (e.g., it produced less static electricity) but one major deficiency relative to nylon that eventually killed the hosiery application. Although PBT had significantly better resiliency than PET, it was still inferior to nylon in that category. After many wearings and washings, PBT pantyhose tended to sag in the knees worse than nylon hose, resulting in a shorter life for the PBT product. Several hundred thousands of dollars and several years of work went down the drain with this failure. The project was well on the way to commercialization before it was killed.

The applications researchers were not dismayed, however. Even before the demise of the hosiery product, they found that PBT made a superior carpet yarn compared to PET. Nylon was the ideal carpet fiber, as it is today, and this was the second major use discovered for nylon polymer. PET polyester had been commercially introduced as a carpet fiber during the late 1960s, and because it was less expensive than nylon, it was rapidly embraced by the tufted carpet manufacturers. After a couple of years in use, however, the polyester carpet fibers became crushed and matted, which led to the eventual failure of the product. Polyester carpets also soiled more readily than nylon carpets and were more difficult to clean.

Celanese introduced PBT-based carpet fiber in 1970, and at least two commercial programs were underway with appreciable retail sales before it was discovered that the PBT fiber was more flammable than PET. Many rolls of carpet were rejected by the mills because they were unable to pass the mandatory flammability tests before shipping. This was another expensive lesson for Celanese.

One could argue that both of these failures were examples of untenable applied research because the deficiencies should have been recognized before the products got so far down the road; of course, these problems are much clearer with hindsight than they were during the development process. The performance limitations in each case could be determined statistically only after many hundreds of goods were manufactured and tested. Such is the nature of new product development and applied research.

Fortunately, there were a few brave souls at Celanese who were ready to come forward with the next potential application for PBT fiber. During the late 1970s, an opportunity presented itself. Stretch garments were becoming increasingly popular in the US. Stretch wovens were in demand in men's and women's trousers, and the denim manufacturers were interested in making fashionable stretch denims. The applications researchers at Celanese found that PBT had unique viscoelastic properties that made it an excellent filling yarn for stretch denims. They did their homework thoroughly this time, and the first successful stretch denim jeans were introduced in the early 1980s. The fabrics were produced by Burlington Industries, and the denims were marketed under the Fortrel ESP trademark name. That material continues to be a successful product.

This case illustrates the challenges of applied research, some of the potential pitfalls, and, in this instance, the protracted time line from the origin of an idea to its ultimate commercial success.

the new director highlighted new product and business opportunities that would result in resumed growth for the company and would recapture market share

A CRITICAL LINK

Applied research can be challenging and time-consuming, and it can result in eventual commercial successes or failures. This chapter examines the methods that work most effectively, the way goals should be set and accomplishments measured, and what resources the applications researcher should cultivate and use to ensure ultimate success.

As stated, the freedom to innovate must be present throughout the entire R&D structure; without it, the organization can become weakened and ineffective. Frequent interaction among R&D personnel and other functions in the organization is critical to ensure a company's success. The applications research group commands a unique position: it is the key link between R&D and sales and marketing and—directly or indirectly—the customer.

PERKING UP THE BUSINESS

Analyzing Product Performance at Magnaflux

It is not unusual for a business to structure its organization around the products it offers its customers. Nor is it unheard of to establish manufacturing cost-reduction programs to increase profitability. And it is not uncommon for businesses to ask their customers what they need and then add the requested features to products. But it is unusual for a company to do all these things at once as the prescription for its rejuvenation and as the means to regain focus on what the customer appreciates most—value.

James E. Doherty

Magnaflux has used a process called product performance analysis to reshape several of its business units into more responsive and productive operations. The case history given here is the experience of one business unit that reduced the interval between the customer's order and the shipment date from 10 weeks to four, decreased the number of employees on the product line by 15%, reduced the number of product parts by 50%, decreased inventory by 35%, and lowered direct cost by 25% during a time when the raw-material cost of a major component increased by more than 100%. In addition, during the 18-month transition period, the main accessory product was revamped, which allowed for a cost reduction of more than 25% and significantly improved customer acceptance because new features were added. At the time of this writing, the business unit is still using the principles of product performance analysis to achieve further improvements.

The operating unit manufactures a broad range of testing equipment. Before the improvement efforts, the product lines were complex, with specific models available for every conceivable customer. Although this diversity of product offerings might be viewed as a competitive advantage, the volume was low for many items, resulting in unacceptably low profit margins. Routine fabrication, which had always been done internally, was outsourced for three years in an attempt to improve efficiency; however, the results were disappointing. Shortages caused by late deliveries, errors, and inconsistent quality plagued the in-house assemblers. To be responsive to customer orders, production control demanded large inventories of materials and parts. Remedial action was obviously required; despite all the recommendations offered in industrial literature and by renowned consultants, however, the best approach was not immediately obvious.

JAMES E. DOHERTY is vice-president of Illinois Tool Works Inc, a division of Magnaflux Corp, Chicago.

A review of the available literature on manufacturing showed that most techniques recommended for performance improvement focus on only one aspect of a business. Depending on their respective points of view, authors variously suggest that businesses do one of the following:

- Establish a mission statement (recommended by industrial psychologists).
- Get close to the customers (recommended by marketers).
- Advertise the cooperative image and promote the product (recommended by salespeople).
- Organize shop floor cells and use just-in-time inventory management (recommended by the production and inventory controllers).
- Implement materials requirements planning (recommended by accountants and production controllers).
- Implement statistical process control techniques (recommended by statisticians in quality control).
- Form quality circles (recommended by industrial psychologists).

What is a manager to do with all this help? How can any manager find the best solution for a particular problem, a solution that works best not only for the manager's own functional area but for the whole business? The answer may be simply to focus on the basics—to be logical and do the obvious.

Magnaflux suspected that more than one approach might be needed to perk up the business, but was surprised by the result of an exercise that identified all the factors that individually affected the quality, cost, and delivery time of the company's products. For the most part, it was found that the same factors had an effect on all three problem areas. This result was interpreted to mean that any approach that focused individually on cost, quality, or lead time would produce improvements in the other two areas. The company quickly recognized, however, that some issues might be understood and dealt with more easily from a particular perspective. Nevertheless, the results of this simple exercise gave Magnaflux the confidence to go back to find the right path for improving the business unit's performance.

TRAVELING THE CHOSEN PATH

The approach selected was simple: evaluate the market, organize to meet its needs, and then measure performance. Although this course of action is not new, that does not make it any less effective. The question remains, however, if the approach is not new and it is effective, why don't more businesses adopt it? Perhaps some look for a quick fix, or they are distracted by a current fad, or the simple routine of hard work is just not exciting enough. It is difficult to say where the problem lies, but perhaps some managers have allowed their businesses to become too complex; it is difficult for them to obtain the perspective they need to perceive the fundamental issues that drive their business to its full potential.

Magnaflux had a complex business, a condition that a succession of managers had allowed to develop. To begin to unravel this complexity, the business's current activities were analyzed to select the ones in which conscientious improvement efforts would provide the greatest benefit in the shortest time.

EVALUATING THE MARKET

To define the core of its business, Magnaflux analyzed current activity using the 80/20 rule (also called the Pareto rule, after the statistician who first proposed it). This axiom states that 80% of a population can be found in 20% of the range. When applied to a business, the 80/20 rule leads to two questions: Which 20% of the products produce 80% of the revenue? and Which 20% of the customers are responsible for 80% of the volume? The answers to these questions define the base activity area or core business.

The 80/20 rule worked for Magnaflux because it forced the company to concentrate on the business areas in which its leverage was greatest. By comparison, energy-depleting, high-risk, glamorous ventures were viewed for what they were: detractors from performance and profitability. By organizing the business's priorities according to the 80/20 rule, the company could be attuned to the market and its strongest product areas. If the principle is applied consistently to every decision that allocates a company's resources and energy, it is possible for the business to succeed regularly.

Exhibit 1 shows the results of an analysis of a product line that was not generating a profitable return. The analysis confirmed the 80/20 rule: of five product offerings, only one (the first) accounts for the bulk of the sales. Further review showed that the products were not well integrated in terms of the features offered and the prices asked. This was clear evidence of an evolution in which specific products were developed to meet special market needs but then were not cut from the product portfolio when that need no longer existed. In addition, the costs in manufacturing the products did not follow a rational order—that is, the products with the highest selling price were not the most expensive to manufacture. When presented with these results, the company realized that it had to restructure the product line and the individual products in order to improve its performance.

Exhibit 1
Analysis of One Product Line Confirming the 80/20 Rule

Product	Sales (%)	Cost ($)	Price ($)
1	75.8	39	72
2	11.2	34	100
3	6.5	25	94
4	5.2	25	100
5	1.3	21	70

ORGANIZING TO MEET MARKET NEEDS

The second stage in the overall business analysis is the product performance analysis. This analysis determines whether the company is delivering what the customer needs in the most efficient manner. The first step of the product performance analysis is to ask the customers what they like and dislike about the products or services offered and to make improvements according to their suggestions. The next step is to find the most efficient way to deliver those products and services. The product performance analysis is based on two well-understood techniques: focus panels and value analysis. It results in delivering a highly valued product to the customer and a high-gross profit to the company.

Although the product performance analysis is straightforward, it works only if the process is followed carefully. Short cuts increase the risk of failure, and the benefits are reduced when the steps are taken in isolation. In addition, the return on investment is poor when the analysis is applied to a poorly selected product, and the success of the process is highly dependent on maintaining the focused attention of highly skilled staff. At Magnaflux, failure has resulted from the inability to keep the appropriate human resources applied, especially during the implementation stage. It is harder than it might seem to stay focused because apparent emergencies continuously tempt managers to temporarily reassign key employees involved in the product performance analysis process. Management must exercise leadership and discipline if product development or improvement programs are to be successful.

Customer Focus Panels

The product performance analysis begins with customer focus panels. To be effective, the information-gathering step must be structured so that the data that customers offer is meaningful. Focus panels serve to organize customer responses. Magnaflux does not rely exclusively on the sales force, however, to obtain customer or product information. Although data from front-line salespeople is informative and helpful, in most cases that information is distorted by the most recent specific customer request. Sales representatives who are worth their keep always know what customers want, but they may not clearly understand what customers need. The point of this distinction is that when customers are forced by pocketbook issues to declare what they need, they are generally decisive. When asked what they want, however, they tend to be expansive and fail to indicate the priority of their desires. In a postmortem on a failed or faltering development project, it is usually found that the company had done a poor job of defining the customers' needs before the project began.

The focus panels used for gathering customer feedback are structured exercises conducted by a third-party facilitator; they are composed of between three and seven customer representatives and a nearly equal number from the company. The company representatives typically include a field service representative, a salesperson, a product design engineer, and a product manager. Representatives from the customers' companies are chosen as much as possible from the same market niche;

if the product must meet the needs of two market segments, separate panels should be used.

Magnaflux strives to include customer representatives with different functions (e.g., operator, supervisor, process engineer) from each of the selected customer companies. It is also desirable for the panel to include customers who own a competitor's products as well as the company's products; in an ideal case, the customer owns both. In essence, the focus panel is composed of individuals who have specific but different knowledge of the product under discussion.

Focus Panels Versus Mail Surveys. A comment often heard from those who have not seen a focus panel in action is that the sample is too small and that a mail survey would be more thorough. Although mail surveys have obvious advantages (e.g., large sample sizes), the problem with them is that survey respondents are self-selected and do not operate from the same reference point. To find out whether this poses a significant problem, Magnaflux conducted a trial survey in a controlled setting.

The experimental survey asked participants to evaluate the relative importance of a group of items known to them. To aid respondents, the company structured the survey to evoke a detailed consideration of specific product attributes. When the results were evaluated, Magnaflux found that the midpoint or norm ranged from 4 to 8 on a scale of 1 to 10, depending on the relatively favorable or unfavorable view of each respondent (see Exhibit 2). In addition, the ends of the range, representing minimum and maximum levels of acceptance, varied.

These variations occurred because a standard of measure had not been established for the survey. Such a universal reference point would have forced participants to gauge their evaluations against a common standard, which is hard to establish

Exhibit 2
Summary of a Mail Survey Showing Variance in the Values Assigned to Product Attributes

	Acceptance Level		
Respondent	**Minimum**	**Midpoint**	**Maximum**
A	6	7	8
B	3	5	6
C	4	7	8
D	5	5	6
E	4	6	9
F	8	8	8
G	5	7	8
H	3	4	5
I	6	6	8

in a mail survey. The need for such a standard was made evident when the survey respondents were brought together to reconcile their evaluations. The group spent considerable effort establishing a standard value for the high end of the acceptance range. After making several attempts to reconcile the rankings assigned to product attributes, the group finally decided on what everyone agreed was the most desirable attribute before ranking the rest.

The same standard value (i.e., the most desirable attribute) was then used in the customer focus panels. The advantage focus panels have over mail surveys is that the common reference point can be defined quickly and early in the evaluation process. Although the question of sample size in focus panels may remain, these panels do produce credible results.

The psychology of the focus group induces individual members to be objective. In addition, all panel members are knowledgeable about the product type; they have the full range of information about the product as well as experience with its use in that market niche. Combined, these factors are conducive to frank discussions among professionals who all seek to discover the essence of a product that does what it should do.

For their part, customers enjoy participating; they are honored to be held in high enough regard to be asked for their opinions. Many participants tell us that no one ever gives them an opportunity to discuss what they need from products directly with the people who can do something about it. For this reason, the product manager and principal design engineer are usually included among the Magnaflux representatives.

Most customers leave the panel saying that it was not what they expected. Possibly they expected a free-form discussion of what they want—an easy topic that requires no effort. They leave the sessions tired and with the feeling that they have done real work—and indeed they have.

The value of the information that can be gained from a focus panel is immense, particularly in light of its modest cost. Magnaflux buys lunch and pays for travel expenses when the customer will allow it. Typically, most of the work can be accomplished in a morning, but some cases require an additional 90 minutes after lunch.

The Mechanics of Conducting a Focus Panel. Focus panels should allow the company to determine the following with respect to a product or product type:

- The chief buying influence.
- The most desirable product features.
- The most desirable product feature.
- A rating of all features against the most desirable one.
- The product attributes most disliked.
- The product attribute most disliked.
- A rating of all the negative attributes against the most disliked one.

In some cases, a free discussion of competitive products that are currently available takes place. This type of exchange must be handled with care by the facilitator because many customers feel uncomfortable discussing specific competitive products in an open forum. The company has finessed this reluctance by displaying competitive products before the panel, ostensibly as a means to refresh its memories about the attributes of products of one type. The subsequent discussion centers on generic attributes. It is sometimes effective to introduce the display between sessions that deal with the desirable and the undesirable product attributes.

most techniques recommended for performance improvement focus on only one aspect of a business; product performance analysis focuses on all aspects

The sequence of events in each panel session is:

1. The group selects by consensus the chief buying influence, the most desirable feature, and the most disliked attribute—These then become standard values for the entire session.
2. The desirable features and the negative attributes are rated against the standard values by vote using cards numbered 1 through 10 (the number 7 is eliminated to discourage fence sitting).
3. Each panel member's vote is recorded along with the range.
4. When there is a large range, a second vote is taken following a discussion in which those participants with the extreme votes explain why they voted as they did.

Much information can be gleaned from the discussions that take place while the panel does its work. The company finds that it is advantageous to have a note-taker who is attuned to the underlying market, sales, product, and manufacturing issues (e.g., the product line manager or the business unit manager). Another advantage comes from having a senior manager present, which lends prestige to the occasion. When a senior manager is present, that person is never an inactive observer but rather participates by welcoming the visitors and setting the stage for the panel by stating the overall objectives of the session. If the senior manager remains for the entire session, a function such as note-taker or assistant to the facilitator is assigned. Under no circumstancese is a senior manager ever a member of the panel.

The Value Analysis

The focus panel is followed by a value analysis program, which directly uses the information obtained from the panel. The objective is not just to find a cheaper way to manufacture the existing product line but to give the customer the best value by investing in the product in direct proportion to what the customer needs. Experience shows that a straightforward cost-reduction program can achieve savings of up to 15%. Magnaflux has greater expectations; along with lower costs, the company wants a product reconfiguration that emphasizes those features the customer values most.

A typical value analysis team at Magnaflux has five members, each with a different perspective on the product to be analyzed: a design engineer, a manufacturing process representative, someone familiar with costs, a representive of the customer's viewpoint, and a rabble-rouser. Each value analysis team is given specific objectives; in this particular case, these were to reduce costs by 35%, reduce inventory to four turns a year, and reduce average lead time to four weeks. These objectives were weighted 60%, 25%, and 15% respectively to help resolve conflicts between actions planned as a result of the analysis.

The value analysis team works for several days through a structured process that produces a function diagram for the product. The team members are asked to use two-word, noun/verb descriptors to keep their thoughts precise and clear. The findings of the focus panel are mapped onto the function diagram, with the focus panel's vote entered alongside each of those functions that produce the features it identified. More than one function can contribute to a feature; in these cases, the vote for that feature is entered more than once.

The cost breakdowns for the product are also mapped onto the function diagram. The team must distribute the cost of each component over those functions involved in its manufacture; the cost is apportioned according to the extent that the feature is produced by that component. Accurate cost information is not always available in the detail that the team needs; in those cases, the team relies on the knowledge of its members and makes educated estimates.

After the target areas have been identified, a brainstorming session follows. Any and all ideas are encouraged. To avoid inhibiting the creative flow, no evaluation of ideas is permitted until the team has produced as many as possible in a period of about two hours. Teams need to be constantly reminded that statements beginning with, That won't work . . . or But . . . are really evaluative statements. Team members are encouraged to offer alternatives that eliminate the flaws they find in someone else's ideas rather than evaluate those ideas aloud. The brainstorming session ends with the team's reviewing the list, discarding the obvious chaff, and selecting a champion for each surviving proposal.

The group reconvenes after 10 days or two weeks, time enough for the champions to have made preliminary evaluations of their proposals. The team members report on potential savings and the level of effort required to implement each of the concepts. More brainstorming, refining, and eliminating of alternative approaches occurs during this session. A similar session is held after another two-week interval.

Supplier day is the next major event. All existing and potential suppliers are invited to hear the team present its ideas. The suppliers are told about the project's objective and are asked to comment on the ideas being presented and to offer their own. After the group session, each supplier speaks individually with each champion. The one-to-one exchange is important to identifying and evaluating all opportunities.

If there are many suppliers, multiple sessions may be required. These sessions can be more than enlightening, particularly if suppliers are encouraged to offer ideas that yield benefits to them; by participating and exchanging information, they can become more efficient and can offer a better price. One supplier said of a particular manufacturing process, "We never understood why you did it that way; it is so expensive and it upsets us. But we never brought it to your attention because you are a valued customer and we thought you must have your reasons." Another said, "You always ask us to provide a quote on each individual part of an assembly instead of the whole assembly. We need to be so careful because we never know which part we well get to make. This costs us more and, in turn, costs you more. If we quote costs for the entire assembly, we don't need to be as careful estimating each part because errors in estimations tend to average out."

the 80/20 rule worked for Magnaflux because it forced the company to concentrate on the business areas in which the company's leverage was greatest

The most important lesson Magnaflux has learned from sessions with suppliers is that to be manufactured efficiently, the product must be designed around the supplier's capabilities. This lesson is particularly important now that material costs far exceed the direct labor content of the products manufactured.

MEASURING PERFORMANCE

The first step in implementing the findings of the value analysis team is to make a formal presentation to management of possible courses of action. Recommendations are classified under three categories:

- Those that can be implemented immediately at no cost.
- Those that require some redesign effort or small investment and can be put in place in less than three months.
- Those requiring major capital investment or at least three months of implementation effort.

In the case under discussion, there were 12 ideas in the first category, 13 in the second, and 2 in the third, all relating to the specific product analyzed. When these ideas were extrapolated to the entire product line, they yielded a four-month payback on investment; when they were implemented, the company's inventory and lead-time goals were realized. To achieve these benefits, the entire product line had to be restructured and streamlined to maximize the use of parts common to two or more products.

This was a bold recommendation for a traditional company; it could not be implemented without full commitment from senior management. A formal project schedule was established, and funds and staff were made available. The business unit manager was appointed project manager to underscore the importance of the effort. Subsequent experience has shown that implementation projects falter without strong management attention.

At the time the product performance analysis was conducted, Magnaflux was also engaged in installing a qualified supplier program in the business unit as a means to combat the problems of late delivery and inconsistent quality. Magnaflux recognized that these two phenomena were not independent but interrelated. The problems arose because the suppliers were organizationally too distant from the assembly unit and the engineering unit where Magnaflux made and designed the product.

It became clear that the poor product performance issues brought up in the analysis were closely related to the business unit's organization. The unit was structured along classic, functional lines. Engineering, procurement, inventory control, and assembly were separate functions, each responsible for a large number of diverse product lines. As part of the implementation program, the company reorganized along product lines to allow for a closer working relationship between the product line production manager, the suppliers, and the other functions associated with or providing support to each line. An additional benefit of this restructuring was that Magnaflux finally knew the real total costs of designing, maintaining, and producing the product line, because no indirect costs had to be allocated to overhead functions.

Organizing around product lines eliminated such broad, functionally oriented operations as shop floor management, inventory control, purchasing, production management, and product engineering that had been supporting several product lines. The new product-line teams got closer to their suppliers, who then became involved directly in product design and accepted just-in-time (JIT) purchasing on an annual basis. Material costs were reduced substantially because the new designs reflected the suppliers' capabilities.

The restructuring produced smaller, focused product performance analysis teams that could communicate and implement change efficiently. The team leader (the product-line production manager) was no longer insulated from suppliers by a purchasing agent and got deliveries on time, and the supplier received immediate feedback on quality issues. The company was forced to change its control system too, and that task proved difficult because the importance of training went unrecognized. Many Magnaflux employees were the product of the company's longstanding functional structure. They had never had to worry about more than one small patch of turf, and the company now expected them to be involved in the total process of producing their product. Training in cost analysis, transaction entries to inventory control, group dynamics, and budgeting would have made the transition less painful.

Magnaflux has come to believe that implementation programs that follow product performance analysis exercises fail when they do not involve the whole organization. When a single functional group (e.g., engineering) is charged with implementation, it can run into resistance from other, less motivated departments or functions in the organization; consequently, the entire project will suffer or fail.

BASKING IN THE GLOW OF SUCCESS . . . AND MORE

As stated, despite many difficulties, at the end of 18 months the number of major products in the product line was reduced from 13 to four, the options offered for each grade of product were selected along more cost-effective, rational lines, and average lead time from customer order to delivery was reduced from 10 weeks to four. At the same time, the number of individuals assigned to the product line was reduced by 15%. (These people were reassigned; job elimination might have discouraged any future employee-driven performance improvement programs). The number of product parts was reduced by 50%, inventory by 35%, and direct costs by 25% in a period when the raw material cost of a major component increased by more than 100%.

data from front-line salespeople is informative and helpful, but in most cases that information is distorted by the most recent specific customer request

Since the company's initial product performance analysis program, the methodology has been applied successfully to a variety of settings. For example, it was used to analyze a whole business unit that was producing custom equipment. The unit was restructured after Magnaflux found that less than 10% of the effort applied to any product was devoted to defining the project scope in detail, particularly the feature about which the customer focus panel felt strongest. The company also found that its effort to define the project scope was spread over too many people, virtually ensuring poor communication.

The product performance analysis methodology can be used to produce significant changes in the performance of a business when it is applied to its core products. In its simplest terms, product performance analysis offers greater perceived value to the customer at more favorable margins and improved efficiencies for the producer. It can be a systematic way to evaluate the market and develop the organizational structure for meeting a market's needs efficiently. Measured results show that the analysis works because it focuses the business on what the customer appreciates most—value.

TECHNOLOGY TRANSFER

Technology transfer covers various activities, including the internal transfer of technology from the R&D or engineering department to the manufacturing department within a company based in one country. It also includes a multinational company's transfer of technology from a laboratory or operations in one country to a laboratory or operations in another country. Finally, it includes the transfer of technology from a research consortium supported by many companies to one of its members. The transfer of technology from a government laboratory to an industrial company, which is another important aspect of this topic, will be addressed in a subsequent publication.

In this section, Keith McKee, Thomas Turner, and Alan Frohman examine technology transfer from an R&D or engineering department to a manufacturing department. The differences in their treatment of the subject are that McKee focuses on the capabilities and needs of the manufacturing department, Turner concentrates on attitudes and procedures needed for technology transfer, and Frohman examines a new model for technology transfer that blurs the issue of transfer by emphasizing parallel efforts. James Key analyzes the transfer of technology from a research consortium to a company, and Tom Thiele looks at international technology transfer.

McKee emphasizes that the design of a product and its manufacturing process must be closely linked. He presents two problems that can occur when the design of a manufacturing process is neglected; he also shows how these problems can be avoided. In addition, McKee examines the particular problems of transferring manufacturing technology to the manufacturing plant. He identifies six factors that are critical in making such a transfer effective.

Turner examines attitudes typical of engineers and manufacturing managers that can create barriers between an engineering department and a manufacturing department. He then presents procedures to help break down these barriers, such as transfer teams, skunk works, producibility engineering, and concurrent engineering. In addition, Turner provides a method that enables managers to diagnose their own problems related to technology transfer.

Frohman looks at internal company technology transfer from quite a different point of view. Frohman states that the linear model of technology transfer—in which R&D hands off to engineering, which hands off to manufacturing, which hands off to marketing and sales—is outmoded. What is needed instead, Frohman suggests, is a new model in which the various functions work in parallel. Frohman then analyzes all the aspects of company operations that must change to ensure successful technology transfer and technological innovation in general. These factors include communication, rewards, and the employees' perceptions of job and career. Finally, Frohman outlines the steps required for implementing this new model of technology transfer.

Key addresses a relatively newer aspect of technology transfer: transfer of technology from a research consortium to a company. He outlines the several stages involved in such a technology transfer, starting from awareness by a company of the technology's value, through a company's efforts to harden a technology to meet its eventual application, to maintenance by a technology support group. Key points out that managers must be wary of such barriers as insufficient communication and wrong timing to succeed in technology transfer.

Thiele examines technology transfer from an international perspective, pointing out that subtle misperceptions can easily occur in international transfers of technology. Although both countries may appear to have similar market needs, there may be subtle differences. Technical standards that appear to be equivalent can have different effects. Other differences can arise, depending on the different uses of a product or a different method of accounting costs. Consequently, great care must be taken in transferring technology internationally to counter not only typical transfer problems but cultural differences.

GROWING PAINS

Transferring New Technology into the Manufacturing Plant

Manufacturing, typically considered in the past an isolated black-box function of a company's production, is becoming increasingly influential in the product development process. To reflect its changing role effectively, manufacturing must become more integrally involved in the early stages of product development. At the same time, companies must carefully examine their attitudes toward changing technology and their use of it.

Keith E. McKee

This chapter addresses both the manufacturing department's role in introducing new products and materials into production, and the process of introducing of new manufacturing technology into the plant. For the purposes of this chapter, technology refers to any scientific or engineering knowledge. Technology can be considered new in either of two situations—when it is being applied for the first time anywhere or when a particular company is first applying it (after other companies have already successfully applied it).

Technology need not be sophisticated or expensive to be of value. The technology involved in changing a fabricated metal case to plastic is simple. Yet this change in technology can make the difference between a product that sells and one that does not. The technology in a roller conveyor is relatively simple and inexpensive compared with the technology in a gantry robot, but in some cases it is the more effective approach.

A MANUFACTURING OVERVIEW

Manufacturing is a process by which material, labor, energy, and equipment are brought together to produce a product that has greater value than the sum of the inputs. This can be shown as a system, as indicated in Exhibit 1. The inputs are material, labor, energy, and capital (or equipment). The output is a product, but there is always some undesirable output—waste and scrap—that should not be forgotten. In addition, such external influences as government actions (e.g., those taken by the Occupational Safety and Health Administration or the Environmental Protection Agency), or natural occurrences (e.g., storms or floods), and the actions of competitors all have effects.

Exhibit 1 presents a broad view of the manufacturing system, based on definitions used by economists and the federal government. The manufacturing plant

KEITH E. McKEE is director of the Manufacturing Productivity Center at the Illinois Institute of Technology Research Institute, Chicago.

Exhibit 1
A Manufacturing System

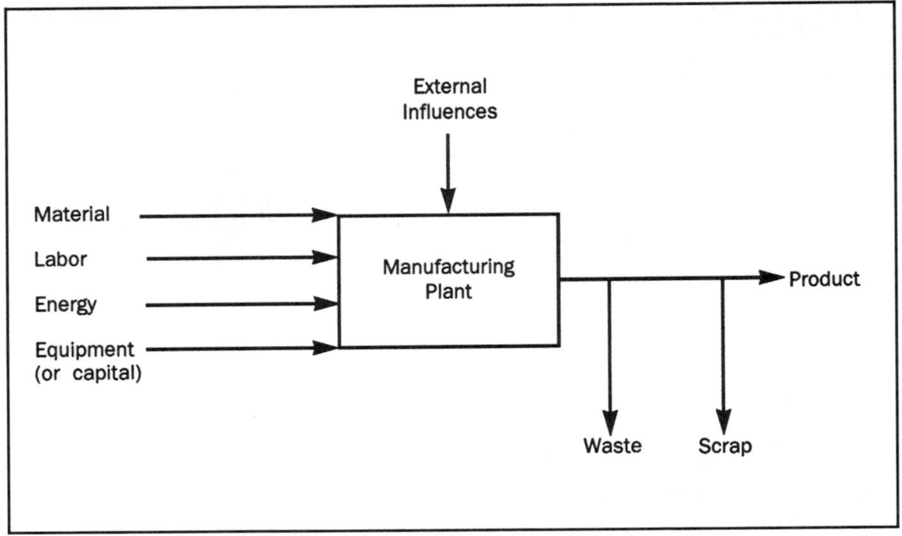

shown in this exhibit could be a steel mill, a machine shop, an oil refinery, an automotive plant, or a textile mill. Those inside manufacturing rarely have such a broad perspective. To them, the production of automobiles, airplanes, refrigerators, and electronics is manufacturing; manufacturing concerns the production of mechanical and electrical parts and assemblies.

This narrow perspective excludes many industries unnecessarily. Therefore, despite the fact that many of the examples in this chapter come from industries within the narrow definition of manufacturing, the chapter itself addresses manufacturing in the broadest sense.

Exhibit 2 presents a view of manufacturing that emphasizes input, output, and feedback. Quality control occurs after manufacturing and provides feedback; feedback also comes from customers. This exhibit illustrates that manufacturing is a closed system that responds to external feedback related to quality and to the quantities that are required.

In both Exhibits 1 and 2, the manufacturing operation is treated as a black box. In reality, this box contains a wide variety of activities, and within the plant there are usually many interrelated activities.

Over the past decades, many more activities have been introduced into manufacturing. Traditionally, manufacturing was based on one process (e.g., machining). A plant produced a limited set of products with a specific type of equipment (e.g., machine tools).

Exhibit 2
Manufacturing Input, Output, and Feedback

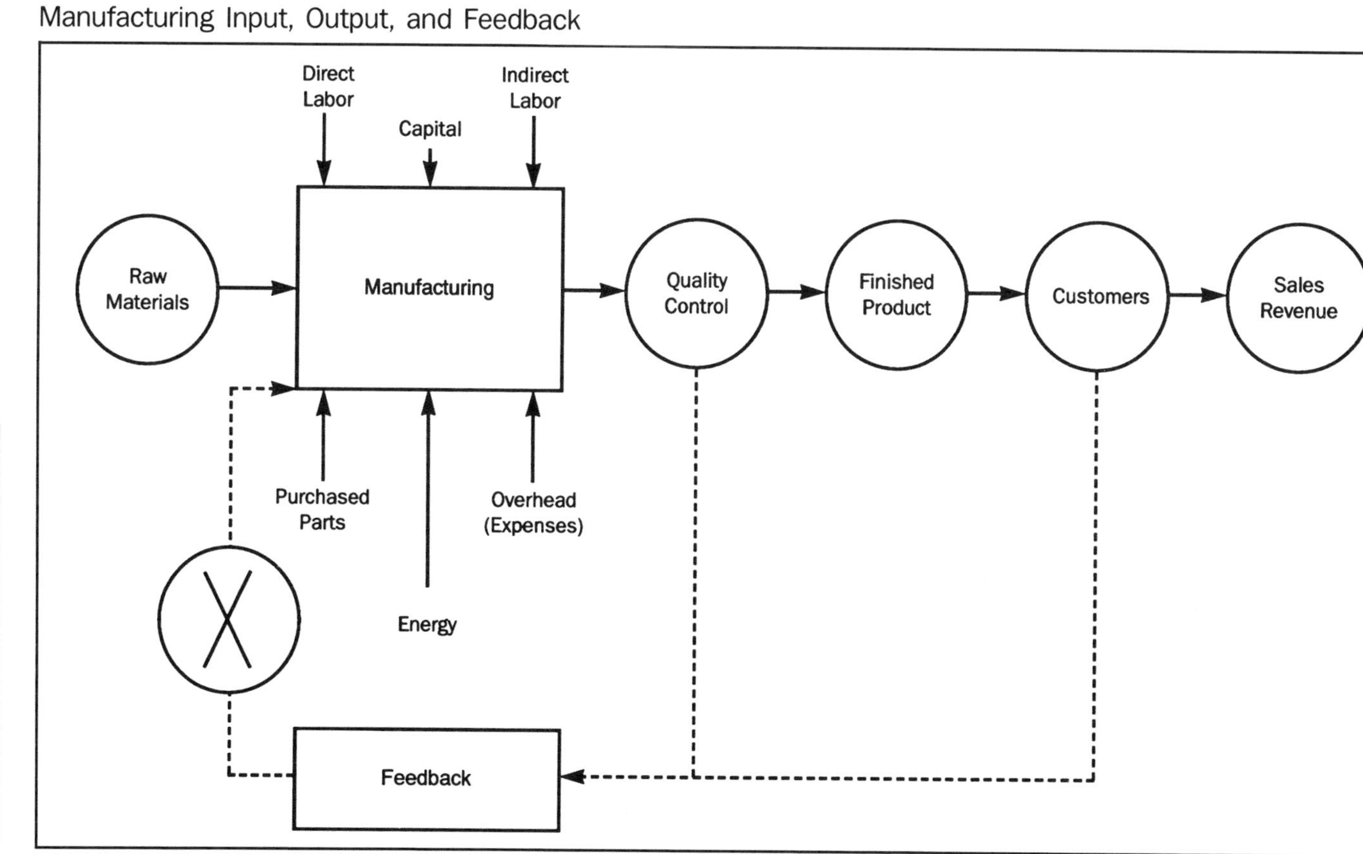

A More Comprehensive View of Manufacturing

Today, however, the typical modern plant consists of many processes. Exhibit 3 shows the processes involved in metal cutting or forming. Besides the principal processes, three other process levels are involved—preprocesses, secondary processes, and finishing.

Preprocesses are operations performed in-house or by vendors to prepare the material for the primary processes. Although these processes could be considered as going all the way back to the mines, the preprocesses in this example are all related to the product being made (i.e., forging, casting, powdering metals, and cleaning).

Secondary processes are performed on the products while the principal processes are going on. Examples are heat treating, welding, and soldering. Secondary processes may be applied by vendors at remote sites or may occur at a remote site within the plant. Whether applied externally or internally within the plant, secondary processes account for an ever-increasing portion of the production cycle.

Finishing refers to those operations performed after the machining. Finishing can include painting or plating, final inspection, assembly, and packaging. All of the operations that are or could be done at the same site as the machining can be included under finishing.

Principal processes are the machining operations themselves. Many such advanced manufacturing systems as numerical control machining centers and cells, and automated tool changers are designed for improving principal processes. Similar developments on tool materials, high speed machining, water jet and laser machin-

Exhibit 3
A More Comprehensive View of Manufacturing

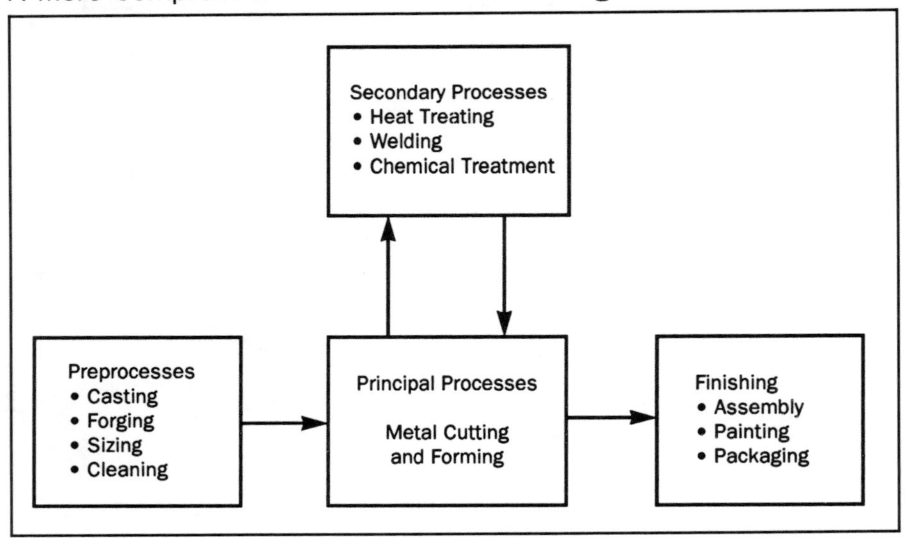

ing, machinery and tool sensors, and improved structures for machine tools are also directed at improving these processes.

to be competitive, manufacturing companies must be adaptable in their ability to use a range of previously-unavailable technologies

Within and connecting each of these processes are such manufacturing systems as material-handling devices (e.g., automatic guided vehicles and robots) and data control systems (e.g., manufacturing resource planning). Improvements are continually being made to these supporting systems, and more improvements are anticipated. Nevertheless, the optimum approach is to minimize the amount of material handling and data control needed—for example, by integrating preprocesses, secondary processes, and finishing operations with the principal processes.

In Exhibit 3 and the preceding discussions, machining operations are considered primary processes. However, one plant's primary processes can be the next preliminary processes. For example, in an operation in which assembly is the primary process, machining could be a preprocess; similarly, in a welding shop, machining is likely to be a secondary process. Whatever the primary process, there are distinct advantages in integrating as many operations as possible.

Manufacturing can no longer be equated with machine tools. In fact, even in the industries in which machine tools perform the principal processes, machine tools represent but a small part of the capital invested and even a smaller portion of the technology involved.

To be competitive, manufacturing companies must be adaptable in their ability to use a range of technologies that were unheard of until recently. CAD, robots, AGV, and automated inspection have all emerged in the past few years; manufacturing materials are also changing. Computers are important in manufacturing, but so are such changes in process technologies as new tools, net shape manufacturing, energy beams, chemical machining, and waterjet material removal.

The Factors Changing Manufacturing

Manufacturing is now in rapid flux. Until recently, the main mission of a manufacturing department was to maintain stable operations. New plants were built simply as modern versions of previous plants. This manufacturing model, however, is no longer appropriate because the following factors are forcing changes in manufacturing:

- Quality—Customers, particularly other companies that are customers, must have parts that are 100% usable in order for them to implement automated assembly and just-in-time inventory and production.
- Flexibility—To be competitive, a supplier must accept product design changes to produce varied quantities and meet flexible delivery times.
- Cycle time—The time from the release of engineering requirements to the delivery of the product must be reduced greatly.
- Inventory—The previously mentioned factors have significantly affected the han-

dling of inventories. Raw material, work-in-process and finished goods inventory must all be reduced.

- Costs—Costs now must be contained and reduced for all elements of manufacturing.
- Competition—Every company, however small and specialized, has or will have competition. Safe niches are no longer available.
- Internationalization—Companies must continuously evaluate the potential for international cooperation and the threat of international competition.

All of these factors require companies to change their approaches to manufacturing. In the past, the policy of company managers in stable companies was to limit the amount of new technology used in manufacturing and to require extensive justification before introducing more. However, technology has now become a critical factor in making a company competitive, and the analytical techniques traditionally used to justify technology are no longer appropriate.

THE EFFECT OF R&D AND DESIGN ON MANUFACTURING

All phases in technology development—research, development, and design (RD&D)—influence manufacturing. Over the past few years there has been an emphasis on improving the manufacturability of product designs. Improvements here can significantly increase the producibility of a product, yet design for manufacturability (DFM) involves only the final phase of the RD&D cycle. Decisions made in R&D can limit the options available during the later design phase. Nonetheless, advantages gained through DFM can be significant.

For example, the Illinois Institute of Technology Research Institute (IITRI) recently conducted a DFM project focusing on a disposable medical product. This product was made from seven parts that were manually assembled; when the product was first designed, the company was producing it in small quantities. The objectives of the IITRI project were to help the company redesign the product and then to develop automatic equipment for assembling the product.

The first phase of this project involved improving the design for manufacturing. The actual design included only three components, which had originally been designed so that they could be assembled easily through automation. When the project was half over, IITRI proudly reported the results and recommended the development of the automated equipment. The customer was pleased with the results but did not immediately authorize the next phase.

A few weeks later the customer announced he had decided to forgo the automated equipment. With the DFM results, he really did not require automation, because manual assembly was now so much easier. Thus, by improving the design for manufacturing, the customer had achieved cost targets that originally had been established for automated assembly.

DFM refers primarily to the assembly of components, but it can also include reviews of manufacturing procedures for making parts and assemblies. For example, a DFM project can involve using a molded plastic part instead of a metal assembly or using a single chip to replace an electronic assembly. With regard to continuous processes, DFM specifies the design processes in such a way that they are insensitive to parametric variations.

typically, it is impossible to select parameters to encompass all anticipated variations throughout production; thus, designs should include process controls that work despite unplanned variations

For example, a company whose manufacturing operations included the extrusion of foamed plastic insulation was encountering frequent defects in its product when continuous processes were operating at the specified design conditions. Using the Taguchi approach, IITRI reviewed the design parameters and redefined them so that almost all defects in the product were eliminated. To do this, IITRI selected the design parameters that would allow the system to tolerate the widest possible range of parametric variations. The goal of a DFM effort for continual processes is to select design parameters that maximize the robustness of a production system.

It is not typically possible to select design parameters that produce a system so robust that all anticipated variations can be tolerated. Consequently, the design should also include process controls that work despite unplanned variations. In addition, as variations occur in one parameter, other parameters should be able to be varied to maintain the integrity of the product.

As an illustration, a pharmaceutical company operated the same process at three plants, where yields ranged from 35% to 70%. Because all three operations were working with the same design parameters, the differences in yield presumably resulted from differences in the knowledge and skill of the operators. As a result of an IITRI study involving all three plants, a procedure was developed for optimizing process control throughout the entire company. When the company implemented this new procedure at the plants, the yields at all three plants exceeded 80%.

Interaction of Product and Process Design

Products and processes must be designed to maximize all aspects of the manufacturing operation, including costs, quality, cycle time, and flexibility. In particular, the design of products and process can significantly affect process control and the manufacturing process.

Exhibit 4 shows how decisions about different manufacturing factors have markedly different amounts of influence over the effectiveness of a manufacturing process. Even though materials constitute one-half of the total product cost, decisions about materials have only 20% of the total influence on the effectiveness of a manufacturing process. On the other hand, cost of design is only 5% of product cost, but decisions about design have 70% of the influence on the effectiveness of a manufacturing process. Thus, focusing attention on manufacturing during the design process can result in cost savings.

Just how can a product's design affect a manufacturing process and how can a manufacturing process be improved through redesign? Two case studies illustrate some answers.

Exhibit 4
Comparison of Four Factors' Influence on Product Cost and Effectiveness of the Manufacturing Process

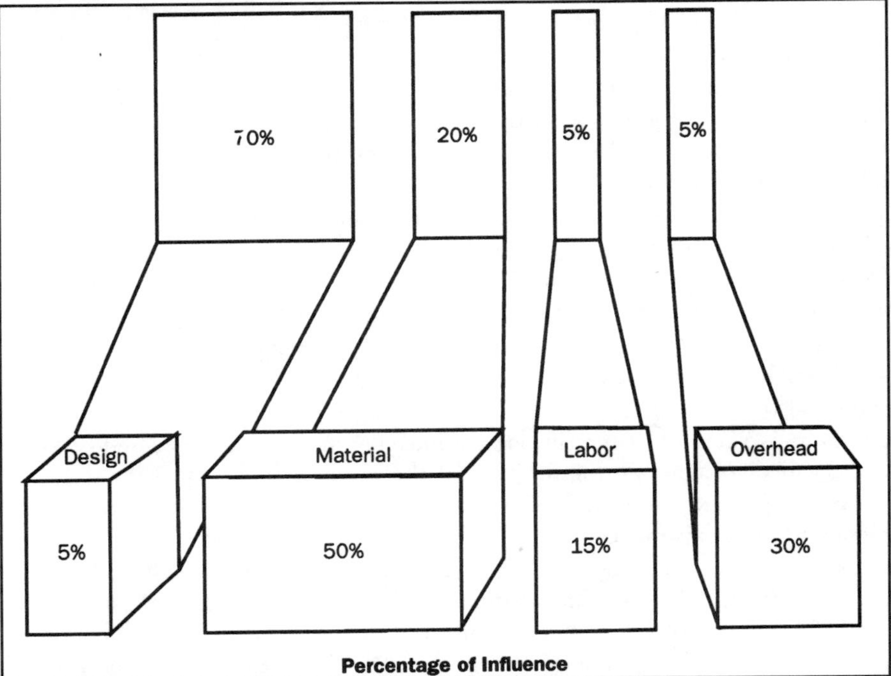

Two Case Studies

The first case concerns an electromechanical product that was designed by a R&D-oriented staff. The company anticipated that it would produce 15,000 of these products over a 5-year period. The electronics components in the product were designed for manual assembly with wave soldering; this design was similar to the one used in developing the company's other products, and the company produced less than 1,000 of these products annually.

The company manufactured the electronic components of the existing products internally, even though it could have had them assembled by companies that specialized in this process. The company also made the circuit boards of these products in-house to maintain control and facilitate repair of boards that needed rework after testing burn-in (more than 80% of them did).

Surprisingly, the company produced the circuit boards of its existing products with almost no process controls, and it did nothing to improve controls when it produced the new product. Interestingly, however, the company did conduct formal design reviews for the new product. The manufacturing department accepted the proposed production approach, presumably because the department was already familiar with it. At least in this instance, therefore, a design and manufacturing review did nothing to improve process controls. The company did not think seriously about process controls for the new or existing products.

too often, personnel choose new materials without considering their decision's potential impact, only to find too late that the material is unavailable

The company designed mechanical components of the new product to handle web materials that could vary in width. The designers of these components had no relevant experience in this area but did have an eminently successful, top-of-the-line design to use as a model. With it, they designed a sophisticated mechanism for relatively simple web handling. To compound the problem, they specified tolerances that were as tight as they could justify. One of the company engineers had found a vendor who said that he could satisfy these tolerances for each of the purchased parts.

Unquestionably, the mechanical system would work, but it was clearly overdesigned in terms of the design approach and the tolerances allowed. The mechanical system cost approximately double what it should. Equally serious, problems in producing this mechanical system were ensured because of the tight tolerances; even the vendors who had claimed that they could produce the parts had problems. Consequently, the vendors' parts were continually out of tolerance, which affected in-house production and ensured problems in assembly. Thus, poor product design can derail manufacturing.

A second case concerns a company that had been producing a thermal sensor for five years with a very poor yield (only 40% of the yield was satisfactory). The sensor was built in accordance with specifications and manufacturing process sheets. After the sensors were cured, they were tested; the sensors that did not pass were scrapped. The company sold 400 sensors per week. To have this many good sensors ready, the company produced 1,000 sensors each week in batches of 100.

When IITRI evaluated sensor tests from the previous two years, it found that the yields had ranged from approximately 20% to 70%. After further examination, IITRI found that these yields were an improvement over the yields of the initial production runs, when the company had considered yields of 20% to 25% to be satisfactory.

This product was a financial disaster for the company. The inconsistent yields generated numerous customer problems. The manufacturing department had worked from the beginning to improve the process controls but had not been successful. Five years after initial production, two manufacturing engineers were assigned full time to review the materials and the manufacturing procedures being used.

The manufacturing department had claimed from the beginning that the design of the product was the root of the problem, but its claim had not been taken seriously. RD&D argued that because the product could be successfully produced at least part of the time, the problem was clearly in the manufacturing process and quality control.

The design consisted of a flat and essentially rigid metal cap that sealed against a machined ring of the same diameter. This metal-to-metal seal was the basis for the design. Presumably, had both metal surfaces been free of nicks, dents, or dirt, this seal would have been adequate. Tight dimensions had been specified for these parts, and inspection was done with considerable care. When the design was reevaluated, two options considered were developing a clean room for the assembly and establishing even tighter tolerances for the mating parts.

When IITRI reevaluated the design, questions arose about design requirements and the design itself. Several alternatives were considered and evaluated. The option selected was to add a sealing material to the cap that would force the metal ring into the seal. The tolerances on the metal ring were reduced, and the edge of the ring was rounded to avoid cutting the seal. The company introduced this design modification within two months, after which the yield rose to 100%.

This case illustrates the intimate interaction between product design and production. Researchers and designers largely establish the parameters in which manufacturing must operate. Manufacturing processes, however well executed, cannot overcome such a deficiency in design as the one just described. The fact that a product can be made with a given manufacturing process does not prove that the process is optimal.

Designers, developers, and researchers all can introduce production problems. Indeed, technologists who in any way influence product decisions affect production. Much of the literature on product and process design interaction suggests that only the designers are involved; this is far from true, because designers are frequently limited by decisions made earlier in the R&D cycle.

Introduction of New Materials

Technologists who develop new materials are usually highly aware of problems that can occur. They are familiar with the many new materials that are developed, characterized, and publicized—but that never become commercially available because the market for them is not sufficiently large. These new materials would become available if there were a sufficiently large market for them, but companies considering using these new materials will do so only if they become comercially available.

Establishing the future availability of new materials can be difficult. Because materials suppliers try to develop markets for their new materials, they tend to be optimistic about future availability and costs. The more research oriented the technologists, the more likely they are to accept these claims; such technologists would naturally like to be the first to use a new material. Seat-of-the-pants designers, on the other hand, are interested in using only materials that have been commercially available for some time and have been successfully used in other products.

The optimal approach, which lies somewhere between these two extremes, is a difficult path to follow. Many products have never appeared or have disappeared because the materials on which they were based did not become available or were withdrawn because there was not a sufficient market for them. On the other hand, using only those materials that are commercially available and have previously been used greatly limits the potential performance or quality that a designer can design into a new product.

Consider the case of the National Aerospace Plane (NASP), which had been proposed as an alternative to the space shuttle. Because NASP is launched from a runway rather than a rocket launch pad, it could have made space travel more practical. Production of NASP, however, depended on the availability of a high-performance material under development. Because this development progressed much less rapidly than had been expected, the NASP concept will not now be pursued. Several major aerospace firms have lost considerable money because of this failure, but in the world of R&D this loss can be considered part of the game.

A similar failure at an individual firm can result in financial disaster. Consequently, companies should generally avoid the risk of using new materials. If a new material is used, material development must be included as a critical path on the PERT chart. Too often, RD&D personnel choose to use new materials without considering the potential impact of this decision on production. Then, when the new product is to be manufactured, the material is either unavailable or available only at a premium price.

In a discussion of the introduction of new materials into manufacturing, the term *materials* can refer to anything that is purchased. For a plant that assembles cars, materials include engines, seats, brakes, or windshields. One company's products, therefore, are the next company's materials. Occasionally a company can be misled with regard to the performance and availability of new materials. Sensors might not work as promised, or a computer program might not run properly. Although this can often happen, the most common problem with materials is that they can be phased out. Castings, motors, electronics, and fabrics that at one time were easily available, frequently turn out to be no longer produced.

Organization of R&D and Manufacturing

Many companies are still organized in a traditional organizational structure, which is illustrated in Exhibit 5. In this structure, the various groups are relatively independent of each other, and interactions between the groups are quite formal.

Exhibit 5
The Traditional Organizational Structure

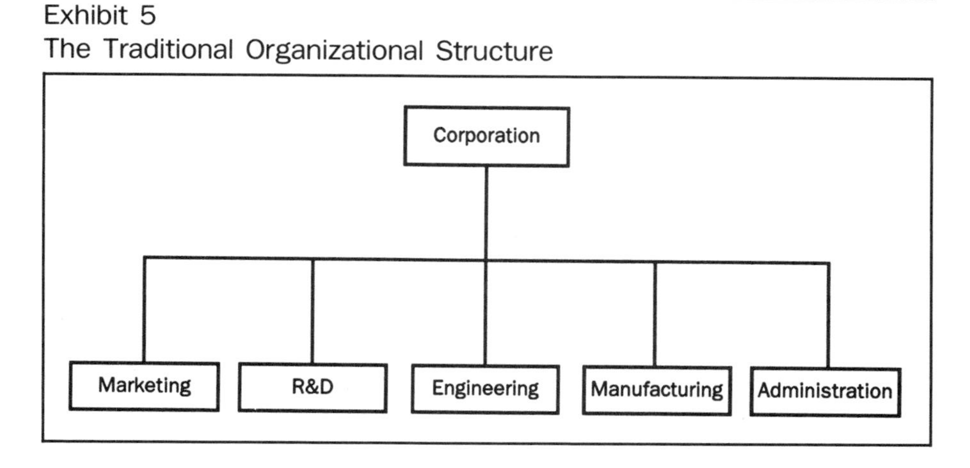

R&D and marketing interact, but they do so almost without regard for manufacturing. The rationale behind this organizational structure assumes that R&D operates more effectively if it is disconnected from day-to-day company operations. In many instances, the R&D department is located far from manufacturing, on the presumption that an R&D department is more effective if it is located in a remote facility.

Occasionally R&D solves one of manufacturing's technical problems, but as a favor, not an assignment. When R&D releases a new product or process, it passes it off to engineering or manufacturing. In most instances, R&D develops the concept for a new product, but engineering develops the product and provides manufacturing any necessary information about it.

When the engineering group in one such company was given a new product to design for manufacturing, it would often find that the R&D department had already built and tested a prototype of the product and that the R&D and marketing departments had already decided what the new product should be like. Often the R&D department released the prototype to engineering, asking it to make the new product available for sale within six months. Sometimes engineering could not reproduce the materials or components used in the prototype because R&D had used materials and components that were not comercially available. The engineering group would then do the best that it could to design an equivalent product using available materials and components. Sometimes the equivalent product did not work because there were too many variations between the R&D prototype and the new product. Then the R&D, engineering, and manufacturing departments would have to work together to design a new product that could be produced. Occasionally, this would take as many as two years to accomplish.

Many alternatives to the traditional organizational structure facilitate interactions among company groups to a greater extent. For example, engineering and manufacturing may report to one person, or the engineering or the manufacturing department may assign someone to work in R&D. Within R&D, a special group can handle manufacturing technology or design for manufacturability.

MANUFACTURING TECHNOLOGY

Manufacturing technology refers to all technologies relating to manufacturing, not just the application of computers to manufacturing operations. Although different industries use different manufacturing technologies, they share many technologies relating to materials handling, inventory, quality control, and plant layout. In fact, more manufacturing technologies are common to all industries than specific to just one.

A recent study by the US Bureau of the Census examined how companies used any of 17 different manufacturing technologies.[1] (Exhibit 6 lists these technologies.) The industries polled were manufacturers of fabricated metal products, industrial machinery and equipment, electronic and electric equipment, transportation equipment, and instruments and related products. Of the companies from these industries, 76% use one or more of these 17 manufacturing technologies; 23%

Exhibit 6
Manufacturing Technologies

Computer-aided design (CAD) or Computer-aided engineering (CAE).
CAD output used to control manufacturing machines (CAD/CAM).
Digital representation of CAD output used in procurement activities.
Flexible manufacturing cells or systems.
Numerically-controlled or computer-numerically-controlled (NC/CNC) machines.
Materials working lasers.
Pick and place robots.
Other robots.
Automatic storage and retrieval systems (AS/RS).
Automatic guided vehicles.
Automated test performed on incoming or in-process materials.
Automated test performed on final product.
Local area network (LAN) for technical data.
Local area network (LAN) for factory use.
Intercompany computer network linking plant to subcontractors, suppliers, or customers.
Programmable controller.
Computers used for control on the factory floor.

use five or more. Most manufacturing technologies are appropriate for a wide range of industrial companies.

Organization of Manufacturing Technology

An enormous variety of organizational structures are used to help companies improve the introduction of manufacturing technology. For example, in some companies, manufacturing technology groups that introduce new technology into manufacturing are located in R&D, engineering, or manufacturing departments. In other companies, manufacturing engineering and design engineering are combined. Several companies have search and development groups that search the world for product and process technologies that could benefit the company.

No organizational structure best facilitates the introduction of manufacturing technology. There is even one company in which a manufacturing technology group is successfully operating as part of the marketing department, on the assumption that new manufacturing processes are associated with new products.

Although organizational structure is not critical to the introduction of manufacturing technology, the following other factors are:

- Capable senior talent—A manufacturing technology group must be staffed with people who are experts in the technologies involved so that they are accepted by vendors and the other members of the company.

- Critical mass—A manufacturing technology group must be large enough that the group members are available to help. Regardless of how capable a group is, it will not be effective if it cannot respond to pressing needs.
- Freedom of action—A manufacturing technology group must be funded so that it can evaluate and develop technologies before they are required. The funds for applying these technologies can be gained from the users.
- Company visibility—A manufacturing technology group must have visibility and be accepted throughout the company. Sometimes this occurs because the leader of manufacturing technology is known and accepted throughout the company. In most instances, a champion within top management must give a manufacturing technology group adequate visibility.
- Facilities and equipment—A manufacturing technology group does not need to have a major laboratory, but it does need space in which to set up equipment. When such a group is introducing new manufacturing equipment, initial setups, if possible, should be situated apart from the production site.
- A broad knowledge of manufacturing—Manufacturing involves a broad range of technologies. Consequently, a manufacturing technology group must continuously learn about other technologies.

One notably effective manufacturing technology group operated as part of the R&D department of a $1 billion company. The group had a staff of 25 and a leader with more than 25 years of experience in manufacturing, engineering, and R&D. The staff was recruited from manufacturing, R&D, and organizations outside of the company (notably vendors of automation equipment). This group started with a three-pronged approach. First, they established a rapport with the operating units to identify their needs. Second, they visited companies, universities, and other organizations capable of revealing the state of the art in various technologies. Third, they set up facilities for developing a computer-aided design and manufacturing system (CAD/CAM). Within two years, this group had designed and developed an effective, state-of-the-art CAD/CAM system and delivered it to the operating units. After the CAD/CAM system was installed, the operating units could manufacture products within 10 minutes of entering data into the system.

Another effective manufacturing search and development group has existed for more than 10 years in a European-based company. This group developed a directory of all the manufacturing technologies that its company needed, and three staff members regularly tour the world looking for new technologies that can help their company.

Much of the success of Japanese companies has been attributed to their searching worldwide for new technologies. Although a few US companies have manufacturing search and development groups, this approach has not been accepted by most US companies.

AUDITS OF MANUFACTURING PLANTS

One of the major challenges facing those who are interested in introducing new manufacturing technology is to identify areas within the plant in which manufacturing technology could be used effectively. Each of the parties in this process has its own perspective. The manufacturing plant staff usually is interested in modern versions of what it already has (e.g., equipment that is faster or has more bells and whistles). Technologists are usually interested in advanced technology, particularly in their area of specialization. Consultants and equipment vendors almost invariably promote the approaches with which they are most familiar. Financial people emphasize return on investment. They usually pay little attention to the long-term impact of technological improvements and focus instead on direct cost reductions.

often, the R&D department released a prototype to engineering, expecting to have the item available for sale within six months

Even in well-organized companies, opportunities for using manufacturing technology are often ill-defined. Management must consider the entire operations of a plant and other issues (e.g., human resources, quality, and strategic plans) in determining whether a manufacturing technology will be beneficial.

At IITRI, we have found that an effective approach for evaluating opportunities for introducing manufacturing technology is to audit the manufacturing plant (see inset). Since 1976, the Manufacturing Productivity Center (MPC) at IITRI has conducted many audits each year. Depending on the company and the focus of a given audit, some facets of the manufacturing operation might get only casual attention. However, assessment of quality strategies, products and processes, materials management, management accounting, and information management is vital for evaluating the status of technology in a plant and for identifying opportunities for introducing new manufacturing technology.

Notes

1. US Department of Commerce, Bureau of the Census, "Manufacturing Technology 1988," (Washington DC).

AUDIT CRITERIA

Each plant audit that MPC conducts is different. The type of audit depends on the nature of the plant and the experiences of the company. In each audit, several areas must be considered methodically.

Quality Strategies

These strategies include the following:

- The company's effective use of:
  - — Labor.
  - — Capital.
  - — Energy.

(con't.)

— Materials
— Appropriate technology to increase quality.

- Management for quality and feedback of information:
  — Responsibility for quality.
  — Quality standards.
- Design verification:
  — Function effectiveness.
  — Reliability.
  — Manufacturability.
  — Testability.
- Manufacturing for quality:
  — Process capability.
  — Statistical process control.
  — Traceability.
  — Diagnostic guidance.
  — Performance prediction.
- Quality Contribution:
  — Inspection and audits.
  — Process control.
  — Design of experiments.
  — Measurements.
- Vendor capability or procurement quality.
- Distribution:
  — Quality perception.
  — Warranty.
  — Repair or replacement.
  — Field service.

Products and Processes

A capable audit should include a comprehensive look at the factors affecting products and processes, including:

- Technology's effect on profits and the workforce.
- Barricades to using technology.
- The extent of manufacturing automation.
- The status of preventive maintenance.
- Machine loading and unloading.
- Machine tools changes.
- How parts are held for machining.
- Machine control processes.
- Shop data collection methods.
- Parts loading and unloading.
- Assembly operations.
- Painting or finishing.

- Feedback controls.
- Parts inspection.
- Monitoring layout of departments.
- Work in process.
- Interdepartmental movement of parts.
- Input and output controls.
- Bill of materials generation.
- Use of CAD/CAM, CIM, FMS, robotics, and group technology.

Materials Management

The audit of materials management must include a thorough assessment of:

- Inventory cost and turnover.
- Systems for managing inventory.
- Order processing.
- Transportation.
- Warehousing.
- Sequence of operations.
- Handling by direct labor.
- Travel distances for materials, equipment, and personnel.
- Opportunities for group technology, product, and process layout.
- Uses of storage.
- Expansion capability.
- Schedule of arrival of out-bound and in-bound carriers.
- The physical environment.

Management Accounting

Areas to be assessed in an audit of management accounting include:

- Management accounting information.
  - Does the information exist?
  - Is it well thought out?
  - Is it used on the shop floor?
  - How well is it used for pricing and for make-or-buy decisions?
- Validity of existing information (e.g., if profitability information is available, how good is it?).
- General questions:
  - Is a standard cost system used?
  - Is it adequate?
  - How are standard costs developed?
  - Does the cost build-up process make sense?
  - What goes into it?
  - At what level are actual costs collected (e.g., factory or cost center)?
  - How is cost information reported?
- Analysis of performance data.
- Budget and consequences of budget performance.

(con't.)

Information Management

Several aspects of the information management system should be scrutinized:

- The adequacy of existing information systems.
- The necessity of reports.
- Speed in obtaining needed information.
- The strategic use of information.
- The cost effectiveness of information systems.

Labor Characteristics and Training

The company's labor environment should be assessed to productively characterize the relationships among workers and management. Important aspects of this environment include:

- The labor-management climate: cooperative or adversarial.
- The degree of idea sharing between hourly workforce and supervisors.
- Employees' perception of management and employment security.
- Employees' feeling of general job satisfaction and morale.
- Employees' understanding of need for quality and productivity improvement.
- The workforce's readiness for work redesign.
- Definition of skill and knowledge needs of company-allocated budget and training cost.
- Adequacy of worker training for current and future work.
- Adequacy of supervisors' training.
- New employee orientation.
- Career development in management, technical, staff, and hourly personnel.

Reward and Incentive Systems

Company methods for encouraging their workers are of importance in the audit. Subjects that should be discussed are:

- Financial and nonfinancial rewards.
- Employee involvement and recognition systems.
- Future plans for further improvement.
- Motivational value of the existing reward and incentive system.
- Opportunity for promotion.

Hazard Control

The workers' exposure to workplace hazards should also be considered in the assessment. Subjects to be addressed include:

- Accident cost analysis.
- Ergonomic considerations of materials handling and repetitive motion.
- Tool, equipment, and workplace design.
- Environmental stressors.

Product Liability

Another area that should be addressed in the audit is product liability. The integrity of the product should be assessed by means of the following criteria:

- Company policy for product safety.
- Safety in product development, in all phases: conceptual, design, manufacturing, transportation, marketing, use, and disposal.
- The degree to which employees are prepared to testify in litigation cases.
- Protection against discovery.

Productivity measurement

Finally, the company's methods of measuring its own productivity should be analyzed in depth, including:

- Factors considered in measuring productivity.
- Measurement data to produce information for:
  - — Labor tracking: time and attendance.
  - — Machine operation monitoring.
  - — Material tracking.
  - — Plant maintenance.
  - — Quality monitoring.
  - — Facilities monitoring.
  - — Job assignment.
  - — Stores control.
  - — Energy control.

BREAK DOWN THE WALLS

Improving the Engineering-Manufacturing Interface

Successful organizations are finding ways to improve the transfer of technology from R&D, product design, and engineering to manufacturing. Such practices as concurrent engineering offer great promise for companies trying to break down the walls that have traditionally separated organizational functions.

Thomas B. Turner

Organizational units, for various reasons, tend to be segregated, or compartmentalized. Although compartmentalization may help each unit perform more efficiently in some ways, it presents problems with the transfer of technology, especially from R&D, product design, and engineering to manufacturing. This lack of integration between product design engineering and manufacturing is ultimately detrimental to new products, but several forces hold the situation in place, in particular, manufacturing managers' traditional conservatism toward new technology.

Fortunately, many large firms have begun working to improve the engineering-manufacturing relationship. These firms include Lockheed, in its historic Skunk Works operation, and General Electric (GE), with its product management method developed for aircraft engine manufacturing. The trend toward improved relations is also evidenced in the recent interest in concurrent engineering.

This chapter discusses problems involved in transferring technology from R&D and product design and engineering to manufacturing and suggests ways of improving the process of technology transfer. The observations presented in this chapter are based on the author's working and consulting experience in firms that manufacture such discrete products as refrigerators, washing machines, electronic equipment (e.g., large radio transmitters), and heavy industrial equipment (e.g., large fans and blowers).

ORGANIZATIONAL COMPARTMENTALIZATION

Each organization, whether a corporation or a unit in a corporation, has a history of successes and failures in its markets. On the basis of these experiences, the leaders of an organization establish what they consider to be the best way of doing things. To encourage the use of such methods and systems, they usually have established a system of informal and formal rewards. In general, employees learn how to advance in such organizations by following the methods of operation and the reward systems established by senior management.

THOMAS B. TURNER, CMC, retired from the Cresap division of Towers Perrin with more than 45 years of experience in industry and management consulting. At the time of his retirement, he was director of the Department of Defense Manufacturing Technology Information Analysis Center.

The hierarchical nature of organizations often results in compartmentalization of functions. This compartmentalization segregates product development from manufacturing and segregates cost accounting from manufacturing as well as from other operations. Ultimately, this compartmentalization is detrimental to the process of technology transfer, and usually the larger and more complex the manufacturing operations and processes, the greater the gulf between the functional elements of the organization.

A Separate Peace

Compartmentalization of functional units often discourages teamwork between the units because it might be construed as interference. During the 1940s, the author was a methods engineer in a large appliance manufacturing organization. A tooling engineer in the same group was designing a large, complex stamping die. This 18-stage progressive die would be used to manufacture a small part for an electrical switch. The part was a hollow, light-tight box, about as big as an adult's little finger, to be made of very thin steel in one pass through the die. When in production, three such dies would be required, each to be used in a coil-fed stamping press and scrap-cutter.

The annual production requirement was for more than 36 million pieces; the schedule was firm, because the part went in a larger appliance: an electric range. The tooling engineer was well qualified, having been a tool and die apprentice in Europe before World War II, but his design and drafting efforts were not going well and he was obviously frustrated.

The tooling engineer started to complain about the work of the design engineers and their product design. When it was suggested that he go back to the design engineers and suggest design improvements that would make the tool design easier, he exclaimed, "I don't tell the engineers how to design products! And the engineers had better not tell me how to manufacture them!" This is an excellent example of the organizational walls that often develop between departments.

Engineers and technologists learn their on-the-job-skills through practical apprenticeship and seek to demonstrate their ability to handle responsibility within a given function (e.g., product design). Their need to be able to market specific skills if they leave the organization influences career choices and limits their opportunity to live and work on the other side of the wall. The lifetime employment practices of the Japanese might afford an advantage here: If engineers know they will never have to work for another employer, they might be more willing to accept assignments in other functional groups on a project-by-project basis.

In the U.S., product design engineers have traditionally concentrated on developing skills, knowledge, and related experience in:

- Materials, processes, and feasibilities, drawing on experimentation and research for innovation.
- Design feasibilities.
- Product performance, life, and durability.
- Cost analysis and competitive analysis.

- Market demands and needs.

Manufacturing personnel generally have different work experience, if not a different technical education, with emphasis on:

the larger and more complex the manufacturing operations and processes, the greater the gulf between the functional elements of the organization

- Industrial engineering, in particular the analysis and measurement of opportunities for improvement in shop operations.
- Tool design and manufacturing engineering (e.g., development and specification of such processes as machining or design of proprietary manufacturing equipment).
- Material requirements planning and inventory control.
- Shop operations.

Technology is breaking down some of the barriers between product engineering and manufacturing, especially in organizations that use CAD/CAM systems. If a part is designed on a computer system and the design transmitted to manufacturing in the form of computer data, it is likely that engineering and manufacturing have previously agreed on acceptable and feasible tolerances and available manufacturing methods.

AUTOMATION COMPLEXITY AND PRODUCT LIFE CYCLES

In some industries, if a part or product is well designed and well accepted, it will be used for many years. In a number of shops, parts or assemblies are still in manufacture though the drawings are more than 50 years old. On the other hand, some products have a short life cycle.

In general, different organizations have different traditions and processes that are a result of the products and technologies involved. Often, the level of technological complexity and the desired life cycle of a product determine the level of integration of the product design and manufacturing functions. For example, at one end of the spectrum is the manufacture of incandescent electric light bulbs. Light-bulb manufacturing is extremely sophisticated and highly automated and has been developed and evaluated over many years.

The common household light bulb does not outwardly appear to have changed much during the last 40 years, except when manufacturers dropped the copper base in favor of an aluminum base. The complexity arises in the coils of tungsten filament wire that are automatically picked up and welded into place on the support wires without human participation. Automatic evacuation and sealing are also sophisticated processes developed many years ago. Few product design changes can be made without careful consultation between the product engineering and manufacturing functions because of the highly automated production processes, which use machinery designed and built solely for that purpose.

Another example of product design that is well integrated with manufacturing is found in the manufacture of ranges and refrigerators, in which the body can progress from the coiled steel to a stamped, folded, and welded body without a human hand touching it. In these industries, engineering-manufacturing relationships have been carefully developed over a period of years to achieve the integration that a high level of automation requires. And the corollary, of course, is that the process cannot be automated without the careful integration of product design into the manufacturing process.

At the other end of the spectrum is the job shop where the product engineer says, "I'm going to design this product so that anyone can manufacture it." Having completed the design work in isolation, the product designer then (figuratively) tosses the drawings over the wall to the manufacturing manager, who says, "I can make anything within reason. If I don't have the process in-house, I will find a subcontractor who can do it for me."

Somewhere in between is the typical manufacturing organization, which is product oriented but may be under substantial competitive pressure. Many manufacturing managers feel they have few luxuries, such as adequate time and budgetary allocation for equipment and staff, because most firms are under great pressure to improve margins. Even such giants as IBM are becoming far more cost conscious.

Of Alligators and Conservatism: Manufacturing Management

Product design and R&D are creative processes, as is the manufacturing industry itself. These creative processes provide some of the great satisfaction of working in industry, and manufacturing managers usually appreciate and respect the application of innovative technology. Manufacturing managers, however, seem to be inherently conservative. Many prefer to learn by experience, not by theory. Many seek reliability and repeatability: Be in command; do not make mistakes.

Manufacturing managers typically live by the calendar, having to make the shipment budget of the week, the month, or the year. They have labor problems, supplier problems, reliability problems, and distribution problems. Product design engineers would do well to try to understand the challenges facing manufacturing managers.

The employees in a large truck plant used to speak of "the alligator" in the plant. "The alligator" was the assembly line, which had to be fed so many axles per hour and so many engines per hour. If the alligator did not get fed, there were real problems.

Is there a simple way around the conflict between the innovative spirit of design engineering and the conservatism of manufacturing? One way may be to eliminate separate departments and place the two functions under a common manager. This approach, according to Keith McKee, director of the Manufacturing Productivity Center at the IIT Research Institute of Chicago, has worked for several of his clients. Regarding compartmentalization, he comments:

> It is interesting to watch the interchange between manufacturing people and the engineers. When new product designs are proposed, manufacturing personnel sometimes act as though they don't know they have the right to refuse the new design. Some of them act as though

they are "second-class citizens." The engineers have a lot of time to think things through and consider alternatives. The drawings are delivered to manufacturing and they have no luxurious amount of time to think things through. They have no quick way to correctly estimate costs and have to make a full set of route sheets in order to come to grips with the estimated costs. Where is the time to stand back and think about it? A technically competent manager can see such difficulties occurring and intervene to manage the situation more effectively.

many manufacturing managers are faced with inadequate time and insufficient budgetary allocations for equipment and staff

MANAGING CHANGE

Although automation could make their lives easier in some ways, most manufacturing managers do not rush to embrace new technology, for a variety of reasons. First, the capital budget may be driven by other matters, such as replacement of depreciated equipment, investment to support new products, or support of manufacturing in other locations. Before accepting new technology, manufacturing managers want to see the technology demonstrated and proved. They usually have little interest in seeking scientific or technological information such as that available from the manufacturing-focused data bases at the Society of Manufacturing Engineers. In addition, overt emphasis on clearing established hurdle rates for return on investment, though not as pervasive as in the past, continues to be a significant impediment to adoption of new technology. Managers are likely to question investments in incremental new technology when returns of 20% to 40% are possible with leveraged buyouts and restructuring.

Despite traditional conservatism about new products and the effect of technology on manufacturing operations, however, most manufacturing managers recognize the potential of continued developments in materials (e.g., plastics, polymers, synthetic fabrics, composites, glass- and metal-matrix materials, ceramics and ceramic-matrix materials, and optoelectronics). In addition, manufacturing managers continue to assess the capabilities and limitations of such methods and technologies as computer-integrated manufacturing, CAD/CAM, just-in-time manufacturing, flexible manufacturing systems, total quality control, Shingo systems, Taguchi systems, and robots.

Although manufacturing managers are faced with rapidly changing products, methods, and systems, many of them think of themselves as conservative and do not always seek change and its attendant risks. On the other hand, R&D departments and product engineers symbolize change to manufacturing managers, and in fact, their function is to create new products and processes with the attendant risks.

TOWARD A SOLUTION: TRANSACTIONS AND NEGOTIATIONS

Several factors affect the nature of transactions and negotiations between the functions in an organization. The first factor to consider is relationships between people—that is, when, how, and between whom negotiations take place. For example, who establishes the timetables, circumstances, and conditions for conducting negotiations between engineering and manufacturing?

If existing facilities and products are involved, negotiations typically take place with existing production and manufacturing engineers; the development process is relatively straightforward. If new facilities and equipment are involved, however, an advanced manufacturing engineering unit might be needed to seek new manufacturing methods. In addition, teaming methods or project management methods might be in order. For example, GE's method, which was pioneered in aircraft engine manufacturing, assigns development personnel and advanced manufacturing engineers to a project team that is responsible for the product throughout its life cycle.

Product or project life may be the key criterion for determining the most effective engineering-manufacturing relationship. If the projected life is short—perhaps a year or two—and requires repeated and frequent work, routine channels should be established and used. On the other hand, if the project is a once-in-10-years or 20-years quantum leap, such methods of organization as teaming and the establishment of an advanced manufacturing engineering unit should be considered.

Whatever the situation, time should always be provided for participation—intensive interchange, debate, and argument. So many times the manufacturing manager is not asked what would work best and is not included in the necessary brainstorming. In addition, the necessary resources should be provided, including allotment for the time and effort required. Key people should be involved, on a full-time basis if necessary, and an account number should be assigned to track costs if necessary.

THE RIGHT WAY AND THE WRONG WAY: TWO CASES IN POINT

A successful example of these principles at work is the Lockheed organization known as the Skunk Works, which developed a long line of advanced aircraft, culminating in the YF-12 and SR-71 aircraft. The YF-12 was a prototype for the SR-71, which was the first production aircraft to fly faster than Mach 3. (Mach 1 is the speed of sound in the atmosphere.)

These all-titanium aircraft were developed and manufactured on an urgent basis by dedicated professionals working in the desert in secrecy. A competitor had already started development of titanium aircraft and agreed to share information with Lockheed. Lockheed started work two years after the competitor and delivered its aircraft two years ahead of the competitor, even though it was dealing with material in which the physical, mechanical, and fabrication standards had not yet been developed and had to be prepared from scratch.

Bob Vaughn was chief manufacturing engineer at the Skunk Works from 1960 to 1970. He cited the following factors as being fundamental to the success of the Skunk Works:

- Competent management, with exceptional technical leadership from senior management.

- Efficiency.
- Accountability.
- Integrity.

He further stated:

overemphasis on clearing hurdle rates for ROI, though not as pervasive as in the past, continues to be a significant impediment to adoption of new technology

> The engineers in this highly disciplined organization were considered to be "scientists with screwdrivers." The organization had only 100 people. The number of people involved was limited "almost viciously." The drawing systems were kept simple, and there was great flexibility in incorporating engineering changes.
>
> Bureaucracy was kept to a minimum. The number of reports required also was kept to a minimum. There was strong emphasis on cost control on funds spent to date, funds committed, and cost to completion. The order of the day was "no surprises."
>
> The climate called for rapid decisions and direct decision in a "special situation." There was strong emphasis on personal responsibility for finding the right answer and doing the job with integrity.

Vaughn went on to become corporate director of productivity for Lockheed Missiles and Space Inc and has since retired. The SR-71 aircraft has just been removed from active service after more than 25 years of outstanding and successful performance. It was flown from Los Angeles to Washington DC, in the record time of one hour and eight minutes on its final trip to enter the Smithsonian Institution. The average speed on this final trip was 2,112.52 miles per hour. This shattered the previous record by two and one half hours.

In other companies, however, there have been recent cases in which excessive pressure to perform caused management to take shortcuts and perform below their usual standards. For example, GE developed a new refrigerator compressor for use in its top-line refrigerators. It failed in 1988, and since then GE has voluntarily replaced 1.1 million compressors. The pretax charge in 1988 was $450 million. This expensive error was thoroughly discussed in an article in *The Wall Street Journal* that appeared on May 7, 1990. The article points out that the new-product failure cost could have been limited if the new product had been phased into production gradually: "Blunders were committed at practically every level. In designing the compressors, engineers made some bad assumptions and then failed to ask the right questions. Managers, eager to cut costs, forced the engineers to accelerate 'life testing' of the compressor, curtailed field testing, and rushed into production."[1]

It is difficult to determine from reading the public accounts whether GE could have avoided making these mistakes and accruing these costs through improved cooperation between manufacturing and product development, though certainly its field failure expense could have been reduced by requiring a more gradual introduction of the new product to the marketplace and consequently allowing the engineering and manufacturing departments more time to perform their work.

LEARNING FROM HISTORY: SOURCES OF INFORMATION

When initiating advanced manufacturing projects, managers need to learn from history and avoid reinventing the wheel. The Department of Defense has addressed many of the issues confronting engineering and manufacturing managers and can be a useful source of information about these issues. For example, the Air Force's system planning methods and IDEF (integrated computer-assisted manufacturing definition) language are useful for planning large projects. Another information source is the Department of Defense Manufacturing Technology Informational Analysis Center—MTIAC—operated by the IIT Research Institute in Chicago.

Literature searches can be conducted at MTIAC or through a commercial data base such as DIALOG—originally developed by Lockheed and now operated by DIALOG Information Services of Palo Alto CA—on such topics as the IDEF language, producibility engineering and planning, or concurrent engineering. Another source is the data base at the Society of Manufacturing Engineers in Detroit.

Managers of job shops may find it useful to study such concepts as producibility engineeering and planning, which ensures a timely and economic transition from development to production. More sophisticated organizations that already use such concepts as advanced manufacturing engineering may wish to consider concurrent engineering, which has been developed and promoted by the Air Force. As described in the November 1989 issue of the journal *Manufacturing Competitiveness Frontiers*, published by IIT Research Institute, concurrent engineering is a systematic approach to integrated concurrent design of products and their related processes, including manufacturing and support. (Another description of concurrent engineering is presented in the article "A Smarter Way to Manufacture," which was published in the April 30, 1990, issue of *Business Week*, pp 110–117.) Such programs call for integrated product and process design teams, designing concurrently rather than sequentially and modeling the design and making trade-off analyses. Dramatic reductions in product development time and cost have been achieved by numerous organizations that have implemented concurrent engineering.

Self Diagnosis

Managers who do not have a preconceived notion of how to approach the matter of engineering-manufacturing relationships might want to consider a diagnostic approach. This method, which is widely used by management consultants, involves the following steps:

1. Setting objectives.
2. Establishing a plan for conducting the assessment.
3. Identifying possible solutions to the problems in advance.
4. Conducting a fact-finding mission to identify the symptoms of ineffective engineering-manufacturing relations:
   — Poor planning and execution of new-product development.
   — Inaccurate product cost estimates.
   — Poor product performance in the field.

— Lack of market penetration.
— Excessive design changes.
— Lack of mutual professional respect.
— Insufficient time or personnel devoted to the essential transactions between the functions.
— Excessively complicated product designs and related processes.
— Improper organization and staffing.
— Unprofitable products.

5. Selecting a course of action.
6. Drafting the conclusions and recommendations and establishing a plan and timetable for implementation of the recommendations.

concurrent engineering holds promise for dramatic reductions in the time and cost required to develop new products and their related processes

CONSIDER SENIOR MANAGEMENT

An attendee at a professional conference commented to the author that when a company selects a senior officer, the last sincere and honest word that person hears is "Congratulations!"

When asked what he thought of senior executives, he responded, "Probably 25% of the chief executives in this country are competent, experienced, capable, educated persons who are well qualified to hold their positions. The balance of the CEOs or general managers are in their positions because of some abnormal drive that forces them to wish to dominate people."

Although this may be overstating the case, these comments point to the importance of understanding senior management's attitude toward improvement efforts. Significant change can occur only with the full understanding and support of the organization's senior executives.

Notes

1. "Chilling Tale: GE Refrigerator Woes Illustrate the Hazards in Changing a Product," *The Wall Street Journal* (May 7, 1990), p 1.

ROUND ROBIN R&D

Adopting the Iterative Model of Technological Innovation

Probably no US company has yet achieved in practice the combination of organizational fluidity and networked information flow that the iterative model prescribes. A considerable amount of change is demanded by this model, and effecting it involves promoting new work habits and forms of communication while retaining the best of the old.

Alan L. Frohman

The prevailing model for technological innovation is linear. The possibility that technology can be used to exploit a market opportunity or to counter a market threat initiates a sequence of design and development steps culminating in the transfer of a new product or process to manufacturing, distribution, sales, and service (see Exhibit 1). This model has had a strong hold on the imagination of US managers for numerous reasons, not the least of them being that it has been shown to work when applied to the management of complex projects. A spectacular example is the success of the management system adopted by NASA that helped to bring about the stunning achievements of the Apollo program.

The linear model was already solidly established well before the 1960s, however, primarily because of its compatibility with the bureaucratic command-and-control paradigm for managing complex organizations. The consequence of applying the linear model has been to civilize the innovation process—that is, to bring it within the frame of reference of corporate management. This has entailed the institution of specialized functions, segmented tasks, and clearly delineated individual responsibilities and accountability.

One consequence of the widespread application of the linear model—and a sign of its deficiencies—is the current preoccupation with interfaces. Perhaps the most apt metaphor for this phenomenon is the baton handoff in a relay race. The literature of technology management is replete with devices to facilitate handoff: liaison functions, interface management teams, and phase reviews as well as various budgeting and financial control procedures whose principal function is psychological rather than financial—(i.e., ensuring that functions share a common interest at the interface). Although the techniques of matrix management have frequently been applied to interface problems, their success has been mixed at best. There is a strong possibility that the failures of matrix management in this context can be attributed to the inappropriate modeling of the innovation process rather than to any inherent weakness in the design or implementation of the matrix structure.

ALAN L. FROHMAN, PhD, is president of Frohman Associates Inc, Lexington MA, a consulting firm specializing in R&D and technology management.

Exhibit 1
The Linear Model of Technological Innovation

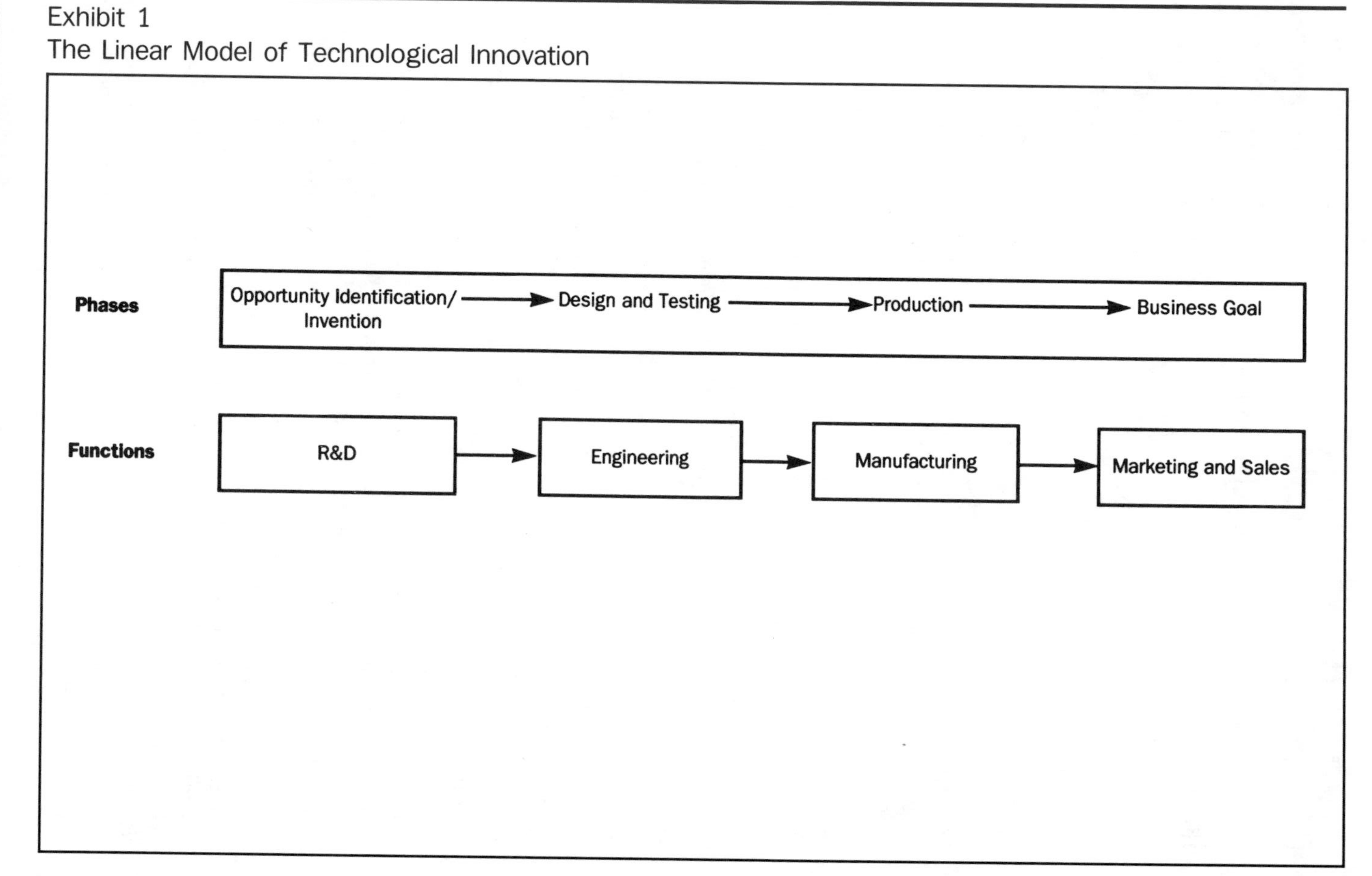

Recently, the linear model has been criticized on several grounds, including:

effective interactive innovation requires the establishment of a network alongside the hierarchical relationships that exist among functional specialists

- Suboptimal resource allocation as a result of overemphasis on a linear progression from scientific discovery to market introduction—The running controversy over the role of the National Science Foundation in funding engineering research illustrates this complaint.
- Slow and costly new product and process introduction caused by excessive caution and cumbersome procedures—The development of the Gillette Sensor shaving system is a recent example.
- Inappropriate program selection and management practices resulting from insufficient emphasis on the iterative and interactive nature of the innovation process—Management focuses on breakthrough programs at the expense of incremental innovations that keep established products in front of the competition.
- Failure to tap into the creativity of employees throughout the organization—Arguably, the growth of Silicon Valley was a consequence of conventionally managed companies' neglect of innovative employees.
- Faulty communication that causes vital information to be overlooked, as in the situation that led to NASA's Challenger disaster.

THE ITERATIVE MODEL

Increasingly, the innovation process is seen as a goal-oriented, iterative learning process (see Exhibit 2). According to this model, innovation teams focus on a collective pool of knowledge and secure and manage the resources needed to generate the innovation. The process requires the establishment of a network alongside the already existing hierarchical relationships among the organization's functional specialists. Identification of a market opportunity or invention of a new product may be initiated within any functional unit of the organization. Resources are then allocated among the innovation teams, and the nucleus of the team is put together. Thereafter, the process depends on the team's functional specialists bringing together the knowledge and resources required to achieve the goal. Their effectiveness depends on their knowledge of and influence on the other parts of the organization. This has an effect analogous to the excitation of a node, which provokes remote elements in a network to cooperative action.

The Networking Process

To be effective, this iterative process requires simultaneous and spontaneous behavior, with individuals acting in parallel in response to their own knowledge of what is required rather than in obedience to a directive from above. This is called networking, and it results in interactive learning not only about technology but also about customer needs, distribution channels, and financial strategies—all the factors involved in completing a product or process innovation.

The differences between the linear and iterative models of innovation can be

Exhibit 2
The Iterative Model of Technological Innovation

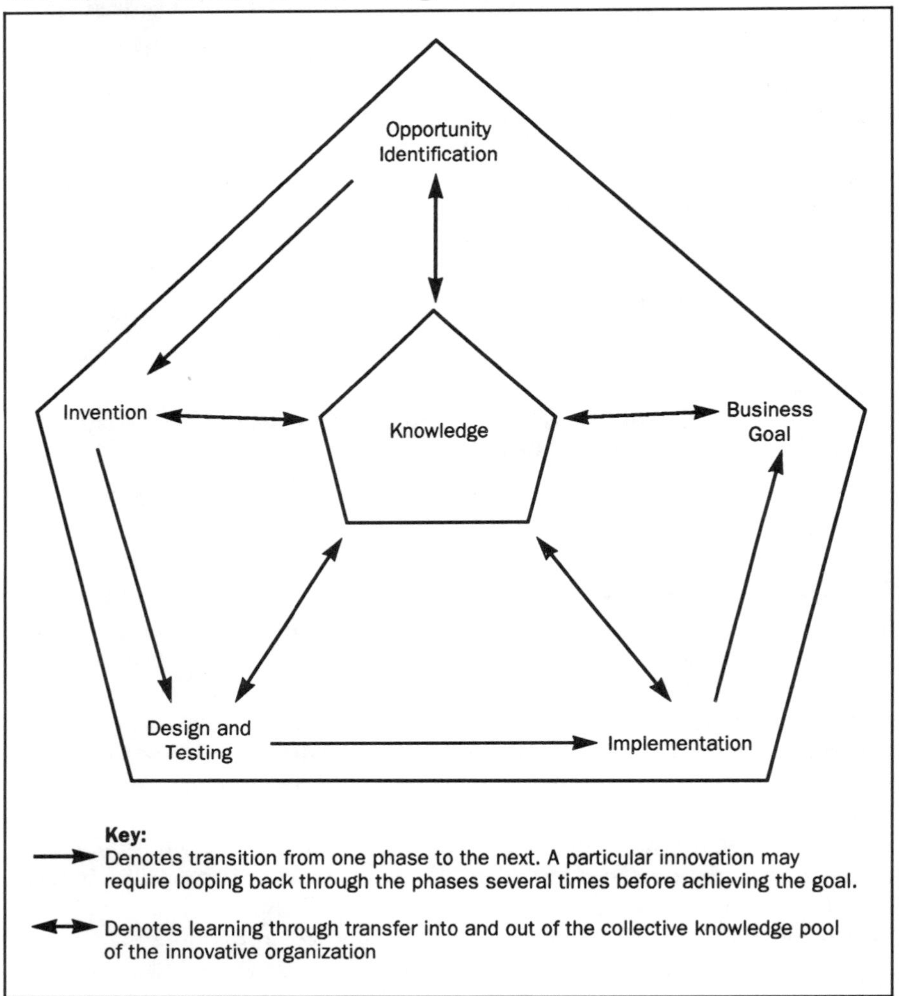

illustrated by the probable responses of management to some typical, generic problems. For example, if the R&D department fails to justify programs in financial terms because technical and financial operations are conducted in isolation, the manager in the linear environment will likely employ new people from outside the company—either technical people with financial backgrounds or financial people with technical backgrounds. Use of the iterative model, on the other hand, requires that technical and financial operations relevant to a particular innovation be conducted within a single team.

Similarly, if a stream of innovations that consistently satisfies customer needs is lacking because marketing does not have the responsibility of guiding technical efforts to that end, the linear model would have management create a new function responsible for analyzing future customer needs. In the iterative model, each innovation team would be charged with matching up a new technology with a market opportunity and so would generate the innovation stream.

successful managers must be able to integrate information that originates not only from team members, but from others inside and outside the organization

The manner in which information is managed is also determined by the innovation model. The linear model leads to a sequential flow (and to the idea of technology transfer). The iterative model leads to a simultaneous flow of information within a network comprising interconnected parts in continual, spotaneous contact. The elements included in such a network may extend beyond the conventional organization itself. For example, as a result of applying just-in-time production methods to inventory control, the boundaries between purchasers and vendors are becoming fuzzy. Similarly, strategic alliances are blurring the demarcation between companies that only five years ago were competitors (and that in some ways remain so, as in the case of GM Corp and Toyota Motor Corp, and Merck & Company Inc and ICI Americas Inc).

In the future, information will need to be managed in two modes. Simultaneous, multidimensional flows will be encouraged for internal and for certain external situations, and sequential flows will be appropriate when crossing the boundaries between nonallied organizations. For the most part, however, management currently remains accustomed to thinking in terms of the linear model of innovation. Although the pervasive effects of this model on assumptions regarding the proper way to manage the information flow may be antithetical to effective management, they cannot be expected to simply disappear. Managers must reeducate themselves to recognize when action consistent with the linear model is justified and when to toss out the old assumptions and behave in a manner that supports networking.

IMPLEMENTING THE NETWORK

An organization adapting to the iterative model of innovation will initially be characterized by a framework of functional experts whose relationships are in a state of flux. For a period of months, team members will associate closely with one another to achieve their goals, and they will be rewarded as a team. The composition of these teams will change constantly, however, to meet the organization's global needs. To succeed in such an environment, managers must be integrators of information originating not only with team members but with others both inside and outside the organization. This is especially true in areas in which technology is changing rapidly. The ability of the technical members of innovation teams to locate and apply relevant information from many sources will be critical to success.

Expectations and Rewards

Employees attempt to please their managers; in exchange, they receive rewards. These rewards may take the form of a raise or bonus, a better assignment, more resources, or approval and recognition. To encourage behavior consistent with the iterative model, managers should encourage networking. Rather than working on narrowly defined problems and remaining within functional boundaries, employees should be asked to coordinate their work with that of other functions, with an eye toward identifying and anticipating problems and developing solutions. In other words, management should direct, encourage, and reward lateral communication.

The problem is that as employees spend more and more time networking, managers may begin to feel less needed, less in control, and less informed. According to the iterative model, however, managers should not be responsible for directing employees. Instead, their responsibility is to encourage the development and effective use of networks and to measure and reward the results.

This is a highly significant shift. Instead of all work, information, and rewards flowing from the boss, the employees direct their own work and work with others to coordinate their results. As a result, the center of communication and decision making can shift to such a degree that management in effect ends up working for its subordinates. Management may be asked to do things for which only the employees, by virtue of their networks, may recognize the need.

All this is not to say that managers should stop defining goals or evaluating performance. Instead, top-down management and networking should be in effect simultaneously. Employees should be rewarded both for accomplishing assigned tasks and for providing or soliciting information or resources outside of these tasks. Individual problem solving should be rewarded as should contributions to the group's efforts.

It is important to maintain a balance between networking and more traditional work. Although concentrating on networking may encourage information flow, the day-to-day work may languish if networking is excessive. By the same token, much work on specialized tasks can result in an overly narrow definition of the problem and its solutions.

Career Paths

If employees are to alter the way in which they perform their jobs, their perception of those jobs and of their long-term career goals must also change. In the linear model, jobs are defined by specialty, and success derives from technical depth rather than breadth. This orientation strongly favors individual contributions and individual credit rather than networking and shared credit. When information flows laterally, technical depth is an important asset in accomplishing the tasks of the team and in establishing and maintaining credibility as a member of the network. Hence, in an iterative environment, the reward for technical depth is not promotion but survival and the chance to keep playing the game.

This attitude can be reinforced by encouraging career paths that lead to a variety of positions and functions. Employees who perceive their futures as being tied to a single function tend to defend it no matter what. Instead, managers should be moved across departmental boundaries, because their identification should be with the business or corporation rather than with the function.

if employees are to alter the way they perform their jobs, their perception of those jobs and of their long-term career goals must also change

Adopting a cross-functional career path means moving from the known and familiar to the unknown and untried. It entails greater dependence on others and an increase in the importance of relationships. For managers, less time is spent telling others what to do and more is spent listening. Some individuals will resist surrendering their professional identity, and they will probably pursue the traditional career path. Although the need for such specialists will continue, it it important for managers to be familiar with both approaches.

OPTIMIZING THE NETWORK

Some people are more oriented to the iterative model than others. The literature refers to them as gatekeepers, but perhaps a more apt name for these promoters of information flow is *networkers*. The transition from a linear to a networked environment can be facilitated if these natural networkers are identified. Research suggests that such individuals are driven by their own interests to keep informed. For example, lawyers dealing with antitrust matters tend to network with other antitrust lawyers. Likewise, scientists working on laser hardening talk to other scientists working in the same area. Effective networkers are distinguished by their inclination to gather information and pass it on; and they tend to be extroverted and social.

Once the natural networkers have been identified, it is important to position them so that the maximum benefit is obtained. Although some networking will happen naturally, a better payoff is likely if management encourages it. First, networkers' jobs should allow them to communicate with others and pass on useful information. Second, networkers should be well informed about company strategy so that they can focus their interests. Third, they should understand the types of information others in the company would find useful.

In addition, networkers should be encouraged to travel and attend conferences and to join societies and relevant associations. They should meet regularly with such sources of information as customers, suppliers, and competitors. They should be in contact with universities and consultants. Several companies, including Merck, have even designated networkers whose job is to keep abreast of relevant fields and communicate their knowledge in a form suited to the needs of those receiving the information. Networkers can be especially helpful in building better partnerships with suppliers. At GM, for example, carefully selected networkers for engineering, purchasing, and disbursements are assigned to serve as contacts with suppliers.

In an R&D organization, technology networkers should develop an awareness of external technological breakthroughs of relevance to the company's ongoing projects. They should maintain multiple contacts with other R&D labs and universities and should make their peers aware of work under way at the company. Networkers should stay as current as possible with new developments through personal contacts and literature and should keep in-house engineers informed of pioneering work.

Transition Policies

The policies and practices of most organizations reinforce the top-down orientation of linear information flow rather than the cross-functional orientation of interactive information flow. Planning, budgeting, and reporting are usually separated by functional unit, and training is given according to function (e.g., senior management usually receives no training). Appraisals and rewards are administered by management with no input from the employee's peers.

To make this framework more amenable to networking, training in teamwork and communication across functional barriers can be useful. Many companies (e.g., General Electric Co, Colgate-Palmolive Co, and Gillette Co) have training classes comprising employees from across the entire organization. Hewlett-Packard Co has modified its planning, budgeting, and reporting systems to cut across functional areas so that the links between such work units as projects, programs, and business are reinforced. GM and Hewlett-Packard have expanded their appraisal and reward systems to include assessments from peers and from others outside the functional group.

It is also important to provide opportunities for employees to be exposed to other functions and disciplines. GE uses rotation programs quite effectively to accomplish this. Other mechanisms used by such companies as Federal Express Corp, Union Carbide Corp, and Grumman Corp include cross-functional task forces, product development teams, and broadened committee assignments.

All these policies can be viewed as steps in the transition to the iterative model of innovation. For them to work, however, more basic changes must have already been made. The reward system must recognize contributions to the network, and employees must understand that part of their job is to make contributions outside their functional areas. When these changes have not occurred, such practices will fall short.

Transition Structures

The process of building lateral links for collaborative networking entails the formation of temporary structures organized around specific problems (e.g., task forces, product development teams, and ad hoc committees). In general, these structures never become business management teams; instead, they are dispersed as each project is completed. The groups have members from different functional units, and their structure tends to be less vertical than that of the traditional organization. In this more organic arrangement, rank is less important than skills or information pertinent to the problem at hand. Authority derives from expertise, and employees with expertise are likely to be found outside of senior management. In fact, leadership may pass from person to person as the project proceeds through different phases. In a product development team, for example, task leadership may pass from designer to engineers to manufacturing personnel. This type of arrangement has the added advantage of encouraging employees to identify with the product or service they are working on rather than with an organizational unit.

Companies that have already incorporated features of the innovation team in their business operations include IBM Corp, E.I. du Pont de Nemours & Co, 3M Corp, Hewlett-Packard, Procter and Gamble Co, and Xerox Corp. One of the most notable examples of its successful use was the development of Ford Motor Co's Taurus and Sable automobile lines. These cars have been popular with consumers, have performed well, and have contributed greatly to the company's income during the past six years.

A Congenial Culture

Although building effective networks can take a long time, they can be destroyed quickly. Traditionally, the insertion of a staff group that analyzed all recommendations for management tended to undermine support for networking. Staff groups inspired defensiveness and the desire to control rather than share information. They led to management by fear and intimidation.

The most essential prerequisite for networking is a culture that tolerates mistakes. If people fear that they will be punished for mistakes, their energy will be devoted to avoiding mistakes and appearing to be in control. The attitude to encourage is that not learning, rather than not knowing, is what should be avoided. Large errors are prevented by learning from small errors, and networking can help in identifying or preventing errors early on.

Companies that freely communicate strategy and business plans to a greater range of employees take another important step in establishing a culture conducive to networking. To build and use networks effectively, people need to understand what is happening in the organization as a whole. The following information should therefore be provided:

- The company's strategic plan and market goals.
- The plans of other departments.
- New product plans from R&D and marketing.
- Analyses of current market trends, the position of competitors, and other factors affecting the business.
- Up-to-date reports on operating costs and profitability.

To foster networking, it is important to recognize that senior management's job is not to solve problems or provide information. Senior management's job is to provide direction and unleash energy, which it should then diffuse throughout the organization. Problems are solved and ideas turned into action by drawing on the knowledge and capabilities of employees across all functional units. For this to occur efficiently, networks must be built. The success of the top-down, hierarchical organization is directly dependent on the effectiveness of lateral connections and the networking of people across personal, organizational, and geographic barriers.

CHAMPIONING CHANGE

To effectively initiate the shift from an organization based on a vertical hierarchy and linear innovation to one based on teams and networking, management should first clearly state what it is trying to do and why. It should inform employees that there is substantial competitive pressure on the company and that change is necessary and inevitable. If the company has a history of initiating new programs only to abandon them later, big, splashy announcements should be avoided. Instead, the necessary steps should be taken quietly and explained along the way to those who will be affected by them. Employees should be continually reminded of what is happening. They should be aware that the transition will take years of effort and that immediate results are not expected. Support for the building of a horizontal organization should be relentless.

Second, changes should be made to exploit the abilities of the company's best networkers. This does not necessarily mean reorganization. It may require only that lateral responsibilities be clarified or that project teams or task forces be formed to get things done. Networkers should be assigned to projects and assignments on which they can be most helpful, and people should be positioned to encourage and facilitate networking. For example, one Johnson & Johnson division changed its product development organization from a discipline-based structure to a customer-based structure, with each customer team headed by an effective networker. At GM Fanuc Robotics, the successful head of the company's US automotive sales unit was assigned to the growing European effort on a part-time basis. Such organizational changes also send a signal to the rest of the organization that real change is occurring, which can itself make it more readily accepted.

Third, it is important to establish a sense of urgency and action. Merely communicating the need for change is not enough. Especially at the beginning, setting a few realistic goals is essential. These may include the establishment of several effective development teams, a cross-functional workshop on a topic of common concern, or increased communication outside a functional unit. At Gillette Co, one of the first steps was to invite personnel from R&D and related disciplines to a two-day workshop on building networks. Many other companies also offer courses and programs in which people can develop their ability to listen to others and understand their perspectives. At GM Fanuc, periodic conferences and workshops facilitate communication and give participants the opportunity to meet new people. At GM, the lateral organization was strengthened by including feedback on networking in performance evaluations. Hewlett-Packard, among others, has also experimented with peer evaluation.

Fourth, the transition should be monitored continuously. Management should meet regularly with teams and task forces to receive their feedback. At companies like GM, GE, and IBM, teams regularly evaluate themselves and use outsiders to periodically measure the effectiveness of horizontal communication. The results of such assessments can reveal unclear goals or misassigned tasks and can indicate where training is needed. Although training in teamwork and lateral networking is a necessary ingredient for success, it must be monitored and adapted as networks develop.

OVERCOMING RESISTANCE

The biggest obstacle to a change from a linear to a parallel mode of communication is the perception that the new way will favor some groups over others. People who suspect that they will lose influence, staff, access to decision makers, and control over funds will resist change. In an operation that values networking, it is clear that R&D, manufacturing, marketing, and the other functions will no longer have absolute authority over resources and activities. R&D will be merely one member of a team whose decisions will determine technology requirements. The same holds true for marketing with respect to such market factors as advertising and for manufacturing with respect to the manufacturing process and suppliers. The team becomes more central at the expense of the functions.

Exactly how power will shift will depend on the previously existing balance of power. In a company such as Polaroid, in which R&D has been extremely powerful, R&D will believe that it has the most to lose. In a company such as Procter & Gamble, in which marketing holds the most power, that department will likely have similar feelings.

Opponents of change can sometimes be won over by allowing them to participate fully in working out the new mode of operation. At the very least, this can force them to justify their position in an open forum. Another approach is to introduce change step by step, by bringing together only two functions at the beginning and then building on that. This has worked well at GM, where the first step toward the development of broader networks involved only product engineering and manufacturing (this was called simultaneous engineering, and it required that product engineers and manufacturing engineers working on the same product be located near one another).

Many view the transition to a more horizontal organization as the creation of an entirely new organization. It is said that Michelangelo saw the form within the stone, and for some the transition has more in common with that perception: all the elements are already there, but they need to be brought out, highlighted, and reinforced. In this view, transforming an organization is more a matter of unleashing than creating.

PASSING THE BATON

Laying the Groundwork for Technology Transfer

Implementing tools and technologies that originate in a research consortium lab is a demanding process. Because both physical and psychological barriers can impede successful technology transfer, all levels of management should be dedicated to overcoming them.

James R. Key

Decreasing an organization's product development cycle through more effective use of technology is a major goal of management, whether the product is generated in the company's research, development, or manufacturing department. By improving the rate and efficacy of technology transfer from external sources, particularly from research consortia, management can enhance company productivity significantly.

Careful selection of new technology and tools can substantially help a company reduce its product development cycles. A company may find the new technology it needs for product development either in commercially available tools or through its own research efforts. To select tools effectively, however, a company must know exactly what is available; staying abreast of every technological development in the world of R&D is a difficult task. As the fruits of the technological age continue to become less expensive and more generally available, companies must work ever harder to stay ahead—or to break even.

The breadth and depth of the research needed for competing in a global marketplace have increased well beyond the resources of any individual company; accordingly, many companies have pooled resources to support research consortia. Cooperative research can achieve a number of objectives beyond the reach of individual companies' efforts. By rapidly attaining critical mass, a consortium avoids needless duplication of efforts in isolated companies. Consortia allow companies to focus more internal resources on applications, precluding replication of generic research results that can be obtained elsewhere. In addition, the amount of technology and knowledge needed for competition are beyond the resources of any individual company's capabilities. These crucial advantages of cooperative research efforts, though recognized by many companies, are understood by few. Consequently, the transfer of technology from external sources is often not as effective as it could and should be.

PRELIMINARY CONSIDERATIONS

Although technology is transferred at the level of technologists, its successful transfer depends on the actions and attitudes of each level of management in each compa-

---

JAMES R. KEY has extensive experience in integrated circuit technology, high-end computers, and communications and is an industry consultant in technology transfer and research consortia.

ny function. In addition, if a technology is to be successfully acquired from an external source, the project must have sufficient attention and commitment from all levels within the corporation.

Technology transfer can be improved by a company's conscious and integrated commitment. For technology transfer to succeed, it must be woven into the fabric of the company's long-range goals and strategic plans by top management. Middle management needs to consciously provide technology support; at this level, a lack of support yields a weak link in the product development chain. Line management must allocate the requisite internal staff and funds for proper operation of the received technology, as well as have sufficient staff and funds to commission and acquire external technology development.

At the same time, technology transfer is still largely an art rather than a scientific discipline. There is no definitive yardstick to gauge the individual steps composing a successful transfer of technology. The flow from concept to production-ready technology is rarely an unbroken, single path. Although research consortia measure dollar leverage, quality and quantity of research output, number of contacts with members, and numbers of reports requested, among other items, there is very little concrete data on the actual transfer of technology. Companies are reluctant to release such information, which might decrease their competitiveness, and engineering departments do not usually document the sources of their ideas and technologies. The technical staff is not rewarded for revealing sources; however, there is potential liability if, when a company claims credit for a development, a university or competitor contends a violation of its intellectual property rights.

BIRTH OF A CONSORTIUM

US corporations first began to look seriously at cooperative research efforts during the early 1980s, when several trends and events forced a paradigm shift in the way integrated circuit and computer manufacturers developed new technologies. A number of events conspired to threaten individual and private research. First, federal research funds from the National Science Foundation (NSF) decreased and Department of Defense (DOD) support shifted from silicon-based technology to focus on more exotic semiconductor materials. The breakup of the Bell Telephone monopoly then effectively restricted outside companies' access to Bell Labs technology. Another factor, the increasing disparity in pay between starting salaries in industry and those for new university professors, made an academic career less and less attractive to job-market entrants, creating a faculty shortage just as computer science and electrical engineering departments experienced massive enrollment increases. At the same time, tightening economic conditions caused companies to significantly reduce support for corporate research programs.

The net result of these trends and events was a dearth of university and private research in the US; concurrently, Pacific Rim and European countries sought to gain market share by investing massively in their national programs to nurture specific industries. US electronics executives, alarmed that their reduced access to research would mean a reduced supply of new technology just as global competitors accelerated their own technology development efforts, began to look for alternative sources of technology. The amount of expertise and technology that companies needed to remain competitive was escalating astronomically. For example, the cost for a new semiconductor fabrication line was projected to rise from $50 million in 1982 to $500 million in 1995 and to $1 billion in 1999. It became clear that no US enterprise, not even such giants as IBM or AT&T, had the intellectual and financial capital to meet all technology needs solely through internal development—particularly when the competition included foreign governments' national programs.

the success or failure of the transfer process is critically dependent on the actions and attitudes of the company's entire management chain

Faced with these threats to their survival, several US electronic companies in late 1982 initiated cooperative research consortia to support research and technology development, sharing rising risks and costs. Having operated over several years now and currently producing steady streams of research and technology, these consortia have proved a successful choice for conducting high-risk, long-range research. A benefit that participants had not anticipated was their new-found forum to set long-range goals for research and to map these goals into their companies' strategic plans.

However, each of the member companies now found itself facing the unexpected difficulties of transferring technology from the research consortium to its own internal development organization. And technology transfer is definitely not a simple matter of acquiring a new piece of equipment, ready to operate; it is a complicated do-it-yourself job, with few or no instructions included.

STAGES OF TECHNOLOGY TRANSFER

Many managers do not understand the processes of technology transfer and technology hardening. To some, technology transfer should be like just-in-time delivery of commodity parts for a high-volume manufacturing operation—simply a matter of transporting the technology (i.e., parts) from the originating lab (i.e., supplier warehouse) to the receiving lab (i.e., assembly line). Unfortunately, technology from a research consortium is rarely in a form that is directly useful to a product developer or manufacturing line. Transport and delivery is only the beginning of the transfer process. In managing technology transfer, a company must undergo eight stages to acquire useful technology and tools:

- Acknowledgment of the recipient's awareness, interest, and tracking of technology.
- Transportation and delivery of the originator's technology to the recipient's lab.
- Confirmation of the recipient's expertise in and understanding of the originator's technology.
- The recipient's evaluation of the technology's suitability for eventual application by product development or manufacturing.
- The recipient's modification and hardening of the technology to meet anticipated needs of product development and manufacturing.

- Demonstration and sale of the recipient's version of the technology to a client.
- Adoption of technology by an end user, who furnishes specific requirements.
- The technology support group's maintenance and modification of the recipient's product in response to the end user's ongoing needs (the recipient usually does not perform this function).

At all stages of technology transfer, the recipient company can encounter company obstacles that menace the success of its project; managers must beware of these barriers during all phases of the transfer.

BARRIERS ON THE HOME FRONT

The first barrier to technology transfer is the would-be recipient's lack of awareness of a technology's existence—still a major problem in many research consortia. Often only a small percentage of the technical staff of sponsoring companies is familiar with the consortium, its goals, and the status of its various research projects. The consortium manager here must assume some responsibility, disseminating information and ensuring that each employee has full access to the research output. Unless this task is part of the manager's performance plan, it will not receive the requisite amount of attention. Company managers should budget time and travel funds sufficient to allow their technology scouts to track external technology developments. The resources for this task, often buried in other projects, tend to get cut early when overall budgets are cut.

A necessary ingredient for the success of a technology transfer is the recipient's faith in the involved technology. A transfer will fail if the recipient does not believe the effort will work or be commercially workable. This uncertainty can also arise if the recipient fears that proprietary company information may be revealed or that a liability may be incurred during the transfer process. If management is not sufficiently confident about the technology transfer, the effort will not attain its potential.

To give the project the essential impetus, individuals must be designated champions of the technology on both sides of the transaction. Champions are needed both to push technology out of the originating organization and to pull new technology into the receiving organization. In addition, the risk and reward structure of the receiving organization must provide sufficiently attractive incentives to encourage its champion to proceed.

Transportation and delivery of technology can be a straightforward process, particularly when the recipient already has the equipment necessary to the technology's immediate operation. If the recipient owns equipment identical to that of the technology's developer, the technology needs no modification. However, the recipient frequently lacks or does not want to purchase the same equipment, and modifications must be made in the recipient's process, tools, or software to accommodate the technology. A common example of this scenario for software occurs when different hardware platforms or operating system versions are involved. This can be a substantial obstacle, and managers need to determine the strategic significance of the project to judge the most efficient allocation of resources.

technology from a research consortium rarely comes to the development lab in a form immediately useful to a product developer or manufacturing line

Before a new technology can be used by a product development group, it must be brought up to industrial strength to meet the group's requirements. The recipient may perform this function and then pass the technology on to an internal tool and technology development organization; external vendors can also accomplish this phase of the transfer.

If the product development group accepts the technology, a technology support group must maintain it in a stable state. Production environments change, and the support group must develop the technology to meet the requirements of production. The development of industrial-strength tools and technologies can be developed by third-party companies, strategic partners, or vendors. This arrangement is insufficient for product development groups, however, which also need the tools and technologies to be supported while they apply them. External support groups tend to be less responsive than internal support groups, and managers must take this lack of response into account when planning for completion of the technology transfer process.

Successful technology and tools go through a life cycle, beginning as fragile, untested concepts, moving onto prototype stages in which they sometimes work, and developing into robust, mature tools whose characteristics are well known for a given class of applications.

PITFALLS IN THE TRANSFER PROCESS

When considering the acquisition of technology from external sources, the manager must be aware of the difficulties involved in the transfer process. As more companies begin to seek such acquisitions from external consortia, the science of technology transfer will likely become much simpler. Currently, however, a number of potential problems still exist in matters of timing, corporate structure and communication, and legal issues, which the manager must take into account before beginning any transfer process.

A Question of Time

New technology, unlike just-in-time deliveries, can have long periods of gestation, taking years to develop to the point of commercial utility after its acquisition by a company. In addition to the time spent advancing the technology to a useful stage, other time factors affect this period of maturation. The attitudes and stability of the company itself can affect technology transfer drastically. For example, companies in the electronics industry frequently reorganize. Managers are moved or leave for other companies; their annual pay structure is based on quarterly and annual results. How do these divergent time scales affect the perspective and actions of managers with respect to technology transfer? What incentives do companies put in place to reward technology transfer that has taken place over several years?

Successful transfer is also very dependent on other types of timing. If it arrives too soon, new technology will likely languish on laboratory shelves; if it arrives too late, it may be unusable. Product development groups that neglect planning do not provide enough lead time to anticipate their own needs; they tend to react quickly, looking for off-the-shelf solutions to their problems of the moment. These organizations look at research and product trends, then jump directly to planning for the next product—considering as scarcely more than an afterthought the tools and technologies that they will need. By that time, crucial decisions have already been made, limiting the range of choices available to the organization and sometimes making the product too costly to develop and manufacture.

Lines of Communication

Communication is an important aspect of a company's operation that cannot be neglected during technology transfer. Attitudes such as not-invented-here often preclude successful transfer: The recipients may be too reluctant to change or invest the necessary effort to adopt the new technology, or the external version may conflict with an internal effort. The external work may not clearly map onto the company strategic plan, either because of incompatibility or lack of a workable strategic plan.

From the initial stages of planning, the manager must ensure that the company employees involved in the transfer are aware of the ramifications of acquiring the technology. Goals and expectations should be communicated clearly, leaving no questions in employees' minds about the propriety of the technology and about possible risks and benefits.

A workable technology transfer strategy must differentiate among invention, basic innovation, and improvement innovation and address the distinctions between them. The strategy must overcome the major barriers of uncertainty in risk, future cost, future performance, and future availability of a proposed technology. Development and manufacturing organizations must correctly anticipate the evolution of a technology that might exist today only in a lab or in a vendor's marketing brochure. Product groups must deal with uncertainty, fear of risks and failure, and the investment needed to get a payback in the scheduled time period.

A Legal Matter

Legal problems can pose a complex challenge to the manager. Antitrust laws, which were originally drafted to protect US companies from unfair competition, actually handicap many research efforts by preventing companies from transferring the very technology that would help them maintain their competitive edge. Recent developments in antitrust legislation have somewhat relaxed the laws' grasp on research consortia.

Before the 1984 National Cooperative Research Act was adopted, it was virtually impossible for companies to jointly fund research in the US. In Europe and Japan, where such restrictions did not apply, many precompetitive cooperative research programs sprang up; some of the best known of these are the Cooperative

VLSI Laboratory and the Fifth Generation Computer Project in Japan and ESPRIT and JESSE in Europe.

in many research consortia, only a few of the companies' technical staff members are familiar with the consortium's goals and the status of their various projects

When Control Data Corp (CDC) sought to follow the lead of the Japanese and European models, attempting in 1982 to form such consortia as the Microelectronics and Computer Technology Corporation (MCC) and the Semiconductor Research Corporation (SRC), many prospective partners cited antitrust fears as reasons against even exploring joining the consortia. Although the Justice Department ruled in 1984 that joint research does not violate the law, US cooperative ventures were still lagging behind those in other industrialized nations.

In 1986, CDC's William Norris and other corporate leaders determined that US firms had a low rate of adoption of research results and technology from universities and consortia. Realizing that the high costs of development and manufacturing were blocking sponsoring firms from aggressively using the results they had funded, Norris and others proposed that antitrust laws be revised to allow companies to share the costs of development and manufacturing. In April 1990, the House Subcommittee on Commercial Law voted to provide antitrust relief for joint production ventures. This legislation, a decided victory for joint ventures, is only slowly starting to visibly affect the joint industry.

Other developments in the legal arena are encouraging cooperative research efforts. More than 700 federal laboratories develop technology for the public good in the fields of defense, agriculture, forestry, and electronics. In the past, legal and administrative barriers have been a bottleneck to the diffusion of this technology into the commercial sector. Through the 1986 Federal Technology Transfer Act, which amended the 1980 Stevenson-Wydler Technology Innovation Act, Congress encouraged access to this large technology resource. CDC was the first company to license software technology through this improved transfer mechanism. Congress recently enacted the National Competitiveness Technology Transfer Act of 1989, which further encourages close alliances between government labs and industry for the purpose of transferring commercially valuable technology.

PREDICTING THE FUTURE

Rarely does a consortium produce results in a form directly applicable by product development groups. The following cases illustrate situations in which company management had difficulties predicting the impact of consortium results on their company's bottom line. Evaluating the value of a technology with undeveloped applications is a difficult task, often through most phases of a project.

Case 1. The research department of an electronics company received software tools for thermal and vibration analysis from the research consortium it had joined recently. This research department approached its company's product development division, offering to apply these tools to the design of a major new product. Responding that such tools had not been needed in the past, the product team declined the offer.

Eighteen months later, the product was late. Caught in an economic downturn, the company badly needed to complete the development of this new product. Despite the fact that budgets were being cut companywide, manufacturing was still begun. After volume production began, field tests revealed a severe vibration problem that a crash engineering effort could not solve. Finally, special shipping arrangements had to be made and special field installations became necessary.

The company learned from this that a research group, at the cost of a few thousand dollars per month, could have been dedicated to solving a problem that eventually grew to cost several million dollars per month in manufacturing. Eighteen months after the research department's offer of new tools had been turned down, the company was experiencing severe cash flow problems. The attendant expense and sales lag far exceeded the total amount that the company had invested in the consortium and research department for those eighteen months.

In its analysis of the problem, management identified the following factors as having prevented the consortium results from being used in a timely manner:

- Insufficient communication between research and product development teams.
- Reluctance of product developers to commit to untried tools from the research department.
- Lack of understanding of the time frames involved for each group.
- The research department's lack of aggressiveness in selling its problem-solving skills to product development.

Case 2. A consortium developed a prototype for an advanced software tool at considerable expense. Having participated in the consortium, the research department of an electronics company decided that this tool was sufficiently robust to be useful. The research group then tried to convince internal product development divisions to try the tool.

After several months, one of the product teams invited the research team to see a new commercial software product from a new company. This software had equivalent performance claims to the tool that the research group was supporting. After much internal discussion, the research department dropped its promotion of the consortium tool, recommending instead that the commercial tool be purchased. Their rationale was that the commercial tool would receive ongoing vendor support, freeing the research team to work on other, less-developed areas of their research.

Critics of the consortium project claimed that the commercial tool would have been developed without the company's investment in the consortium. Consortium supporters claimed that the consortium's work and subsequent publicity actually spurred the formation of the new company and its product. Furthermore, these supporters maintained, the familiarity that the research group gained with the consortium's tool provided an in-depth knowledge of the subject, allowing the researchers to make a quick and informed decision on the purchase of the commercial product. Researchers pointed out other occasions in which the company had purchased untried commercial software that had not performed as advertised, leading to schedule slippages and budget overruns.

When analyzing the situation, management concluded that the knowledge gained had justified the company's investment in the consortium. The goal was to obtain software tools with the right performance characteristics. It was unnecessary that the consortium provide the actual tools that were to be used in the company's production environment.

> **to be useful, new technology must be understood, adapted, hardened, and supported by internal groups**

SUMMARY

Although various mechanisms and processes can provide equitable access to research results for companies, many participants do not invest enough internally to absorb the output from research consortia; nor do industrial members deal adequately with the output of consortia at all levels of management. Management can improve the productivity of technology transfer through several actions:

- Conducting a thorough evaluation of the role of tools and technologies in the company's strategic plan, and integrating the requirements for external tools and technologies into the plan, rather than handling them on an ad hoc basis.
- Reexamining the company's process for bringing in technology, particularly those that need modifications to meet any unique company specifications.
- Reexamining the significance of technology and tool development in each division. The roles of research, product development, and manufacturing are well understood; the roles of technology and tool support are not, and this factor can have a severe impact on budgets and schedules.
- Consciously identifying champions for bringing in technology. Company pruning and restructuring has seriously depleted the reservoir of individuals who carry out this function.
- Recognizing the considerable amount of time needed to bring a new technology up to speed, and differentiating between technology development cycles and product development cycles.
- Recognizing that significant resources are required for the receiving organization both to receive and to harden technology. To be useful, new technology and tools must be understood, adapted, hardened, and supported by internal groups, even if third-party developers and vendors must be used.

Ignoring or undervaluing the knowledge that dedicated research consortia offer is a dangerous mistake that can hamper company operations. In both of the preceding case studies, the research consortia's findings could have been advantageous to their sponsoring companies. If consortia results are either applied or dismissed too quickly, however, a company can easily lose the advantage it would gain from participating in a consortium. In both cases, the important precept is long-range thinking. Severe myopia—staring too closely at research proceedings and findings—can often result in a companywide headache.

UNLOCKING THE US-EUROPEAN TECHNOLOGY TRANSFER DOOR

Ensuring Joint Development Success

European designs of technical products are renowned for their ergonomic benefits and sleek look and feel. International joint ventures are one way of bringing high-quality, successful European designs to the US while cutting product development time. Such ventures have many pitfalls, however, and managers must pay careful attention to communicating, planning, and finding common ground in the global marketplace.

Tom N. Thiele

With product life cycles shortening and competition growing internationally, it is increasingly important for technology managers to ensure that the technology development process in their organizations does not reinvent the wheel. Although in-house development efforts are essential for productive growth and innovation, every option involving access to technology should be considered, including licensing, alliances, and transfer among subsidiaries. Using available technology and designs may seem much more cost- and time-effective in comparison with in-house development; still, other challenges and problems can occur that need to be carefully managed. This case study focuses on the lessons learned during the author's several decades of involvement in technology development with several major manufacturing companies, much of which included either joint-venture activities or the international transfer of design technology.

FORMING A JOINT VENTURE

Technology and successful product design availability were two of the driving forces behind a major joint venture between a large European company and a US electrical equipment manufacturer. The companies saw significant advantages in the opportunity to bring to the US completed product designs—used in successful European products—from which new products could be manufactured in the US with little or no modification. The designs were already successful in Europe, and the European company had an outstanding reputation for the quality and thoroughness of its designs.

A joint venture was formed with the expectation that the development process in the US could be shortened and that development costs could be reduced if the domestic producer used these well-designed products to meet the needs of the US

TOM N. THIELE is a registered professional engineer and formerly was the director of technology application at Siemens Corporate Research, Princeton NJ.

market. These efforts were undertaken for several products, with varying degrees of success. Some important lessons were learned, as described in the following sections.

Different Market Needs Affect Product Design

In the particular European markets involved, product design is, in general, highly technology driven. Because of the dominant position of this large European partner, as well as the tendency of large European technical corporations to serve as expert consultants to the customer, product designs are driven less by specific customer preferences and more by the supplier's interpretation of customer needs. In contrast, customer specifications are much more important in the US market because:

- The US partner is in a third-tier competitive position (i.e., it competes directly against two much larger companies).
- The competitive environment is much tougher.
- Independent consultants play a much stronger role in determining bid requirements.

For example, one of the products of the joint venture—a programmable logic controller (PLC)—is used for the control of industrial equipment. A very sophisticated and highly flexible design had been completed and successfully sold in the European market. The European product's performance record was excellent, and the product had gained a reputation as one of the best of its kind.

When this PLC was first introduced to US customers, their reactions were positive with regard to function, flexibility, and versatility. However, the users were required to learn a complicated programming language to achieve the desired performance. The user manual for the device provided several hundred pages of detailed explanation. In contrast, leading US manufacturers of similar devices had designed their products to be easily understandable by plant operators who had limited formal education and were unwilling because of time constraints to learn a complex new programming language. The US instruction books were perhaps a dozen pages in length and made generous use of pictorial illustrations to explain the equipment's operation and programming.

The European users were usually engineers, and the design was oriented to their skill level. This design orientation was not easily modified to meet the very different use patterns and market needs in the US. The ultimate solution to this problem was to redesign the product to ensure that the operator interface and instruction manuals would meet US needs. To achieve this solution, however, the business managers in Europe who were responsible for the design had to be made aware that US market needs differed significantly. In addition, they had to be convinced that the product would not succeed in the US market unless it was changed to meet those needs.

Standards Affect Product Design Decisions

European business managers must understand that their product will not succeed in the US unless it is changed to meet the needs of the market

When adapting European electrical products to fit US market requirements, certain obvious changes are needed in voltage and frequency. Three-phase industrial power is typically distributed at 380 volts AC, 50 Hz in Europe but at 240 or 480 volts, 60 Hz in the US. There are more subtle differences, however, that can force major design modifications or make an existing, optimized design inapplicable.

Consider the example of an industrial variable-speed DC drive. Standard practice in Europe is to rate these drives at various power levels in kilowatts (kw), corresponding to the continuous output capability of the system. In the US, a corresponding power rating in horsepower (HP) is used. Although it might seem simple to convert a 54-kw drive (a standard design size in Europe) to 72.4 HP, it is not. Unfortunately, the US standard National Electrical Manufacturers Association (NEMA) ratings are stated in integral numbers (e.g., 50, 60, 75, 100), and customer orders typically specify only these standard values.

Furthermore, a 54-kw drive (designed for a nominal 500 volts used in Europe) would be downrated to 66.6 HP at 460 volts and would have to be sold as a 60-HP drive, with no design changes. The comparison is even more difficult to make, because the NEMA rating is based on continuous duty at the nominal rating and a 1-minute rating at 150% of continuous duty. In contrast, the European ratings are continuous duty and have no overload provisions. If a European 500-volt design is applied with no changes to meet the US NEMA ratings at 50 HP (460 volts), a 60.8-kw continuous-duty rating is needed; because the next larger available unit size is 75 kw, a further downrating occurs. Because both the cost and the size of the units increase with continuous-duty ratings, it is not competitive to apply an existing design to this product type.

The comparison is further complicated by the fact that the European rating standard basis is 35° ambient and the US rating standard basis is 40° ambient. Another area of concern relates to the differences between electrical clearances and between testing standards. The European norm is typically established by the International Electrotechnical Commission (IEC); the US follows a combination of standards from the American National Standards Institute (ANSI), the Institute of Electrical and Electronic Engineers (IEEE), and Underwriters Laboratories (UL). The only practical method for meeting a standard that requires a physical dimension or relationship is for a company to consider it in advance (i.e., during the design process). Although the current trend is toward global standards, complete standardization has far to go in many areas, particularly in implementation.

In the example of the variable-speed DC drive, some power and control circuit dimensions also conflicted with US market requirements. After the US company had completed a detailed evaluation of product capabilities, it found an effective solution: it would design the power and interface circuits in the US to meet the US-specific requirements, and it would use an optimized digital-control module that would be identical for the European and US markets.

Physical Form Can Be Adapted to Meet Market Needs

Some products may have a function for which the requirements are the same in both the European and the US markets, particularly when the product in question is a subassembly to a higher-level product. Even if the technical requirements are sufficiently alike that the products can be functionally equivalent, however, the packaging and ergonomic needs of the overall application may dictate differences. The difference may be as simple as determining whether to place an off button at the lower left and color it red for US customers or to make the off button green and place it on the upper right side of the product, as is standard in Europe.

More serious concerns are the volume and shape of the products and components for the functions required; the layout of components must often differ for the specific end use. As mentioned, devising a common mechanical format for the international market is further complicated by the need to support variations in standards and specifications.

One compromise is for the companies involved in the joint venture to seek a common solution that meets different ergonomic needs. If there are still differences—for example, in the front-panel layout of a consumer product or in the repackaging and relayout of printed-circuit boards—the receiving company should make the changes required. Both partners will still be able to apply much of the basic design in repackaged form. However, technology managers should not underestimate the task of retesting and qualifying the repackaged version. In the process they may find that the packaging design ideas of foreign designers provide a fresh, new perspective that challenges the way they had always solved the problem in the past.

Product Configuration Flexibility Affects Design

In a product that has many possible applications or that can be used in a variety of configurations, the question often arises as to how much flexibility needs to be built into the basic design. For example, the IBM PC uses the open-box approach—that is, the computer's design includes several card slots for user options. In contrast, the original Apple Macintosh was self-contained and closed; there was no simple way of adding features. Another example comes from the automotive industry, in which some manufacturers' products have options that allow for customization, whereas other manufacturers prepackage certain combinations and offer few choices to the customer.

These examples illustrate a dichotomy in design philosophy that is affected by two fundamental questions:

- Is the product function well defined at the outset and will it remain basically the same over the product life, or will the application evolve and allow the user to benefit from product modifications and upgrades?
- Is there a substantial customer base that requires a fixed-function product (with attendant quantity and cost benefits), or must the base product be configurable for a variety of customer needs and uses?

The answer for any particular situation depends on several factors, including the length of the product life cycle. For example, if the product will be replaced in three to five years, the problem of upgradability and flexibility can be resolved by offering a new version for purchase. On the other hand, if the product needs to last 10 years or more, the ability to upgrade becomes more important.

devising a common mechanical format for the international market is further complicated by the need to support variations in standards and specifications

Another factor is the product's configuration flexibility. If the market is fragmented, with customers each expecting a product tailored to their particular needs, a product that is easily configured for different functions is important. Alternatively, if a sufficient market segment exists for a specific configuration, the company may obtain a cost advantage by designing different versions for each application.

For example, in one product, the technology to be used was acquired from an overseas company and was embodied in a flexible design that allowed for various plug-in circuit cards to be incorporated on a mix-and-match basis to achieve a wide variety of capabilities. In the US market, however, 80% of the customers wanted a specific combination of features. By eliminating the sockets, multiple-wire harnesses, and connections and by combining all the critical functions on a single printed-circuit board, the company developed a design for less than half the cost of the more flexible version. Moreover, because the resulting product had fewer assemblies and interconnections, product reliability was enhanced. Yet the fundamental design—the product technology—was similar even though the packages were quite different, each optimized for different market needs.

Cost-Accounting Systems Affect Technology Transfer Decisions

The evaluation of product design approaches and the transfer of technology raise questions concerning alternative designs and their respective costs. Technology managers may assume that it is necessary simply to calculate the costs, convert to a common currency at the current rate, and compare the numbers; this is only part of the picture. This approach to calculating costs is ineffective because the cost-accounting structure may vary considerably between the countries and the companies involved.

For example, consider the traditional comparisons of direct product manufacturing costs made on the basis of material, labor, and burden. To compare the differences in actual cost between two or more designs in this way, the technology manager should know that the standard accounting system in one company may include a materials handling and inventory charge that is added to direct material cost and is considered part of material cost. In addition, the cost of development may be amortized over the expected production quantity and added to the direct manufacturing burden. The fact that the same terms are used to collect various parts of the cost structure does not mean that their definitions are the same; the technology manager should investigate whether the cost-accounting structure is the same. To ensure that an applicable comparison is made, it is useful to calculate direct material cost plus labor hours for product assembly and testing. Local labor rates and shipping costs for any special materials should then be applied, taking into account all the factors of assembly labor.

SHARING THE COST AND RESULTS OF JOINT PRODUCT DEVELOPMENT

Joint product development provides an opportunity to build on the strengths of two organizations to obtain a better result than either might accomplish alone. However, some important caveats apply. The organizations involved are likely to have different motivations and objectives; they should try to discuss these during the preliminary stages of development so that there will be no unpleasant surprises later. The companies should also ensure that their technical managers have approved the joint venture. A product that is feasible and desirable technically may be inappropriate from a marketing, manufacturing, or financial viewpoint. All of these factors are further complicated when the cultural differences of the corporations and the countries are taken into account.

For example, a US company and a European company wanted to develop a product that would serve both markets with the identical technology. The companies followed all the steps outlined in previous sections. They prepared a detailed specification. The product, built around a microprocessor programmed to perform electrical control functions, was to have two versions (of both hardware and software). Different mechanical packages would be used to accommodate the differences in the physical constraints of the associated products being controlled.

Because the same microprocessor was used and because most of the software functions were identical, the software development task was common to both. After the specification was agreed to, however, the technical involvement of the US company dwindled to quarterly progress reviews. When the project neared completion, the designing partner (the European company) performed the qualification testing on its version and provided samples for testing in the US. Because of differences in the US hardware design, however, certain software functions didn't perform as required, although the original specification had been met. In addition, corrections were difficult to make because the European designers had had no problem with their version and the US development group possessed no detailed knowledge of the software structure and code.

This problem is becoming increasingly important because operating software is often built into a product at the time the hardware design has been frozen. At this stage, even a simple change to one line of code becomes expensive, because a large quantity of manufactured parts must be scrapped or the company must find a way to correct the problem without tearing down each piece.

Companies can minimize the likelihood that this type of problem will occur by maintaining technical involvement throughout the project, even if one partner does most of the detailed design work. Not only will that partner gain invaluable inside knowledge of the product's design, but the relationship between the partners will ensure that operating experiences and problems that may occur later will be shared more openly. This is especially important when joint development requires cooperation between units of a multinational company or involves an alliance that will last beyond the present project. The relationships and success achieved should make future efforts even more effective.

the fact that the same terms are used to collect various parts of the cost structure does not mean that their definitions are the same

COMMUNICATING DURING THE JOINT PRODUCT DEVELOPMENT EFFORT

In one example of joint product development, no management agreement was obtained and no basis for a common design was established. The project ended up as a competition between the two design groups, who withheld from one another design details that might have simplified the other's design tasks. Furthermore, the adversarial relationship greatly reduced the likelihood of future successful cooperation between the organizations involved. In an environment in which reduced product lead time is essential, this example illustrates the importance of open communication in joint development projects.

Technology managers may conclude that most of the problems would have easily disappeared or at least have been resolved quickly if only there had been better communication between all parties involved. However true this is, communication between people with different backgrounds, cultures, and native languages requires a formal methodology if it is to be effective.

Consider the supposedly simple process of holding a meeting, exchanging information, and making mutually satisfactory decisions. One US senior manager involved in this project spent three days in product meetings in Europe and returned satisfied that everything he intended had been accomplished. He wrote meeting minutes and sent them to his counterparts in Europe. To his chagrin, he was quickly informed by his European colleagues that what he thought had been accomplished had been understood by them in a completely different way. Their response completely reversed what he thought the meeting had accomplished and built mistrust and ill will into the relationship.

A much more effective communication method is to prepare written minutes on site before an important meeting has been concluded, thus providing an opportunity for all parties to review and approve the meeting's accomplishments before departing. Although this process may seem time-consuming and unnecessary, it is an effective way to avoid the misunderstandings that can seriously damage relationships.

Another important requisite of successful international cooperation is to learn some of the language and customs of the desired overseas partners. It is especially important for US managers to make this effort. Although complete mastery of a foreign language may not be practical, it will certainly prove worthwhile for managers to learn enough to show that they have a sincere interest in a project's success and to prevent social impropriety. In particular, US managers should know that most Europeans are required to study six years of English as part of their formal schooling, whereas foreign language study in the US is not generally encouraged, particularly in the technical curriculum. This difference alone puts US technical managers at a disadvantage.

Finally, managers should learn from their mistakes as well as from the mistakes of others. They should review a project after it has been completed and determine why it failed or how it could have been more successful. In addition, they should share this information with others in their company and with other companies, including the partner if appropriate.

SECTION

6

NURTURING INNOVATION

Small and large companies alike should be concerned with nurturing innovation. In this section, Russell Horres and Yuv Mehra describe innovation in small, start-up companies. Louis Mischkind analyzes what is required to nurture innovation in a large, established company, and Robert Tuite discusses a venture program to encourage innovation.

Horres, who has worked at both small and large companies, contrasts the speed and agility of a small, start-up company with the slowness and established procedures of a large company. According to Horres, the sluggishness of large companies stems from the policies established by each of the company's functions. Individuals in a small start-up company, on the other hand, can dedicate their energies to developing new products. Generally, they are highly motivated, their attention is not diverted to existing products, and they do not feel compelled to analyze every option thoroughly before acting. Therefore, Horres writes, they are able to overcome the barriers to innovation more quickly than are their large-company counterparts.

Mehra, who worked for a large company before starting a small one, relates the history of his new company. He first explains the reasons for spinning off a large-company technology to form the new company; he then traces the steps involved in the company's launch: gaining financial support, addressing legal issues, setting up an office, staffing the new company, and bringing the new product to the marketplace. At each step, Mehra faced challenges that technical managers who work in a laboratory usually do not encounter.

Mischkind analyzes the challenges of nurturing innovation in a large, established company. Drawing from the testimony of 200 R&D professionals, Mischkind sums up their and his own conclusions about innovation. Interestingly, many of the same characteristics that Horres attributes to small companies are the same attributes that Mischkind emphasizes as requirements for innovation in a large company: entrepreneurship, teamwork, flexible organizational structure, and appropriate incentives. Mischkind concludes that, within the more structured climate that it needs to maintain existing businesses, a larger company should nurture a small-company climate of innovation.

Finally, Tuite analyzes the efforts of large corporations to develop new businesses by means of new-venture programs. Tuite explains how companies establish new-venture programs, how they work, and what the results have been. Tuite also relates the lessons companies have learned from running new-venture programs—for example, how to deal with the risks and rewards for the company and for the venture management team.

TOO MANY COOKS . . .

Managing Innovation in Start-Up Versus Established Environments

Russell Horres

Managers in established corporations have been repeatedly surprised by the appearance of technologically innovative product concepts from the design teams of newly formed companies. This study contrasts the dynamics of developing innovative new products in start-up organizations with developing them in established corporations.

Many factors account for the difficulties found in managing innovation in an established corporation's product development environment. The most significant factor is management's attempt to balance the complex interrelationships among functional departments, each of which in its own way adds requirements to the product attributes, development schedule, design procedures, and manufacturing processes. Individually, each requirement can be rationalized and justified as essential to the success of the project. Any corporation with more than 10 years of development experience can cite ample reasons for the establishment of checks and balances on the innovation process, and functional managers are all too eager to refer to the corporate history.

The first department to enter the process is R&D, which often can cite reasons for restrictive development guidelines, extensive design reviews, and documentation as the right way to develop products. An equivalent factor in a market-driven organization is the product planning function, which reacts to the length of development projects by striving to include every conceivable new feature and by requiring interrelationships with existing product lines from each new product. Product planning seldom appreciates the added complexity and attendant schedule commitments of such requirements. The purchasing function becomes deeply involved in establishing a vendor base that meets a predetermined list of requirements and controlling the sourcing of vendors by requiring conformance of parts to established bills of materials. Operations with established, capital-intensive manufacturing processes and well-defined manufacturing procedures generate additional constraints. Service functions, having experienced difficulties maintaining existing products, develop an additional set of product requirements. The quality assurance and reliability engineering functions, often independent of product development teams, establish product performance standards and corporate quality goals. The cost accounting function, capital approval processes, corporate legal counsel, and, depending on the nature of the business, the regulatory affairs function can all add requirements for conforming to established procedures.

Convention is a powerful force, and no well-schooled project manager is going to commit heresy by attacking the very fabric of business practices. Although many

RUSSELL HORRES is the vice-president of R&D at Ivac Corp.

participants bemoan lengthy development cycles, functional areas provide only the appearance of improving the flow of innovative products, each—according to its own perception—finding value-added contributions to the integration of the new-product design effort with existing processes. Although an individual contribution may seem essential, taken altogether, numerous contributions can actually cripple a project, just as the Lilliputians of *Gulliver's Travels* subdued the giant with a myriad of tiny threads.

IN THE BEGINNING

At its inception, all of a start-up corporation's efforts are devoted to the creation of an innovative product. There are no existing products to build, market, or sell; all efforts are focused on design. The concept of departmentalization does not exist, and at a minimum, there is a recognition of overlapping roles among the participants. The acquisition and management of capital to fund these efforts are paramount and supported by all founding members.

In the face of limited resources, much effort must be contracted out during this initial phase. Designers have carte blanche to meet future customers' needs with only broad targets for features, manufacturability, serviceability, and product costs. The development team sources all vendors, procures parts, and defines quality controls and specific manufacturing processes down to the shipping container. Very often, design-team members are also involved in the manufacturing start-up and initial marketing of the concept.

When the new product becomes accepted, R&D funding is cut back drastically as efforts turn to marketing and production. The new company experiences phenomenal growth, tiers of management develop, and departmentalization ensues. Departments establish policies and controls, innovation slows, and the founding entrepreneurs frequently become disenchanted with the pace and leave the Lilliputians to their work.

FUNCTIONAL BARRIERS

The major participants in new product development are marketing, R&D, and manufacturing. Although these roles may be recognized in the new enterprise, the concept of departments and individualized functions do not have to be dealt with. In the established corporation, functional areas create significant barriers to innovative product development. These barriers are manifest in the communication difficulties among the functional groups and their appreciation of the corporate goals and initiatives. Interdepartmental barriers are often deeply rooted, with attitudes passing from generation to generation like a virus.

Most contemporary corporations have recognized the limitations of sequential product development and have instituted product development teams with representatives from the participating functions. Concurrent engineering—that is, developing manufacturing processes simultaneously with the product design—is also becoming widely employed. These management processes are intended to help ensure that the resulting product will meet customers' needs, optimize a company's competitive position, achieve the company's predetermined cost and quality objectives, and be integrated smoothly and rapidly into the manufacturing process. Unless care is taken to ensure teamwork and effective communication, these processes will only establish a platform on which to exercise functional imperatives and thereby subvert the very purpose for their existence. Creating the processes is easy; actualizing their promise is a genuine challenge.

at its inception, all of a start-up corporation's efforts are devoted to the creation of an innovative product

Providing a forum for frequent exchange is seldom an answer to reducing barriers. The answer lies in developing an understanding of each function's capabilities, motivation, jargon, and roles and in building respect for the value each function brings to the development process. Individuals representing the various functions must understand the value of teamwork; the team needs to nurture nonconformists to exploit the creativity that such individuals can bring to problem solving.

LEADERSHIP STYLES

Much has been said about the role of leadership styles in managing innovation. Although no right or wrong style can be proved to stimulate innovation, the start-up environment seldom permits the luxury of experimentation. Most often, a directive, controlling style complemented by the charisma and vision of the innovator produces highly effective leadership. The intense focus of the development activity and small size of the start-up corporation result in total alignment of purpose and minimize the role of communication and people skills.

To complement the leadership of the new venture, the start-up organization's team membership must be composed of carefully selected individuals with similar values. Each member must understand his or her role in the venture and be prepared to dedicate a major portion of time to the success of the endeavor. Rigid perceptions about job descriptions are unlikely to be successful in the start-up environment.

RISK TAKERS

By its very nature, a new enterprise attracts more results-oriented individuals. There is little gratification for process-oriented individuals seeking to expand their management abilities by coordinating large numbers of staff members in the new enterprise. On the other hand, the ability to accomplish tasks, make decisions, and reach goals in a highly efficient manner presents significant opportunity for gratification for both types of individuals.

Individuals who choose to associate with a start-up, particularly those who leave the comfort of the established organization, are also risk takers. Contrary to the notion that these individuals are not team players, there are close working relationships within the founding group of many new ventures. Indeed, the pressures to attain profitability and the enormous time commitment often induce stresses in the staff as the team replaces family in its demand for time. Not infrequently, this imbalance results in the loss of founding members. This closeness and peer pressure for commitment often overwhelm even mature managers who have aspired to participate in a new enterprise for years.

The established corporation that pursues innovative product development does not benefit from the personality traits that are characteristic of people who elect to join a new venture. In a more mature company, employees are conditioned to be process oriented and averse to risk, and as team members, they are functionally oriented and career focused. Although the commitment a company gains from these individuals can be great, progress occurs largely on a business-as-usual basis.

MOTIVATION

The subject of motivation cannot be overlooked in examining performance differences between the start-up and the mature corporation. In the start-up, equity participation and the possibility for significant returns are incentives to make a major commitment in time and creativity. Stories of stockroom clerks and secretaries becoming wealthy during the acquisition or public sale of a new venture are legendary and are frequently discussed among team members in start-up organizations. The potential personal reward also provides the rationale to these individuals' families to understand the extraordinary commitment of time and energy that new ventures demand.

The drive for fairness and equity across corporate functions inhibits the established corporation from wielding this power tool. Incentives are limited and seldom effective. For the majority of the employees in these more mature organizations, success means they will be assigned to another team.

FOCUS

A fundamental leverage that the start-up enjoys is focus and dedication to the new product development process. Without the attendant distraction of ongoing operations, each team member can devote substantial effort to the success of the endeavor. There are certainly fewer resources dedicated to the effort, but each resource is more effective because it is not diverted to competing activities.

By contrast, the project development team in the established corporation must be concerned with ongoing production, marketing, and sales. Because the R&D representatives are largely focused on development, they consider requests for assistance on field problems or for customization of existing products a distraction. Other team members face even more significant distractions because the major emphasis of their functions is maintaining the ongoing business of the corporation. Personnel development and training programs are additional time commitments. The impact that these competing activities have on the ability of team members to focus on the project is seldom appreciated.

ANALYSIS PARALYSIS

by its very nature, a new enterprise attracts more results-oriented individuals

No single factor has more relevance to the established corporation's lack of responsiveness than its predilection for overanalysis and reanalysis of options. In most cases, the innovation with which the start-up blindsides the established competition has been under consideration for years by a strategic planning group in an established organization but has been viewed as too risky or radical for the market. Each successive pass at an innovative concept seems to become increasingly conservative until the organization finally convinces itself that the product won't be successful.

The most expedient product development actions taken by large corporations are those that respond to competitive product introductions. These initiatives are destined to be late in the marketplace and will result in, at best, adequate second-generation products. A compounding factor is the dilemma a market leader faces in any strategic decision to render an existing, successful product line obsolete. As difficult as this decision is, it will ultimately be mandated by external factors if the company is to remain a leader in the face of competitive innovation.

A variant on the theme of overanalysis is the reevaluation of a project once it is under way. At the first signs of difficulty or schedule delay, risk aversion takes over, more analysis is done—sometimes by a new forecasting team—and the project is canceled. Indeed, the continual turnover of team members is a major cause of schedule delays and goal inconsistencies. US industry's habit of rotating people through assignments without regard to the goal of the assignment must be broken.

POSSIBLE ALTERNATIVES

It is not feasible or necessarily desirable to attempt to simulate the dynamics of managing innovation in start-ups in corporations with established markets. Risks must be managed more carefully when losses can be significant and product introductions in mature markets demand extensive coordination between corporate functions.

This does not mean that managers must accept an overly bureaucratic process. In addition to systematically evaluating the steps in the innovation process that are not value-added, managers can structure some portion of corporate resources to focus on the technology base of new markets, and entrepreneurial dynamics can be encouraged in new product innovations for these sectors. The most widely recognized and successful structure for these enterprises is the dedicated, freestanding venture organization that has the freedom of a start-up and the financial resources of the established corporation. Although other structures might be successful, few will argue that the new ventures group is not a powerful source for driving innovation.

UNLEASHING CREATIVITY

This chapter illustrates the complex nature of corporate dynamics surrounding the management of product development activities. The need to reduce product development life cycles and increase worldwide competition is placing significant pressure on the product development process. Unless drastic steps are taken to change corporate thinking, policy, and action, the established corporation will continue to be outmaneuvered in the area of innovative product development.

Major US corporations have the capacity to be creative but have locked out this creativity from within through bureaucratic, risk-averse, conservative corporate policies. The areas in which a change of culture would have the greatest impact include product planning, team focus, team member consistency, incentives and rewards for risk takers, tolerance of nonconformity, and reduction of functional barriers.

TECHNOLOGY SPIN-OFF

Surmounting the Obstacles of a Start-Up Venture

This chapter details the author's experiences with technology innovation in a multibillion-dollar company and the ensuing spin-off of a start-up company with a charter to develop and market the innovative technology to the oil and gas industry. These events occurred at a time when oil cost $10 a barrel and economic conditions throughout the industry were at their lowest level in several years.

Yuv R. Mehra

During years of growth and prosperity, the oil and gas industry went through several stages of technological development. In the course of this development, the oil and gas processing plants became increasingly larger to meet stringent product specifications. The industry's rapid growth resulted in a supply of raw materials that exceeded demand. With a shift in the supply/demand ratio, operating flexibility became important to maintain profitability.

The flexibility to meet market demand was almost nonexistent in most of the processing plants. Most of the facilities were operated using systems that were established before the availability of microcomputers, which contributed to the lack of flexibility. It was apparent that an order-of-magnitude change in operations was needed to meet the profitability challenge.

THE INNOVATION PROCESS: A FRESH LOOK AT ESTABLISHED OPERATIONS

Although the last significant technological innovation in light hydrocarbon separation occurred approximately 25 years ago, it was anticipated that the challenge posed by shrinking operating margins would encourage the industry to accept innovations. However, the lack of recent innovations indicated an industrywide reluctance to try new concepts. The general view was that to be successful, any innovation had to offer a simple solution.

Before working to develop a desired solution, it was necessary to understand the salient features of commercially available technologies already in use in the industry. A review of the technical capabilities and limitations of these technologies led to the development of a hybrid concept that capitalized on the positive aspects of a few technologies and minimized their drawbacks. The process basically came down to taking the time to understand proven industry practices and to sort through a maze of choices to achieve a useful innovation.

YUV R. MEHRA is president and chief operating officer of Advanced Extraction Technologies Inc, an energy-related development company in Houston.

The concept finally proposed had several applications in related yet independent segments of the hydrocarbon processing industry—specifically, natural gas, petrochemical, and petroleum refining. The versatility of the concept was attractive from a business standpoint because soft economic conditions in one industry segment could be offset by favorable conditions in other segments of the industry.

Intellectual Property Issues

Even though this innovation occurred in a large organization, the thought of owning, developing, and marketing intellectual property was relatively new to senior management of the company. Further, because of the simplicity of the concept, which incorporated proven principles in a novel manner, research into the state of the art in this and related technologies became quite extensive. Because the success of any technology business relies heavily on the strength of its patents, however, the company could not afford to bypass an exhaustive patent search, despite the substantial cost. Therefore, on the recommendation of a technology licensing consultant, an independent patent consulting firm was retained to provide comments on the patent applications on file and to ensure the integrity of the patent portfolio.

An additional benefit of hiring outside counsel was that the technical group obtained expert guidance on what constituted allowable disclosures to internal and third parties. New guidelines had to be learned over time to help company personnel distinguish between information that pertained to patents, know-how, trade secrets, and proprietary versus confidential issues.

The patent consulting firm engaged in this capacity was also instrumental in helping the company sort through the complexities of patent filings. Although patent filing procedures are challenging in themselves, even trickier issues are selecting the countries outside the US where patent protection will be sought and determining the right time to make these filings, especially when the full potential of the invention is not clearly defined.

It is tempting for inventors to seek worldwide patent coverage, but the reality is that patent filing, prosecution, and maintenance costs are substantially higher abroad than in the US. Unlike the US, which grants de facto patents on a first-to-invent basis, most countries follow the first-to-file system, which means that if a patent is not initially filed in a particular country but is desired in that country at a later date, it may be too late to file for and obtain a patent. For this and other reasons, it is an absolute necessity to analyze carefully the potential of foreign markets both from geographic and political viewpoints. Part of that analysis should address the enforceability of patent rights in those countries.

Technology Tests and Demonstrations

While the patent filings and their protection were being taken care of, a simultaneous effort was necessary to demonstrate the key features of the proposed technology. Because even a small pilot-scale facility requires a substantial initial investment, the selection of a proper location is always a time-consuming and challenging task.

Given the nature of the proposed concept and senior management's lack of familiarity with the nature of research, there was at first a strong incentive to develop an additional commercial production source that used the new technology. After several alternatives were explored, however, it was decided that a small pilot unit would be built in one of the existing plants operated by the company to reduce the initial research and development investment.

because the success of any technology business relies on the strength of its patents, the company could not afford to bypass an exhaustive patent search

The portion of the existing facility that was to be used to conduct the pilot production run based on the new concept was isolated from the rest of the operating plant. The available equipment was inspected and retrofitted according to the requirements of the new process. The constraints of the available resources necessitated that the pilot tests focus on only the key features of the technology. A team of dedicated professionals from the engineering and operations departments of the company was charged with carrying out the necessary testing.

After four months of construction and testing, the plant was successfully started up. Within 20 minutes of feedstock (i.e., raw material) input, the product was obtained. Achieving this objective (i.e., the process innovation) was an experience shared by the engineering and operations teams, and the satisfaction felt by all participants is now quite difficult to express in words. From that happy moment and for the next 11 months, data was collected continually to define and document the capabilities, scope, and potential of the new technology.

After the salient features of the proposed technology had been demonstrated and patent applications were filed in the US and 22 foreign countries, the next step was to build a commercial demonstration facility. This facility was a necessity not only because it was a strategy for overcoming the industry's resistance to innovation but also because it would be the basis for a potential licensing business. To carry out the licensing activities successfully, it was projected that an additional investment of $11 million would be required to cover operating expenses for the next three years and the cost of constructing and operating the commercial demonstration facility.

A Change in Course

The technology innovation process took a turn in another direction when in 1983 and 1984 the parent company was purchased by another major corporation that wanted to expand its business into the energy sector. As expected, a reorganization plan was put into action to consolidate the operations of the two large companies. Given the oil and gas industry's state of economic decline and stagnation in 1986, when oil cost less than $10 a barrel, it could have been a long time before a commercial-scale facility incorporating the new concept would be built.

Even though the group handling this technology innovation in the acquired company did not have a counterpart to merge with in the acquiring company, the opportunity arose to spin off the technology. Contrary to the general public perception, it has been this author's personal experience that in the energy industry, innovation and its acceptance are achieved by the operating companies and not by the engineering contractors. Consistent with this idea, it was preferred that the technology stay with the operating company that had championed it.

Recognizing the different priorities on which the new corporation was focusing its efforts, it was decided that the technology would be divested. Among the potential purchasers of this technology, the following candidates were identified:

- Other operating companies (as opposed to engineering contractors) in the hydrocarbon industry.
- Engineering contractors offering services to the hydrocarbon industry.
- Private individuals interested in establishing an organization with the charter to develop and market the technology to the hydrocarbon industry.

As expected, there was little response from the operating companies, who were facing the same or similar economic conditions that were primarily responsible for the originating company's decision to divest the technology. Interest in acquisition of the technology was expressed by some of the major engineering contractors—provided that the originating company commit itself to building the first commercial plant during the following year (1987). This requirement was inconsistent with the corporation's directives and was unacceptable.

The lack of interest or acceptable terms from potential buyers was not the only problem the originating company faced with possible divestiture; the matter of patent rights—which permit the owner to exclude others from making, using, or selling the invention—also had to be addressed. After the company expended considerable resources to bring the technology to its present state, senior management concluded that a new patent owner would develop the technology fully but would withhold it from the rest of the industry. To ensure that the innovation would be made available industrywide, senior management decided that a small company, independent of the parent corporation, should be formed using acquired venture capital.

THE START-UP COMPANY: FINANCING FIRST

The timing for starting up the new company was not auspicious. In 1986, venture capitalists were in no mood to even talk about financing a high-risk energy-related start-up, and the banks were of no help either. In addition, the new technology—which existed only on paper—was not projected to produce any revenues for the company during the first three years of its operation.

All was not doom and gloom, however. One factor weighing in the start-up company's favor was the overall strengthening of patent enforcement efforts, which made start-ups based on new technology patents more attractive to investors. The patent applications that the originating company had filed had the potential to produce a portfolio of 17 patents in the US, with corresponding coverage in 14 additional countries. A highly respected patent law firm was consulted, and it confirmed the strength of the proposed patents. It was apparent by then that the strategy for the new company had to be based on the potential of the patent portfolio.

Financing for the start-up company had to come from individuals who had both a vision and the desire to get into the energy business. It was decided that the most effective way to locate and gain access to such individual investors was to seek out an intermediary. This intermediary would need a strong financial orientation as well as the legal background necessary to be able to create the means to offer company shares through private placement in exchange for capital to cover operating expenses over the projected lengthy sales cycle.

after several alternatives were explored, it was decided to build a small pilot unit in one of the existing plants

Searching for an Intermediary

After numerous discussions with company managers and independent consultants over a six-month period, a strategy to obtain financing was in place. The company that owned the technology would put forth the terms it was willing to accept; provided these terms were agreeable to a prospective intermediary, the company would offer the management of the start-up corporation to that intermediary. Through business contacts, such a candidate was identified and a meeting was set.

The company's strategy made possible a deal whereby a new corporation was formed to acquire the technology, and the inventor was transferred to the start-up to assume responsibility for daily operations. To minimize potential conflicts of interest, this individual refrained from participating actively in the negotiations. His immediate role was to provide information to both parties and, in the process, strike a balance between maximum return for the current owner and a final agreement that all parties could accept after the technology was transferred to the new organization.

The agreement also provided for sharing future royalties received through a minimum payment over a specified period of time. Although the six-month negotiation period had its exciting and frustrating moments, it was ultimately rewarding because an agreement was reached that all parties were enthusiastic about.

There was much apprehension on all sides regarding the prospects of the start-up. All parties approached the new venture in a positive manner, with self-interest taking a back seat to a new spirit of cooperation dedicated to achieving success. Each party made a conscientious effort to show utmost respect for the others' abilities. If a lawyer does not engineer projects and the engineer does not practice law, there is reason to believe that the relationship can grow—provided that both parties nurture one another, meet each other's needs, and accept their own limitations at all times.

Raising the Necessary Capital

Acquiring the technology merely gave the new organization the go-ahead to capitalize the company; raising the capital was a new and very real challenge. Previous informal discussions with prospective individual investors needed to be formalized. Given the uncertainty of the company's actual future performance as compared to estimates of revenues and returns on investment, it was also essential to analyze these projections carefully. The final numbers could not be so conservative as to jeopardize financing, especially at a time when investment money was so tight. On the other hand, the figures also could not be so optimistic as to mislead the investors.

Having given up all hope of raising $11 million, which would have allowed the new corporation to capitalize fully and to control its own destiny, the plan now was to raise $3 million to cover at least the first three years of operating expenses. Because no revenues were projected during that period, the offering was not attractive to many potential investors, and only $2 million was actually raised. Undercapitalization put significant pressure on the team to seek a commercial demonstration facility as quickly as possible and to keep expenses to a minimum.

Setting Up Shop

The next step was to establish the business in new quarters. The team soon learned that the $2 million obtained in initial financing and the potential of the acquired technology were not enough to obtain an office lease in Houston in 1986. Abundant office space was available to those with a corporate umbrella; however, because the company was a start-up, prospective creditors demanded answers to the usual barrage of questions, including:

- Who are you?
- How do we know you will pay your bills?
- Will the former owner sign a collateral agreement?
- How about personal guarantees from company officers?
- What is your company's credit history?

Personal guarantees were out of the question, and clearly no credit history was available. Despite substantial obstacles, the team was eventually able to get an office space that suited the company's needs.

Likewise, the team was completely unprepared for the complexities of setting up office, including obtaining telephones, computers, and accounting services and purchasing the necessary furniture and supplies needed to equip a business. Team members' experience working in a big company was of little help, because these support functions were often delegated to others on the staff.

Hiring Staff. To say that it is not easy to find qualified staff for a start-up business is an extreme understatement. It requires nothing short of convincing creditable professionals to leave organizations for which they have worked for several years, perhaps even with an acceptable level of job satisfaction, to join a company that did not exist yesterday. In addition, one of the former mentors of the technology recommended seeking only employed candidates and ignoring those who were actively seeking jobs. That additional criterion certainly did not make the task any easier.

The company's greatest advantage in the staffing issue was the number of years team members had had in the industry and, consequently, the number of personal contacts they had made, many of whom were personal friends. Believing that the team offered an opportunity to share in the success of a new venture, it turned to those individuals with whom its members shared a mutual understanding and with whom they were compatible, recognizing the risk that the pressures of a start-up effort could sour those relationships. By focusing on the task at hand and by using a commonsense approach, the team felt certain that it would be able to overcome any obstacles.

the plan was now to raise $3 million to cover at least the first three years of operating expenses

Marketing. The prospect of marketing the new technology was daunting. It was a small start-up operation selling new technology that required a substantial investment from the large, bureaucratic organizations in the industry. The initial obstacle was the not-invented-here syndrome—that is, experts reacting unfavorably to receiving advice from a small outfit. The other major obstacle concerned the size and the nature of the companies approached. The long planning cycles in the large organizations are tantamount to stagnation from a small company's standpoint. By the time large companies consider taking an interest in a proposal, the small company could be out of business. This difference in operating speed was anticipated when financing was sought for the start-up.

These problems notwithstanding, however, the start-up company's single largest obstacle has remained the lack of a commercial demonstration unit. Well-qualified and experienced engineers in companies to which the technology has been marketed have been unable to give acceptable reasons why the new technology would not work. Given this universal agreement that the technology has merit, it is difficult to explain why nobody wants to be the first to prove it.

Engineers who have seen presentations about the proposed technology understand its inner workings and concur that the risks are economic and not technical in nature, but internal directives from their management say not to try anything new. Most of the managers responsible for making project decisions are not technically trained and therefore do not understand why their technical people are frustrated. The start-up company tried on one occasion to help close the gap between the managerial and technical parties in the same organization but was advised that it was interfering with the company's internal politics and procedures.

As an alternative form of introduction for the proposed technology, the start-up company applied for research funding from the Small Business Innovation Research program sponsored by the US Department of Energy (DOE) for two consecutive years. Both applications were denied on the grounds that the concept, which is based on sound principles of chemical engineering, is quite simple and therefore does not deserve a research grant. It is this author's belief that the DOE really means that the project's probability of success is so high that government funding is unnecessary.

AT THE END OF THE TUNNEL

Through faith and perseverance and thanks to the continuing support from investors and industry associates as well as the addition of resources from some other small ventures, a project has been identified that will conclusively demonstrate the technology in 1991 under the cosponsorship of an industry-funded organization. During the past four years, the company has been able to recapitalize using resources provided by the company's initial investors, and it foresees the day when it will eventually taste the fruits of a long market development cycle.

Perhaps the most important lesson to be learned from this experience is that all parties involved in a start-up must understand that there are no quick returns. To achieve high returns on their investment, they must be in for the long haul.

AGING TO PERFECTION

Managing Innovation in Large Middle-Aged Companies

Innovation is the lifeblood of competitive organizations. It is rarely constrained by lack of creative talent or limits to change based on principle (e.g., the so-called laws of nature). Problems concerning innovation reside with organizational expectations, processes, and structures. In a sense, this is good news—"We have found the enemy and it is us."

Louis A. Mischkind

Innovation is the application of new ideas to create new processes, technologies, or products. The success of an innovation is measured by how much it contributes, directly or indirectly, to revenues or profits. To ensure success, innovations in product development or technology must be coupled with creative initiatives in manufacturing, marketing, finance, service, administration, and human resources. Magnificently engineered devices with no markets or profit margins are simply unacceptable to sources of capital. This does not mean, however, that failed innovations should be ignored forever. On the contrary, resurrecting creative ideas that are technically sound but were formerly found impractical is an innovative venture of the highest order. Complex enterprises can benefit from archiving fundamentally sound ideas whose time had not yet arrived.

The fountainhead of the creative initiative in organizations is dedicated and intelligent people; they are the basis for any attempt to foster improvements in innovation in product and technology development. Although an individual's character and intelligence are developed long before corporate affiliation, the organization has the power to encourage or inhibit the full measure of these qualities.

Exemplary management and progressive corporate cultures tend to attract and retain vital and committed employees. A creative climate, however, often requires actions that elude otherwise well-run companies. This is especially true as companies become successful and their cultures reach middle age.

It is difficult to determine when an organization is no longer young. Significant development cycles last somewhere between two years to five years. After several cycles have occurred, most organizations begin to develop a past, along with some indelible habits. After twenty years, middle age is inevitable. Conscious rejuvenation then becomes imperative if the organization is to maintain a competitive innovative advantage.

To intensify the problem, periods occur when a considerable increase in technical and product innovation is absolutely essential for survival. (The information

---

LOUIS A. MISCHKIND, PhD, conducts research and installs solutions in Fortune 500 companies on such subjects as communication, innovation, entrepreneurial action, technical vitality, morale, and executive selection and development. His focus has been on high-technology organizations. He received a PhD in organization psychology from New York University.

and semiconductor industries, among others, are currently in such stages.) In addition, organizations may have cultures in which technical personnel perceive themselves as second-class citizens compared to employees in sales, financial, and legal departments, which further exacerbates the situation.

Under the combined circumstances of middle age, intense turbulence, and perceived lack of appreciation for technical creativity, improving innovation becomes urgently necessary. There is no time for vacillation; extraordinary improvements are essential, from the plant floor to the executive suit.

This chapter focuses only on product and technology development—two critical areas of development that, it is commonly agreed, must be afforded the highest priority if an industry is to survive and prosper in the 20th century and into the 21st century. This chapter therefore identifies the major challenges facing senior management, points out ways to develop a creative organizational climate, and describes four simple processes that can lead to increases in innovation within well-established and complex environments—that is, middle-aged companies.

PROBLEMS AND CHALLENGES IN MANAGING INNOVATION IN MIDDLE-AGED COMPANIES

As an organization matures, originality gradually erodes. This is the disease of organizational aging. Age is not really the issue; rather, it is the cultural characteristics that frequently arise as an organization matures. These cultural characteristics (i.e., problems) are described in the following paragraphs.

One problem stems from a tradition of successful innovations. After experiencing success, an organization tends to repeat the precise sequence of actions deemed responsible for the success. Inevitably, a time comes when these actions are no longer appropriate. Inertia, brought on by habit, vested interest, and superstition, seriously hampers innovation.

Another problem stems from failed innovations. After experiencing failure, an organization tends to try to avoid future fiascoes by installing formal and informal rules. These controls are frequently predicated on faulty logic and half-truths and, more important, are maintained long beyond their usefulness. Unexamined constraints are the bane of innovation.

A third problem is that a value system ultimately takes hold that narrowly defines which innovations are possible and broadly construes which are impossible. This mind-set results in a self-generated imprisonment in the past. Yet another important problem occurs when an organization paradoxically demands instant spontaneity. Organizations are often frustrated because innovations do not occur on schedule. It must be understood that management's function is to foster creativity, create an environment conducive to innovation, and be patient.

Furthermore, an effective operational organization (i.e., the core business) operates on premises that can conflict with innovation teams. After all, the operational organization is constructed to produce low-cost, high-quality products in quantity. This goal is admirable: without success in this regard, there is no capital for innovation. That there should be conflict between these sides of the business is ironic—

innovations are produced to fuel the operational organization. Nonetheless, there is no escaping the fact that innovation must be treated as a separate activity.

An organization's core structure can be compared to an elephant that tramples every innovative mouse in its path. The operational organization can virtually guarantee a marginal return for every dollar spent on its activities. The innovation organization can only promise to sustain a future revenue stream (if it is successful). This is not such a powerful argument when management is measured on a yearly or, more often, quarterly basis. Innovation is always vulnerable when its activities are embedded in the operational organization.

an atmosphere favoring innovation must be created by the words and actions of senior management

SEVEN CHARACTERISTICS OF SUCCESSFUL INNOVATION

A review of the problems of innovation teams in large, middle-aged organizations reveals seven characteristics of successful innovation:

- Senior management sponsorship and support.
- Clear and appropriate goals.
- Team selection and balance.
- Intimacy with the market.
- Flexible organizational structure.
- Flexible controls and measurements.
- Appropriate incentives.

These characteristics of success are described in detail in the following sections.

Senior Management Sponsorship and Support

An atmosphere favoring innovation must be created by the words and actions of senior management. Through words (i.e., official company communications), senior management tells employees what it wants and expects. Through such actions as direct, intensive sponsorship of innovative activities (e.g., the Manhattan Project, the Apollo program), senior management demonstrates its true intent.

Clear and Appropriate Goals

Although senior management can create a favorable environment and sponsor specific programs, practical innovation must spring from sources much deeper in the organizational hierarchy. Nothing can happen until teams focus on specific issues requiring creative output.

These goals must be stated in specific technical, financial, and market-oriented terms. Broadly demanding something new, exciting, and revolutionary is likely to lead to frustration and wasted effort. The first step in launching innovation is a hard-nosed investigation of what needs to be done. This inquiry must be properly executed by interfunctional surveillance teams of innovative individuals from the technical and business communities. The use of consultants is also advisable during the early stages.

Successful innovations are occasionally big breakthroughs. More often, they are such evolutionary, piecemeal changes as significantly improved versions of existing products and processes. Whether radical or basic, innovation goals must be expressed in precise technical and market-relevant terms.

Team Selection and Balance

Invariably, new product development managers say the team is the key to successful innovation. They emphasize that successful teams have intense cohesion and high levels of mutual respect. Personality conflicts do occur, but this is not unhealthy; they can be managed if each individual values the roles others are playing and if there is minimal duplication of effort and creative interpersonal leadership. Successful innovation teams require two major roles: technical and entrepreneurial.

Technical Team Members. Technical talent comes in two varieties. In both cases, there is a mastery of fundamentals. The first variety consists of technical specialists who understand history in terms of past failures and successes. They are street-smart, with extensive informal networks. Most often, this type is associated with long tenure.

The second variety consist of technical specialists filled with ideas about how to do things differently; the force of precedent does not weigh heavily on their shoulders. Their knowledge is leading edge. They are usually new to the business, having recently been involved in university training. They are usually short-tenured and young. (Age, of course, is not the criterion. Some people are disposed to creative expression for the length of their careers.) Some team members show characteristics of both varieties.

Entrepreneurial Team Members. A cliche in venture capital circles is that it is better to back a first-rate entrepreneur with a second-rate idea than a great idea tied to someone who can't pull it off. First-rate entrepreneurs are pragmatic visionaries who infect those around them with a sense of urgency. The identification, cultivation, and use of this talent is the most important condition for successful innovation. Entrepreneurs fill an innovation-enabling role—they create an environment of possibilities, which is the real mother of invention.

Achieving a Functional Balance Among Team Members. A mixture of functional skills is critically important for achieving short cycle times and ultimate success in the market. Innovative products and technologies must pass through several functional organizations before they reach the consumer. To facilitate this passage, innovation teams must include product designers, manufacturing, financial, marketing, and service personnel from the very beginning of the project. Each of these groups will become more active as the development cycle unfolds; however, they must start together and share a common team dedication.

In short, team balance involves the proper mixture of creativity, experience, leadership, and the range of skills needed to move an enterprise from idea to product shipment. It is possible (and desirable) for a single team member to fulfill several of these requirements. Nonetheless, the era of the one-man band is over. The situational demands of modern innovation have become so complex that team formation must be a conscious endeavor.

first-rate entrepreneurs are pragmatic visionaries who infect those around them with a sense of urgency

Intimacy with the Market

Innovators yearn for more intimate association with the end user. The reason is obvious: for a product to succeed, it must appeal to a willing buyer. The relative number of buyers is another name for market share, which is one of the prime measurements of business success.

Customers have access to numerous competitive offerings in the marketplace. As a market matures, buyers become more sophisticated and develop a clearer understanding of what they want in terms of such product features as function, price, aesthetics, and ease of use.

Without direct contact with customers, innovators are at a tremendous disadvantage. Often, no information is available to them except what they read in trade journals or hear through the grapevine. Even if they are fortunate enough to have access to the second-hand accounts of salespeople (which happens rarely), much can be lost in translation. Even direct contact with customers (which is an even rarer occurrence) can be of limited value because customers may not be able to verbalize what they want in technical terms.

Useful technical understanding comes from first-hand observation of and conversation with customers to discover the kind of product or modification that will make a competitive difference. Successful innovators must be adept at translating customers' needs into technical design. If they are merely astute technically, they must be trained as to develop such translation skills.

Removing the Barrier to Understanding Customer Needs. What blocks the developer from this urgently needed intimate customer contact? As organizations become more successful and complex, the distance between innovators and customers increases. Marketing and customer service often shield the innovator from the user. These groups sometimes justify their actions by asserting that technical people lack savoir faire and may therefore upset the sales relationship by saying the wrong thing to the wrong person. Salespeople sometimes fear that technical types will embarrass the company. Briefings between the two groups could minimize this possibility; any remaining risk is more than balanced by the advantages.

At the other extreme, innovators have been accused of avoiding users, preferring to live in their sequestered laboratory settings. This is a myth. Successful innovators (particularly from leading-edge companies) have said that their failed innovations lacked only a nuance of customer understanding and that, if they had another chance, they would want a deeper appreciation of customer needs.

Flexible Organizational Structure

As mentioned, innovation must be treated as an activity separate from the organization's core business. Therefore, innovation teams must be structured somewhat differently from core business units—especially in terms of team size and in the availability of resources. Typically, when an operational organization focuses on a key issue, enormous resources are deployed. In contrast, terms that have been successful in delivering innovations are almost always lean and mean. They thrive on daily face-to-face contact and simple structure.

Of course, the size and complexity of the team depends on the type of innovation. For example, a revolutionary technical development may involve as few as 5 to 10 individuals. A significant modification to a main-line product, however, may need as many as 100 team members. In any case, innovations are produced by organizations that are streamlined relative to the impact of their results.

With respect to resource availability, the positioning of the team (physically and organizationally) should permit easy access to such key services as information processing, model shops, purchasing, and human resources. This access could be accomplished through vertical integration with direct resource ownership; this is undesirable however, because it leads to organizational obesity.

A much more effective solution is for senior management to issue some type of document of understanding to all employees, stating that the team's activities deserve high priority. The entrepreneurial spirit will take the team the rest of the way. Entrepreneurs are experts in creating informal networks and obtaining the information and resources they need. Whatever structure is adopted, the team should also be empowered to rapidly design, prototype, test, and experiment with its creation in actual customer environments.

Flexible Controls and Measurements

An organization's core business uses measurements and controls to ensure efficiency and thus maximize ROI. The innovation team, however, requires breathing space and time for incubation.

Schedules and reviews in the operational organization are highly structured and not easily amended. In contrast, the innovation team must be loosely bound by these demands, especially during the early stages of innovation. For a specific time period, senior management should adopt a hands-off policy. This moratorium on close scrutiny should probably not be less than three months or more than twelve months—in any case, the timing should be agreed upon in advance.

Many innovations fail because senior management suddenly demands immediate and unequivocal proof of progress. The inability to satisfy this demand usually leads to a scaling down of resources when they are most needed and an ultimate downward spiral to the innovation program.

Senior managers often describe innovators as prima donnas who expect to be pampered and shown infinite patience. On the contrary, successful innovators welcome measurements and evaluations. After all, they believe their offering to be clearly superior to other available products. They do not feel they should be indefinitely free from hard-nosed scrutiny; they only insist that mangement adhere to the timing of reviews that was originally stipulated.

a large organization that wants to pursue innovation must cope with devastating disadvantages

Appropriate Incentives

Some R&D professionals maintain that such special incentives as bonuses, awards, stock options, and profit sharing spur innovative output. Others state that the opportunity for self-expression is the prime motivator for innovators.

Creative people seek situations in which they can be totally immersed in their work. They are often passionate, focused, difficult to manage, even monomaniacal zealots. In their world, monetary reward is welcomed but is perceived more as a way of keeping score than as a prime motivator.

Potential innovators take their lead from the prevailing culture. If winning is treated as business-as-usual, and losing is viewed as a calamity (which is too often the case), many creative people will play it safe. Innovators are creative but not necessarily heroic. Therefore, the incentive arrangement for innovators should closely match that of an ordinary entrepreneurial venture. At the very least, this means that the downside risks for failure (in terms of career advancement and financial compensation) should be balanced with rewards for achievement.

FOUR PROCESSES OF MANAGING INNOVATION

The seven characteristics of successful innovation must be converted to processes that can be implemented. Each process should specify short-term, tactical actions that will expedite innovation. In addition, the organization should describe strategic steps that will imprint an innovative mentality on the corporate culture favorable to sustaining improvements in the long term. In this endeavor, various levels of management must play their part.

Focus on the Importance of Innovation and Specific Challenges

This approach establishes innovation as a top priority. It also involves the systematic selection of specific challenges to attack. Strategically, management must be philosophically and emotionally dedicated to the future. It must ask why not instead of why. A culture must be created in which the concepts of *surpass* and *leap* are equal to those of *maintain* and *reap*. Senior management should allocate money, time, and other resources for the future because it is convinced that the organization's survival is at stake. Tactically, the organization can implement the specific, immediate actions described in the following paragraphs.

Introduce Media Events Featuring Innovation. Suggested events include articles in company publications and videos. The content would review and celebrate past achievements and declare the organization's intentions for continued focus on innovation.

Establish Corporate Directives and Instructions on Innovation. These would ask management to show progressive improvement in the use of staff resources for creative purposes. Such instructions have had healthy effects in such areas as quality and service. The key is to spell out methods and measures of compliance.

Identify Areas in Need of Innovative Action. Every year or so, senior management should identify general targets to attack. This activity should usually highlight industries or market segments, infrequently a specific product, and never a method. At the next level of the management hierarchy, every major line of business should identify significant and specific gaps in the product line or technology. This endeavor should be carried out by interdisciplinary surveillance teams. In turn, innovation teams should develop specific details and plans. If one strategic gap could be correctly identified every two years by each business area and then successfully addressed, corporate performance would improve enormously.

Allocate and Protect Financial and Staff Resources for Targeted Innovation Programs. The actual figures allocated would be a small proportion of the total budget and headcount because innovation teams operate most effectively when they are relatively constrained. It is of critical importance, however, not to withdraw resources that have been earmarked for innovation and shift them to the core business during a crisis or because unforeseen problems develop in other areas of the business.

Strategically, the organization must develop a method of accountability for the general management of innovation activities. Performance evaluations of executives must include innovation responsibilities. For example, the 3M Corp provides measurements that ensure accountability for innovation. One measurement is that 25% of revenues must come from products developed during the past five years.

Select and Educate Employees for Innovation Projects

Using this process, management recruits and identifies those individuals who are best suited for innovative activities and then develops them for practical action. Strategically, employee selection must begin with initial recruiting of new hires, and development must be a continual process that extends from new hires (i.e., trainees) to executives.

The character traits of innovators often include impatience, irreverence, high spirits, and considerable self-confidence. Although these traits are certainly not universal, management must be prepared to deal with unconventional people with big egos. Because they will be relatively more difficult to manage, the challenge is to assimilate innovators into the prevailing organizational culture. Whether they like or dislike it, innovators must understand that the shortest distance from idea conception to market share is rarely a straight line. The company recruiting logo should read: "Wanted: Talented Troublemakers Who Are Willing to Learn How to Make a Difference." Tactical actions that can be instantly implemented are described in the following sections.

Assign Innovators to Recruit New Hires. This approach is based on the maxim, It takes one to know one. It is not enough to set educational grade-point standards or focus recruitment on prestigious schools if recruiters are incapable of distinguishing good students from innovators. Instead, the recruiters should have technical sophistication and an appreciation of the difference between mundane and creative mind-sets. Direct conversations with recruits along with examples of their output frequently provides enough information to allow innovators to spot promising candidates.

the skills needed for a specific effort will always cross functional lines and will usually cross business lines

Identify Innovative Technical and Management Potential. The organization must distinguish between innovative and operational types within the existing employee population. This should be systematically accomplished through programs that assess an employee's executive and technical abilities. In the short term, much can be achieved simply by asking managers and peers to answer the following questions:

- Who are the people you seek when you are technically stuck?
- Who do you seek when looking for new approaches?
- Who do you want on your team when it's crunch time?
- Who knows best how to handle bright bucking broncos?

The answers to these questions should then be reviewed to identify the employees most frequently mentioned.

Share Technical and Managerial Personnel to Serve on Innovation Teams. The talent search must not be confined. The proper skills needed for a specific effort will always cross functional lines and will usually cross business area lines. All departments are loath to let their best people go (even if they are not currently fully utilized). This dilemma calls for organizations to establish an innovation team broker at the highest level of the organization. This individual must have an extensive understanding of talent and competent interpersonal skills.

Establish Broadening Programs to Develop Innovative Technical and Management Talent. The organization should institute programs to accelerate the development of business acumen in the pool of identified innovators. Although some formal training should be considered, the focus should be on experience. Components of an accelerated business experience program for technical individuals may include:

- Assignment to a marketing organization—Experience with marketing is invaluable. The assignment should be long enough for the innovator to develop:
  - A deep understanding of the customers who use (or will use) the developer's products.

- An appreciation of the sales mentality.
- An interpersonal network that will keep the developer in continual contact with the marketplace.

- Assignment to a manufacturing organization—The goal here is to deepen an appreciation of the challenges of product manufacturability and quality. Again, it should last long enough for the innovator to develop real knowledge and informal contacts.
- Assignments or training in finance—Many technical innovators know almost nothing about how organizations are measured. This ignorance can considerably ruin their credibility. Financial training should provide skills in pricing, competitive analysis, forecasting, cost analysis, and other business measurements.
- Assignments as assistants to executives—These jobs alert the young developer to the issues and demands of managing the big picture. Mentor relationships could be established with value that lasts long beyond the assignment.

Tactically, these assignments could be offered on an ad hoc basis. Strategically, the organization should institute a systematic program encompassing all of the assignments described.

Organize and Use the Innovation Team

This process deals with getting the job done. The goal is swift development cycles and high product success rates in terms of customer acceptance. This happens when the right people are put to work on the most important programs within a conducive organizational environment. On a continuing basis, senior management in each business area must put carefully selected teams in place to exploit significant technological and product opportunities.

The seven characteristics of successful innovation point to some of the tactical actions that should be required of the innovation team. These are described in the following sections.

The Innovation Team Should Consist of Motivated Individuals Who Are Innovation Oriented. The best people from all corners of the organization should be made available, and innovation team membership should be voluntary.

The Innovation Team Should Be Small, Frugally Funded, and on a Demanding Schedule. The budget and schedule should be developed in a participative manner.

The Innovation Team Should Be Interdisciplinary. For example, at inception, designers and marketers are likely to be more active than manufacturing and financial personnel. Nonetheless, the presence of all is important.

The Innovation Team Must Be Afforded Certain Freedoms. Specifically, the team should be free to translate the focus area into technical and business plans and specifications. The result may differ to some extent from the vision of the original surveillance team; still, this is appropriate. It is a waste of creative talent to demand strict adherence to early direction and to impose rigid constraints on the development process.

the innovation team should be subject to informal but hard-nosed critiques of its plans and assumptions

The Innovation Team Should Be Given Ready Access to Required Services and Audiences. The services mentioned (e.g., model shops and tools) allow for unfettered opportunities to produce prototypes. Access to audiences—for example, leading-edge users—allows the product to be tested for real-world feasibility.

The Innovation Team Should Conduct Several Cooperative Reviews. Such reviews subject the innovation team to informal but hard-nosed critiques of its plans and assumptions.

The first step in the review process is the selection of experts who are not associated with the program. They should receive information about the project in advance and then spend one or two days listening to the project team present its case. The experts should subsequently offer suggestions and evaluate the probability of the product's success. Ultimately, the team decides whether or not to act on this advice.

A team should plan to conduct a review as soon as it has a firm idea of its technical and market direction. As the team passes through critical early checkpoints, the review process should be repeated. This technique has been used by the 3M Corp for more than 20 years; 3M believes it is one of the cornerstones of the company's success.

The Innovation Team Should Periodically Focus on Team Development. This activity involves a two- or three-day retreat to build work relationships, share ideas, and develop intensive plans. At first, the use of a facilitator is recommended. A team development retreat should be held immediately after the team is formed and perhaps at other times when cooperation, morale, and fundamental plans need to be reexamined.

Promote Operational Harmony in the Innovation Team

This process has two goals: to ensure that innovative developments are effectively transferred to the core business, and to maintain the morale of innovation team members as they return to the core business and as they face the potential apprehension of colleagues who were not on innovation teams. The following sections provide guidelines for meeting these goals.

Transferring Products to the Organization's Core Business. The interdisciplinary composition of the innovation team is proved valuable when the product is ready to be transferred to the operational organization. For example, when a successful hardware innovation is transferred to manufacturing, the manufacturing members of the innovation team will be responsible for the transition. Simultaneously, the inclusion of team members from marketing should lead to more accurate product availability dates and smoother delivery to the customer.

Maintaining the Morale of Innovation Team Members. Management must be careful that a feeling of elitism does not develop among members of the innovation team. Management must also ensure that operational personnel do not perceive themselves to have second-class status. For example, special incentives for innovators must be handled with care. One method is to distribute the gains of corporate entrepreneurs into three portions: one-third to the parent organization, one-third directly to the corporate entrepreneurs, and the final one-third to an intracapital account to be used by corporate entrepreneurs to fund future ventures. Although this unique proposal may have merit in limited circumstances, its effect would be profoundly negative on the employees not involved with innovation, making it especially difficult for innovators to reenter the core business. Therefore, this method is not recommended.

To keep up the morale of innovators and to minimize any negative impact on the rest of the employee population, the following tactical actions are recommended:

- Select team members who have previously demonstrated innovative output—When individuals are widely respected by their peers, their presence on the team should produce little negative reaction.
- Select team members on the basis of informed consent—Team members must be fully cognizant of the demands and risks of the assignment. Such risks or inconveniences may include early termination of the project, long hours and hard work, and dealing with the inevitable reentry into the general employee population.
- Provide incentives for achievement, including one or more of the following:
  - Corporate awards—If possible, these should be conferred on the team rather than on an individual member.
  - Promotions and salary increases.
  - Stock options and awards.
  - Continued challenging assignments—Success makes team members prime candidates for future ventures.
  - Creative gain-sharing—One interesting technique is to issue token stock in the organization. Additional shares could be issued for exemplary individual and group performance. At some point in the future, the token stock could be given real dollar value and redeemed in cash by the bearer.

THE BEST MIX

Organizations approach innovation crises in a variety of ways. One is to establish an office that receives new venture ideas from the general employee population. A technical review panel sifts through the submissions. Those that pass initial screening are converted into business plans and are either submitted to potential sponsors or nominally funded for further development. Although this approach is highly egalitarian, it is esentially hit or miss. The promising ideas often yield a not-invented-here response from prospective sponsors. On the other hand, when organizational cultures are primed to accept innovation ideas from unlikely sources, this approach can be successful.

the operational organization must spearhead and nurture innovative projects as offshoots of the organization's core business

Another approach is to establish new venture organizations (also called independent business units) that are completely separate from the core business. These are vertically integrated business units that own multifunctional support groups. This approach, however, has had a spotty track record. It also generates the not-invented-here resistance; it is costly and unwieldy; and it often forces organizations into areas in which they have little experience or understanding.

The processes recommended in this chapter require the operational organization to spearhead and nurture innovative projects as buds off the main stem—that is, as offshoots of the organization's core business. This is certainly not a novel idea. Blue-ribbon task forces have existed for centuries. For the most part, however, these groups are formed in crisis mode to correct the mistakes and omissions of the core business unit. Instead, such groups should be created as a first choice rather than as part of crisis management.

The days are numbered for large, permanent, advanced product development groups. A large organization that wants to pursue innovation must cope with devastating disadvantages. The characteristics of successful innovation call for teams that are small, temporary, highly focused, meticulously selected, flexibly controlled, and creatively managed. In the past, such teams have formed the cutting edge of corporate growth and they continue to promise such growth.

Finally, and most important, these teams should be under the umbrella of the general manager of the core business unit. This individual must be held accountable for both current and future profitability. In truth, these responsibilities define the term general manager; the steps presented in this chapter show how these responsibilities can be fulfilled.

WHY NOT INVENTED HERE?

Establishing a Corporate New-Venture Program

Despite several distinct advantages over independent, venture capital-supported entrepreneurs, large corporations have not achieved comparable returns from their investments in business proposals developed and supported through internal innovation and new-venture programs.

Robert J. Tuite

The publication of Gifford Pinchot's book, *Intrapreneurship: Why You Don't Have to Leave the Corporation to Become an Entrepreneur* (New York: Harper & Row, 1985), signaled a rebirth in the interest among large corporations in finding a way to internalize the venture capital process for new-business development. During the 1980s, several large corporations experimented with entrepreneurial approaches to technology-based new-business development, with mixed results. For example, in one large, highly diversified corporation, a corporate venture board installed a worldwide, bottom-up innovation program that, by the end of 1990, resulted in the use or commercialization of more than 140 new processes, products, or businesses. The program has resulted in more than 4,500 proposals—almost 1,000 in its peak year—and nearly 90% of the more than 140 adopted proposals were funded by and implemented in an existing business unit or another organization within the corporation (e.g., manufacturing or research). In addition, the venture board itself sponsored 14 internal start-up ventures that were managed, at least at inception, by their "intrapreneurial" founders.

On the other hand, with annual R&D expenditures approaching or exceeding $1 billion, large corporations often find that their size and complexity make technology-based diversification outside their established businesses extremely difficult. Despite seemingly overwhelming advantages in technological capability, marketing expertise, channels of distribution, and human and financial resources, large, technology-rich companies have been consistently outperformed in the development of new businesses by small, venture capital-backed start-up companies. The experiences of corporations that have tried a free-enterprise, business incubation approach to internally generated new-business development offer many valuable insights for other companies contemplating an innovation and internal venture program.

FOSTERING INNOVATION

The origins of many large corporations can be traced to lone entrepreneurs with a single product based on a core technology. In most cases, R&D in these corpo-

---

ROBERT J. TUITE, PhD, has more than 30 years of experience in research, technology management, and new-venture development, both within a large Fortune 50 corporation and in the external venture capital community. He is currently president of RJ Tuite Associates, a consulting firm specializing in innovation, technology management, and entrepreneurial corporate business development; Tuite is also on the board of directors and is director of investment evaluation for Tech Ventures, Inc, a venture capital firm that specializes in seed investments in technology-based start-up companies.

rations began primarily as a defensive effort aimed at maintaining and expanding the original product line. From this modest beginning, however, R&D eventually evolves into a corporate institution with a large annual budget and must meet the demand for increasingly sophisticated products that provide opportunities for business expansion. A company's investment in research eventually yields a large body of scientific knowledge and an extensive, applications-oriented technology base. Viewed in this context, a corporation's commercialized products actually represent only the tip of the technological iceberg, beneath which lies a much broader base of knowledge and technology.

Large corporations are responsible for most of the technology currently developed in the US. Of the $150 billion invested in R&D during 1990, more than 70%—$108 billion—was targeted for R&D performed in and by industry, and the percentage is even higher for applied research and development.[1]

Despite this overwhelming advantage in technological assets (and perhaps even greater advantages in human, financial, and informational resources as well as marketing expertise and access to distribution channels), large corporations have generally been outperformed by small companies in the process of forming new businesses. Small businesses (i.e., those having less than 500 employees), often backed by venture capital investors, have produced more innovations per employee, moved products to market faster and at lower cost, and recovered their initial investments more quickly than large corporations.[2]

As a company grows larger and more complex, it becomes increasingly difficult to exploit the company's vast intellectual property assets by developing new products or serving new customers, particularly in markets beyond its core businesses. Although new products derived from the company's technology base may have exceptional market potential, many are never commercialized because they do not conform with the company's strategic direction, marketing approaches, distribution channels, willingness to take risks, or corporate image. In addition, the corporation may not have, or be willing to dedicate, the human and financial resources required to exploit the new opportunities, especially if they come at the expense of support for its existing businesses. Thus, the gap between R&D output and its use as the basis of new operating businesses (i.e., the role of the new business development function in a typical corporation) grows wider and becomes more difficult to bridge.

The Venture Capital Approach

During the 1980s, venture capitalists were extremely successful in commercializing new technology by financing and providing management assistance to entrepreneurial start-up companies. Although returns on such investments vary, the success of many venture capitalists has demonstrated that the combination of a committed, energetic, and talented entrepreneurial team, a group of operationally experienced and well-connected investors, shared objectives, and the potential for exceptional wealth can be a powerful formula for developing new businesses.

Although the traditional image of these embryonic companies has been of individual scientists, engineers, and inventors working out of a garage, increasingly they are well-researched technology spin-offs from universities or the R&D laboratores of established companies. Typically, a technologist leaves the parent organization and seeks financial backing for a new company on the basis of specific technological expertise. Rather than fight this phenomenon, many universities and corporations are now actively supporting the process, negotiating directly with venture capital firms for a combination of equity in the new firm and follow-on royalties in exchange for the rights to and transfer of the technology.

a corporation's commercialized products are only the tip of a technological iceberg, beneath which lies a broader base of knowledge and technology

Several large corporations have sought to emulate the venture capitalists' method—both as a means of corporate renewal as well as for diversification—but generally with limited success. Managers of these companies have tried various organizational approaches to stimulating, nurturing, selecting, and managing new ventures; these approaches range from a total corporate commitment to new business development to empowering small, dedicated, multifunctional project teams (i.e., the matrix management approach, in which team members come and go depending on the current need for their particular expertise). In most cases, the parent corporation retains at least partial equity in the spin-off company. Regardless of the approach, management's objective is to foster a spirit of intrapreneurship, in which the corporation attempts to simulate internally the entrepreneurial environment characteristic of small, start-up companies.

The Keys to Innovation

The key variables in these various approaches seem to be the processes whereby the corporation develops and manages new businesses: To a great extent, long-term success depends on the degree to which the corporation values innovation. A successful new-business development process requires a corporate management that adopts an opportunistic approach and is willing to experiment and accept the uncertainties, false starts, and failures that are characteristic of start-up ventures. In addition, careful consideration must be given to the degree of control the corporation will exercise over the new venture's evolving strategy after the new business is operational. Corporate management must define the degree of autonomy and accountability its venture managers will have for making independent tactical decisions that may differ from the parent company's original assumptions and objectives in launching the venture.

Because large corporations are generally subdivided into profit centers or strategic business units, senior managers have long sought methods for using the corporation's assets to create new businesses that do not fit the individual charters of the existing businesses or sectors but fit the overall charter of the corporation. Emphasis on the challenges of operating and enhancing curent businesses, however, coupled with the risk-averse nature of most corporations, tends to preclude the entrepreneurial approach to new business development that characterizes the free-enterprise system that exists outside the corporation.

CORPORATE ENTREPRENEURSHIP PROGRAMS

The optimal approach to stimulating innovation is one in which the concept is interwoven throughout the corporate culture. One company, famous for its innovative culture, uses such mechanisms and reward systems as:

- Dedicating 15% of its R&D budget to opportunistic new-business development. This monetary commitment encourages grass roots experimentation, innovation, and entrepreneurship.
- Using a portion of its travel budget to allow technologists to meet with customers and discuss customer needs.
- Establishing a management guideline that 25% of the revenue from each business unit must come from products less than five years old.
- Conducting an annual technology fair that simulates a free market for new technologies and innovation at an early stage in their development.
- Celebrating when a new business surpasses $2 million in sales and is profitable.

Another large corporation has divided senior management responsibility into operations, or existing businesses, and new business, which includes R&D, business research, licensing, and a portfolio of small start-up businesses. Several corporations have formed senior-level venture boards to facilitate new business formation in areas outside their traditional businesses, and to facilitate approval of corporate spin-off ventures. In one corporation, the venture board is the corporate approval mechanism for channeling nonstrategic technology projects into a special corporate venture capital department that was created to provide seed capital and business plan development and to locate external sources of equity capital.

Generally staffed by a small group of senior managers in order to provide both credibility and rapid decision making, corporate-level venture boards have also played a subtler role in stimulating a more entrepreneurial climate within the corporation. In one large corporation, the venture board adopted a threefold mission that involved creating new businesses, establishing an environment that fosters innovation and entrepreneurship, and providing career opportunities that would help the corporation attract, develop, and retain entrepreneurial talent.

This venture board created and managed a modest venture investment pool that was to be funded by the corporation for no more than five years. After that, the fund was to be self-sustaining, with the proceeds of internal sales and external divestments providing the capital for future ventures. The venture board then provided a nurturing environment for the champions of new ideas during the early stages of development and established a venture capital approach to funding and operational management for more mature proposals that were not picked up by an existing business unit. In addition, the corporation created three new organizations to facilitate the process of developing and managing new ventures by:

- Finding, shaping, evaluating, and funding new-business proposals and directing the proposals to the most appropriate potential sponsors.
- Monitoring and guiding venture board-sponsored businesses, once they are funded and operational.

- Creating a venture advisory panel, consisting of selected line and functional managers, to provide the new ventures with access to required resources while minimizing disruption of ongoing operations in existing units.

senior managers have long sought methods for using the corporation's assets to create new businesses that do not fit the individual charters of existing divisions

These nurturing organizations were created to provide an efficient, acceptable decision-making mechanism for such issues as matters of corporate policy and allocation of limited resources. They also played an advisory role by guiding the numerous small, temporary, semiautonomous, flexible, and responsive businesses or projects operating in the corporation until those ventures obtained full sponsorship.

THE INNOVATION MANAGEMENT PROCESS

Approaches to new business development can be classified as strategic or opportunistic. Large companies have traditionally approached new business development from a strategic, or top-down, perspective. A staff group tries to determine which new businesses to enter—either with internal resources or through acquisitions—by matching internal strengths with external opportunities. The opportunistic approach, however, takes a bottom-up perspective by emphasizing the creation of a free-enterprise environment and the necessary infrastructure to facilitate, guide, and control the development process. The opportunistic approach relies on the ingenuity of employees to discover the perceived customer needs that can be addressed by existing corporate capabilities and thereby provide the impetus for new businesses. All employees involved in the process must believe that the rewards for successful implementation outweigh the penalties for failure.

Studies have shown that most significant innovations have evolved through a bottom-up process. For example, a study by R. Barth, ("How Innovators Work, and What Stops Them," *Financial Times*, March 6, 1986) found that 62% of 73 successful innovations originated from a bottom-up process and generally were implemented despite opposition from senior management. When allocating resources to new projects, therefore, progressive growth-oriented companies generally try to balance a predominantly top-down strategic planning approach with a mechanism for responding to proposals that are generated from a bottom-up perspective.

In one corporation, senior management encouraged employees to submit their new-business proposals, promising a nurturing environment in which ideas would be directed to the most appropriate unit in the corporation. To improve the corporate climate for innovation and to create awareness and stimulate employee participation, this corporation also sponsored such activities as:

- Publishing a quarterly innovation newsletter highlighting successful innovations and innovators.
- Issuing bulletins detailing newly recognized needs and discoveries.

- Holding an annual technology fair to promote applications of technologies that were ready for commercialization but were not yet sponsored by any existing business unit.
- Sponsoring a speaker's series featuring experts in the fields of creativity, innovation, and entrepreneurship.
- Conducting educational seminars on various aspects of the start-up process (e.g., marketing, finance, and negotiating).

Because of the high risk, unpredictable nature, and high attrition rate inherent in innovation, the innovation management process must follow a set of well-defined phases and gates. During each phase, resources are added, with the investment directed toward producing information or evidence to augment pertinent knowledge and reduce project risk. At the same time, less-promising and riskier innovations are discontinued or delayed. Properly managed, most of the venture investment pool is thereby channeled into those proposals with the greatest potential for success.

The innovation management process comprises three phases—each with its own set of substages—that correlate well both with the external venture capital process and with the range of skills typically found in a large corporation. The three phases of the innovation management process are:

- The concept development phase, in which an idea is developed and refined to formulate a supportable business proposal (i.e., the idea has sufficient substance and focus to attract the first seed investment).
- The business development phase, which takes the proposal from the first seed investment to a product prototype and identification of target customers (i.e., the point at which the typical venture capitalist first shows significant investment interest).
- The start-up phase, in which the embryonic business begins production and market development, with a goal of becoming a commercially sustainable and operationally stable company (i.e., reaching the point at which the business typically would go public or be acquired).

Portfolio Managers

Each phase should be managed by a group of portfolio managers who are recruited because they have the skills required for a given phase of the business development process. With a primary goal of creating strategic value by investing in opportunistic innovation, the portfolio managers are involved in all phases of the business development process. Portfolio managers review proposals at an early stage and, working through their proposals' advocates or champions, help develop and shape the proposals to fit the strategic vision of some business unit within the corporation. The portfolio managers also provide access to ad hoc resources within the corporation, lend implementation help as required, and, at the appropriate point in the development process, help sell the project to an internal customer.

Because the ultimate customer for any innovation is always an established unit within the corporation, the innovation management process must be designed so that all proposals are handled in the same way, with the only differences being the point of entry into the process, which depends on the maturity of the proposal, and the point of exit (e.g., as a well-researched proposal, a working prototype with identified customers, or a fully operating business). During all three phases of the process, proposals are routinely evaluated for business feasibility, sustainable competitive advantage, and consistency with corporate vision and strategy. Therefore, portfolio managers make essential contributions during all three phases of the innovation management process by providing:

the venture board provides a nurturing environment for champions of new ideas during the early stages of development

- A temporary home and a fair hearing for all new ideas and proposals.
- Dedicated assistance in proposal evaluation, development, enrichment, guidance, and packaging for internal sale.
- A corporatewide network to access information, complementary expertise, and ad hoc resources.
- Early stage empowerment and, when appropriate, financing for proposal evaluation and development.
- A nurturing business incubator environment, with its supporting infrastructure.
- Contacts with potential sponsors (i.e., the ultimate internal customers of the proposals, capabilities, or new businesses).

Business Incubators

The business incubator has become a popular and accepted mechanism to provide an environment that increases the probability of success for start-up businesses.[3] During the business development and start-up phases, incubators offer flexible, inexpensive office space, on-site business planning and management assistance (e.g., personnel recruiting, accounting, and access to other resources), and various support services (e.g., conference rooms, shipping and receiving facilities, copiers, fax machines, clerical help, telephones, computers, furniture). Many corporations have also used skunk works to encourage innovation and fast-paced product development.

The Concept Development Phase

In the concept development phase, portfolio management is handled in one corporation by a group of innovation facilitators who operate as part of a companywide organization called the Office of Innovation.[4] This network has branches throughout the corporation to support the development of employees' proposals. To provide a supportive environment, innovation facilitators are chosen for their interpersonal skills, inherently positive attitude, and aptitude for possibility thinking. Because they are familiar with the objectives and strategies of the existing business units, innovation facilitators can help the idea champions gather and analyze information from published sources and internal consultants, shape proposals to address business unit needs, and establish contacts with potential internal sponsors. When appropriate, innovation facilitators help consolidate similar or compementary proposals into a single proposal. They also help originators recognize cases in which their proposals lack substance or do not fit the company's strategic plans. To avoid conflict with the originator's line management, only proposals that fall outside the current charter of the originator's immediate organization are accepted by the Office of Innovation. A properly managed Office of Innovation can operate with minimum budget, relying mainly on its ability to bootleg resources to accomplish its mission.

In this corporation, approximately 8% to 10% of all proposals survive the concept development phase and obtain internal seed sponsorship. Of these, approximately 85% are sponsored in existing organizations. Proposals sponsored by existing business units or other established corporate organizations continue to be monitored for statistical purposes but no longer fall under the responsibility of the venture board.

The Business Development Phase

In the seed, or business development, phase, proposals that do not fit the mission of any existing business unit can be considered for their potential as stand-alone businesses. After confirming both the lack of a major objection from any existing organization and some level of support from at least one organization, the seed portfolio managers work with the would-be entrepreneurs to develop a business proposal, obtain technical and human resources, develop a prototype of the product or service, and identify potential target markets and early customers. Seed financing is provided in stages, and at the appropriate point in the process, a temporary and partial release from current job responsibilities is negotiated with the would-be entrepreneur's management. Generally, the entrepreneurs devote about 20% of their time to their projects at the start of the seed phase; successful ventures eventually require a full-time commitment.

The business planning process typically requires an assessment of the industry structure and dynamics and of the proposed venture's competitive advantage. This assessment is followed by the development of alternative strategies, evaluation of associated risks, agreement on a strategy, and preparation of a staged operating plan supported by financial projections. During the business development phase, proposed ventures beyond the scope of existing business units eventually must be sponsored by a corporate officer who can provide assurance that the new business will be of strategic corporate interest if it succeeds. The seed sponsorship phase is complete when the new product or business is announced or, in the case of a new process, is used internally. For a start-up venture, the management team is in place, the business plan is completed, a working prototype has been produced, and targeted customers with a willingness to buy have been identified.

In this corporation, approximately 30% to 40% of seeded proposals survive the business development phase—either inside an existing business unit or under the direction of the venture board—of these, nearly 90% are commercialized or adopted by an existing business unit or other established organization; the remainder receive direct venture board sponsorship.

Seed portfolio managers are recruited from both the technical and marketing ranks. They are selected for their business acumen, project management experience, and ability to assemble the various elements necessary for a stand-alone business.

properly managed, most of the venture investment pool is channeled into proposals with the greatest potential for success

The Start-Up Phase

During the operational, or start-up, phase, projects sponsored by existing organizations are managed within those organizations and are monitored only for statistical purposes by the venture board. New business ventures managed by the venture board are expected to grow, become profitable, and stabilize their operations within three to five years. This process involves such tasks as developing manufacturing capacity, establishing distribution channels, developing marketing programs, and building a sales force. New businesses are expected to continue testing the validity of such strategic assumptions as the needs and segmentation of the customer base, pricing, and the length of the sales cycle. A key indicator during this phase is how well the new business has anticipated and can respond to competitive challenges and how quickly it can adapt its product to meet these challenges.

In this phase, venture portfolio managers serve as the key link between the venture board and the individual ventures. Venture portfolio managers are selected for their experience in managing a start-up or small business. Like the seed portfolio managers, they are recruited from technical (or manufacturing) and marketing or business backgrounds.

PROBLEMS ENCOUNTERED AND LESSONS LEARNED

Managers face many challenges in implementing and sustaining a broad-based innovation and internal venture process within a large corporation. During the concept development phase and the early stages of the business development phase, managers and innovation facilitators must balance the needs of clients (i.e., proposal generators) with those of customers (i.e., potential sponsors). At this point, investments are relatively small and any problems can be handled within the corporation. In subsequent stages of the overall process, however, new and relatively inexperienced venture managers begin to communicate with potential customers outside the corporation, and a new set of issues must be addressed.

An internal venture capital operation within a large, diversified, technology-rich, and vertically-integrated corporation should have several distinct advantages over an independent venture capital company:

- A broader, deeper technology base on which to build new businesses.
- Access to company personnel with a broader range of expertise and skills.
- Economies of scale in such areas as research, manufacturing, purchasing, and distribution channels.

- The ability to access internal organizations and to create natural intracompany alliances without concern for inadvertent disclosure of confidential information.
- Established distribution channels.
- More accessible and efficient means of obtaining financing (executives of typical start-up companies often spend as much as one-third of their time trying to raise money).
- A stable image, through association with the parent company, which lends instant credibility to the new business, both with suppliers and potential customers.

Despite these inherent advantages, large companies still find internal venturing extremely difficult. Although these companies seem to have received considerable value from such programs, most of the benefits have been achieved during the concept development and early business development phases, which are relatively inexpensive to implement and pose little external risk to the corporation.

Even with survival statistics that would be acceptable to a venture capitalist, however, the operational or start-up phase poses risks that management considers disproportionate to the potential benefits. Although the long-term economic performance of such programs is not easily quantified, it is unlikely that large corporations will ever attain venture capital-like returns on their investment in the new ventures. There are many reasons—some related to the structure and others to the execution of internal new-venture programs—but the difficulties of internal venturing can be traced to one issue: the risk-versus-reward decision. Both corporate management and the venture management team often believe they have more to lose than to gain.

Corporate Risks and Rewards

The potential damage to a corporation from problems with new ventures far exceeds the potential benefits.[5] Damage may be in the form of litigation (or the threat of litigation), harm to the corporation's public image, or any action that falls short of meeting exacting corporate ethics standards. In addition, corporate management typically views even the most successful ventures as too small and taking too long to have any real impact on a large corporation. Furthermore, the turmoil associated with failed or faltering ventures places disproportionate demands on scarce management time relative to the ventures' contribution to short-term corporate performance.

Management Risks and Rewards

Few, if any, large corporations have been willing to offer managers of internal new ventures rewards comparable to those they would earn from successful external ventures, in which compensation depends more on the level of value created than on the number of resources controlled. Any corporate constraints on the maximum compensation of the venture team tend to limit the caliber of venture managers that can be attracted to internal venture programs. On the other hand, large corporations have traditionally provided some security for entrepreneurs who fail, if only because management retains a sense of obligation to an employee who took a risk.

Successful venture managers must share in the financial rewards, or the new-venture process must be integrated into the company's career development program. New business formation, like R&D, is a high risk enterprise; like R&D employees who move from project to project, entrepreneurial employees should be viewed as valuable resources that can be developed through training and experience and can successfully launch more than one business. Corporations have experimented with equity compensation plans and with using venture management jobs as career development positions, and generally have found the career development approach to be more consistent with the corporate culture.

proposals sponsored by existing business units continue to be monitored for statistical purposes but no longer fall under the responsibility of the venture board

Executing the Plan

In addition to the risk-versus-reward decision, corporate management must address several important issues related to the execution of internal new-venture programs. These issues are described in the following sections.

Avoiding Unreasonable Expectations. Management expectations regarding the size and growth rate of new businesses are often unrealistic. In addition, both the individual ventures and their managers are often prematurely afforded a stature beyond that which is warranted, which creates the wrong image within the new ventures and hurts working relationships with managers of the established businesses. The perceived pressure to grow rapidly has led many ventures to build excessive staff and infrastructure in anticipation of revenue that is slow to materialize. Continued failure to meet objectives, not only in the individual ventures, but at the venture portfolio level, has diminished the credibility of many venture programs.

Establishing Clear, Balanced Objectives. Any program aimed at improving the corporate climate for innovation tends to have an array of objectives, both tangible (e.g., generating new businesses with strong cash flow and earnings) and intangible (e.g., changing the corporate culture). Corporate management often fails to establish ahead of time the desired balance between the tangible and intangible objectives, which are more difficult to measure but must be accommodated to achieve the overall corporate mission. The hierarchy of objectives must be clearly articulated and documented by senior management.

Defining Management Controls and Exit Plans. Relatively inexperienced yet highly autonomous managers, inadequate cash management mechanisms, and the use of the corporation's name in a new venture's advertising all combine to create potentially significant management control problems. Entrepreneurs, realizing that their target market is more difficult to penetrate than originally assumed, often lose focus and try to find new markets. A parent corporation's continued willingness to finance a faltering venture despite revenue shortfalls results in cash overruns, which deplete the venture fund and deprive subsequent venture candidates of needed seed and start-up capital. Exiting a failed venture that has been identified as part of a well-known parent corporation is more complicated after the venture has developed loyal customers. A key requirement in any business plan is an exit plan that defines the steps to be taken if the venture fails to meet its goals within a reasonable period of time or level of investment.

Ensuring Management Competency. With self-selected entrepreneurs as managers, corporations soon discover that the skills and characteristics that are effective in working through the corporate bureaucracy to start a new business differ from those needed to manage the business once it is operational. With limited incentives to attract managers with start-up experience for subsequent phases of the business, management turnover has become a time-consuming and demotivating problem. Corporate management must recognize that competent venture management at each stage of the business development process is the most important factor in new business success.

Coaching Venture Management. The functional management skills developed in a large company often differ from those needed to coach and nurture fledgling entrepreneurs. A new management style must be developed in corporations that are commited to internal new-venture programs. For example, one corporation required each venture to have a board of directors that included members from outside the corporation with start-up or industry experience. Corporate management also needs to place greater emphasis on training and development courses and seminars for new-venture managers.

Completing the Due-Diligence Process. Before investing, any experienced investor will conduct a due-diligence investigation to study the assertions and assumptions in a business plan and to check the references of the new venture's management team. To avoid the analysis paralysis typical of large corporations, venture programs often rely on the milestone-based, staged financing process used by venture capitalists to achieve management control. Although this process works reasonably well through the stages prior to market entry, it is difficult to sustain once the venture has established a customer base. Because customers view the venture as part of a large corporation rather than as a start-up company, the parent corporation must place greater emphasis on early due diligence, thereby removing many of the advantages of speed of execution that were originally sought.

Balancing Marketing and Technology Orientations. Many of the ventures launched in internal new-venture programs are primarily technology driven and lack comprehensive marketing plans. Proponents of proposed ventures must obtain input from potential customers before assembling a business plan. This approach shifts the onus of rejecting a new-business proposal away from the portfolio manager and onto the potential customer.

an internal venture capital operation within a large, diversified, technology-rich corporation should have several advantages over an independent venture capital company

Venturing Strategically. To avoid building a collection of strategically unrelated businesses, internal venturing must be strongly coupled to a corporate development effort, with corporate vision, intent, and strategy as its guides. Early in a venture program, corporations often mistakenly rely on verbal assurances of strategic fit and continued support for the venture. Even with successful ventures, a formal process for strategic sponsorship is essential. Without it, faltering ventures quickly lose their appeal, and successful ventures come to be viewed as encroaching on the charters and distribution channels of established business units.

Demonstrating Support. With the number of conflicting objectives and the multitude of stakeholders in a typical large corporation, the most important success factor for an innovation and internal venturing program is senior management involvement. Start-up companies are the riskiest form of new business development, especially when driven by a new technological capability, and venture managers must believe that senior management has a stake in their success, rather than feeling they are merely on trial.

THE STATE OF INTRAPRENEURSHIP

After a period when intrapreneurship took on a fad appeal, many corporations have abandoned the internal venture approach to technology-based new business development. Diversification has given way to a stick-to-the-knitting mentality, in which cost control and focus on core businesses has taken primacy over revitalization and new business growth. In this climate, all future new businesses that evolve from a corporation's technology base must fit the strategies of one or more of the corporation's sectors or business units. Most corporate managers believe these parameters provide adequate flexibility for pursuing new business opportunities. Most large corporations have highly focused business units, each with the management competency to develop and operate businesses within their sphere of activity. In this climate, any new bottom-up business proposal must ultimately be sponsored and implemented by an existing business unit, with little opportunity to launch a new business at the corporate level.

In one corporation only a scaled-down version of the Office of Innovation has survived, with a corporate-level charter to develop innovation proposals that cross organizational boundaries. However, other organizations within the corporation have taken advantage of internal entrepreneurial skills and networking capabilities to achieve targeted innovation objectives. In this corporation, the new-business development function in the business planning department of each business unit provides a receptive audience for well-formulated innovation proposals. Facilitators can continue to nurture and play the role of champion for a proposal, with the due-diligence responsibility falling on the sponsoring business unit that must implement the proposal if adopted.

Many corporations have recently embraced the concept of value-based management, in which the value of each business in a company's portfolio is judged on the same criteria as those used by the external capital markets. Rather than judge an internal business strictly on its operating earnings, a warranted equity value is calculated on the basis of the net present value of its projected cash flows, using the business's cost of capital (i.e., a measure of the risk inherent in those projections) as the discount factor. Depending on its growth potential, a business might be valued at 10 to 20 times its annual earnings, which reflects the fact that an annuity of that size would be required to duplicate the business's projected cash flow. Theoretically, this methodology puts new business valuation on the same basis inside a corporation as in the venture capital markets.

Because valuation of a start-up company with little or no history is based more on assumptions than on supportable projections from past performance, the cost of capital is much higher, especially in the early stages, when only a relatively small amount of money is at stake. When funding early-stage ventures, venture capitalists typically require projected returns of approximately 60%, or roughly 10 times their money in five years. To the extent the new venture remains viable, the cost of capital decreases during subsequent rounds of financing, when the need for capital generally grows to fuel the venture's success. As a business matures, the cost of capital gradually approaches that of an established business; therefore, this methodology is philosophically consistent with comparing start-up companies to established businesses and should lead to better assessment of the relative value of the new business component of a corporation's total business portfolio.

CONCLUSIONS

Innovation and new-business development are the lifeblood of any corporation, and somehow must be integrated into corporate strategy. Large corporations can reap substantial benefits by adopting internally a more market-driven, free enterprise approach to expanding their business base.

When a small change in the core business can overwhelm even the most optimistic projections for a new business, however, established companies are forced to place a higher priority on protecting their existing businesses than on developing new businesses. To remain competitive in their core businesses, corporations must set objectives, develop plans, and focus their resources on meeting those objectives; therefore, unanticipated opportunities are often viewed as unwelcome distractions.

Compensation and career growth also favor managing the established business over the more speculative start-up business; corporate incentives typically are based on the extent to which operating plans are met, which places a greater premium on predictability than on experimentation. When combined with the fact that corporate stature and compensation are determined by resources controlled rather than value added, the overall result is an environment that actually inhibits innovation.

the perceived pressure for rapid growth has prompted many ventures to build excessive staff and infrastructure in anticipation of revenue that is slow to materialize

Large corporations have already met a similar challenge, however, in their technology development operations. Because a separate R&D function is created, people are removed from the environment of the current business to develop the knowledge, technologies, products, and processes for future businesses. Because the source and the timing of the next discovery cannot be predicted with any degree of certainty, R&D management has created an environment conducive to experimentation and can accommodate frequent failure as part of the learning process. Separate strategies, goals, and budgets are often developed for fundamental research, applied research, and product development. In recognition of R&D's low yield, incentives in R&D are also geared toward meeting scientific and technological objectives, independent of subsequent commercial success, and people are moved from project to project throughout their career, rather than being tied to the commercial success or failure of any one project. Despite the low yield of commercial successes, morale is maintained in the R&D environment because the results of employees' efforts may still provide the basis for conference presentations, publications, and internal reports, all of which demonstrate the value of the individual scientist's or engineer's contributions.

By providing a temporary home for ideas and businesses that have not yet been adopted by established business units, the large corporation can provide a comparable environment for new business development. A small amount of seed money and a business incubator environment, managed by a dedicated, appropriately compensated innovation and venture organization of portfolio managers that are skilled in nurturing innovators and innovations, can help both the innovators and the potential customers for their innovations. The efforts of this organization should be focused first on information gathering, then on packaging the proposal to support the appeal for financing, and finally on staged implementation with limited business objectives at each stage. The business incubator also allows a potential sponsor time to accommodate a new project outside the annual budget cycle and avoids the not-nvented-here syndrome by virtue of the legitimacy afforded the incubator environment.

Employee morale is enhanced because all ideas get a fair and informed hearing; good ideas are not killed by premature business scrutiny, and bad ideas die before costs become excessive. Free exchange of ideas—especially among ad hoc, self-selected, multidisciplinary teams—provides synergy and cross-fertilization between disciplines and businesses and generally leads to better proposals. In addition, it contributes to the education and development of professionals who might otherwise begin and end their careers as functional specialists. The larger, more diverse, and more technologically rich a company is, the more it should benefit from such a corporate-level nurturing organization.

Rewards and Reinforcements

As with technical innovation, business innovation must be rewarded for its own sake. With more than 90% of the proposals never reaching the seed investment phase, and a similar attrition rate among those that do achieve seed sponsorship and commercialization, there must be some form of corporate acknowledgment or reward for achieving milestones short of commercial success. Corporations that have nurtured innovation and new ventures have found that many of the best proposals have been from an employee's third or fourth try. By making the significance of the reward proportional to the degree of commercial success or internal impact, management can emphasize the importance of successful innovation over mere creative activity.

Without the full support and cooperation of the established business units, a corporate innovation and new-venture program cannot survive. Therefore, the new-venture organization must invest the time necessary to understand the strategies and needs of the business units, develop credibility, and build the necessary personal relationships. Often, it must defer to the existing organizations on such issues as allocation of scarce resources. The proper attitude is to view the business units as the ultimate customers of the new-venture organization. Innovation is a buyer's market, and the business units will all have competing projects that did not receive budget allocations. Even a director of research will find it difficult to gain funding for new technology in this type of environment. The use of idea champions and advisory boards as political intermediaries, as well as such auxiliary programs as technology fairs, speakers, and newsletters, all can help communications and engender business unit support.

A small seed fund can be useful during assessment of the feasibility of potential new businesses and evaluation of the capabilities of would-be entrepreneurs. However, the entire range of value-creation options (i.e., licensing, partnerships, establishing a start-up company) should be considered. Although all seeded proposals should have the support of one or more business units, the new-venture organization also should have the ability to support a project until it is placed in one or more target business units. The organization also must have the option of divestiture should a proposal fail to attract business unit interest. Seed-stage businesses can provide a low-risk proving ground for potential entrepreneurial managers in a large corporation, though it is often difficult and time-consuming to bring would-be entrepreneurs' objectives into alignment with those of the corporation.

Any new business must go through an entrepreneurial phase, and a corporate business incubator can provide an ideal environment for developing small entrepreneurial ventures separate from the operating environment of the established businesses. Start-up companies face problems and challenges that differ from those of the typical established operating unit. Until their size and stability warrant a more permanent home in a parent business unit, new businesses can benefit from being grouped into a portfolio under management skilled in start-up operations, with the ultimate owner of the venture having a seat on the venture's board of directors.

If there is no appropriate business unit for a new venture, there needs to be a corporate development process for the formation of new business units. Although some new businesses may spin off from existing units, others may be completely

without a formal process for strategic sponsorship, faltering ventures quickly lose their appeal

new because they are based on a new capability in research, manufacturing, or technical service. The key issue is strategic fit, and the process must designate how that is decided, by whom, and at what stage in the business development process. Again, divestiture must remain an option for ventures that have developed to that point in the process and then fail to meet the increasingly difficult test of strategic fit.

The internal new-venture process is most often used as a corporate development tool for new businesses that are viewed as business experiments with only a possible future strategic fit or those that fail the strategic-fit test outright and are spin-off or divestiture candidates. Unless there is a corporate career development path that emphasizes start-up experience, the corporate spin-off route is more advantageous to the entrepreneur and the corporation. The entrepreneur is freed from the constraints of the corporation and is able to compete on equal terms with other firms, both for customers and for capital. Although financing becomes a major issue, the entrepreneur shares fully in both the potential risks and rewards. In addition, the corporation eliminates the resource demands associated with the start-up effort, avoids the expense of a truncated project, and retains the advantages and limited legal liability of a minority shareholder or limited partner. Finally, should the spin-off company become a commercial success with obvious strategic value, the corporation can repurchase the stock of the venture at the higher price, but from the vantage point of an already significant shareholder.

Acknowledgments

This chapter was adapted from a paper originally presented at the International Forum on Technology Management in Brussels, July 1989. The author gratefully acknowledges the assistance of W. Gerald Norton, Allison F. Dolan, and Nancy Ferrell, who extracted and interpreted information from the Office of Innovation data base; Pier A. Abetti and Jack W. Savidge, who reviewed an early draft of the paper and offered a number of helpful suggestions; and Hollister B. Sykes, Ernest Sternberg, Donald Hoag, Robert Szakonyi, and Alan Fusfeld, who reviewed and critiqued the presentation manuscript.

Notes

1. *National Patterns of R&D Resources: 1990*, National Science Foundation Report nsf 90-316 (Washington DC, 1990).
2. National Science Board, *Science and Engineering Indicators—1989* (Washington DC: US Government Printing Office, 1989), p 141 and references cited therein.
3. R.W. Smilor and M.D. Gill, Jr., *The New Business Incubator* (Lexington MA: Lexington Books, 1986); and P.A. Abetti, C.W. LeMaistre, and M.H. Wacholder, "Technopolis Development in a Mature Industrial Area: The Role of Rensselaer Polytechnic Institute," in R.W. Smilor, G. Kozmetsky, and D.V. Gibson, *Creating the Technopolis: Linking Technology Commercialization and Economic Development* (Cambridge MA: Ballinger Publishing Co, 1988).

4. R. Rosenfeld and J.C. Servo, *The Futurist* (August 1984), p 21; R. Rosenfeld and J.C. Servo, *Intrapreneurial Excellence* (November 1985), p 4; and C.H. Chandler, *Journal of Business Strategy* (Summer 1986), p 5.
5. H. Geneen, "Risking the Company for One Random Jewel," *Venture Magazine* (January 1985), p 46.

INDEX

F

G

I

T

V

W